T0315291

Ecological Challenges and Conservation Conundrums

Ecological Challenges and Conservation Conundrums

Essays and Reflections For A Changing World

John A. Wiens

Winthrop Research Professor
University of Western Australia
and
Emeritus Distinguished Professor
Colorado State University

Registered office: John Wiley & Sons, Ltd, The Atrium, Southern Gate, Chichester, West Sussex, PO19 8SQ, UK

Editorial offices: 9600 Garsington Road, Oxford, OX4 2DQ, UK
The Atrium, Southern Gate, Chichester, West Sussex, PO19 8SQ, UK
111 River Street, Hoboken, NJ 07030-5774, USA

For details of our global editorial offices, for customer services and for information about how to apply for permission to reuse the copyright material in this book please see our website at www.wiley.com/wiley-blackwell.

Library of Congress Cataloging-in-Publication Data

Names: Wiens, John A., author.
Title: Ecological challenges and conservation conundrums : essays and
reflections for a changing world / John A. Wiens.
Description: Chichester, UK ; Hoboken, NJ : John Wiley & Sons, 2016. |
Includes bibliographical references and index.
Identifiers: LCCN 2015044916 | ISBN 9781118895108 (cloth)
Subjects: LCSH: Ecological assessment (Biology) | Conservation of natural resources.
Classification: LCC QH541.15.E22 W54 2016 | DDC 333.72–dc23 LC record available at
http://lccn.loc.gov/2015044916

A catalogue record for this book is available from the British Library.

Wiley also publishes its books in a variety of electronic formats. Some content that appears in print may not be available in electronic books.

Cover image: Ruffed Grouse (Bonasa umbellus) feather on the forest floor, Vilas County, Wisconsin.
Photo: John Wiens

Set in 9/12pt, MeridienLTStd by SPi Global, Chennai, India.

1 2016

To my grandchildren, Madeleine, Henry, and Annika.
May you always enjoy Nature's beauty.

Contents

Preamble: Why this book?

Many books have been written about ecology and conservation in the changing world that is fast upon us. This book is an idiosyncratic one. It's not a traditional science book, dealing with facts, figures, explanations, and theories, although it is a book about science. Neither is it a traditional conservation book about protecting biodiversity in the face of the looming environmental crisis or the perils of climate change, although environmental change is the underlying theme. Rather, it's a personal accounting of some of the challenges and conundrums I see confronting ecology and conservation as we move into an uncertain world. Fresh thinking is needed; my hope is to prompt some of that thinking.

The book is built around an eclectic collection of essays that I wrote over the past decade for the *Bulletin of the British Ecological Society*, plus a few older ones. The *Bulletin* is distributed to all members of the Society—an appreciative but somewhat circumscribed group of professional ecologists and students, mostly based in the United Kingdom. The essays are presented here as they were published, along with some new prefacing material to explain why I wrote the essay or to update the topic.

Essays, however, are usually written as stand-alone pieces. To tie them together, or at least to keep them from floating off as untethered balloons, I've included some accompanying text to provide context for the essays and offer my thoughts on some topics and issues in ecology and conservation. These writings are really reflections rather than reviews or syntheses. Those who want a textbook, richly annotated with literature citations, will need to look elsewhere, but those who want some fodder for thinking about the challenges facing ecology and conservation will, I hope, find this book rewarding.

I said that this book is not a science book, and that's immediately evident in the writing style. By striving for objectivity, scientific writing tends to avoid opinions and personality. Essays, however, aren't like that. Essays express opinions. They are intended to be stimulating, sometimes controversial, somewhat personal, and even (shudder) entertaining. They are meant to be read and enjoyed. I've kept the accompanying chapters short, so each can be read in a single, relaxed sitting. I've always been interested in why ecologists and conservationists ask the questions or address the issues they do, so I explore the history of ideas—a historiography of sorts—more than is customary. I use examples that I'm familiar with, so there's a lot about California, the American West, and Australia. I've retained the illustrations that originally accompanied the essays, but I've otherwise used pictures and graphs sparingly, hoping that the words will speak for themselves. I haven't gone into technical details and have tried to avoid mystery by limiting the use of jargon. I use footnotes to offer asides and direct readers to some relevant literature, but I relegate the full citations to the end of the book.

This is a book I've wanted to write for some time. I want others—ecologists, conservationists, environmentalists, students, my professional colleagues, resource managers, policy makers, even philosophers; indeed, anyone with an interest in conservation and the natural world—to feel my excitement and share my concerns by thinking about the issues I consider in the essays and chapters. My hope is that, after reading an essay and the writings, you'll nod in agreement—or perhaps you'll nod more vigorously in disagreement (but hopefully not nod off)—and will pause to think, and enjoy the thinking.

The book is organized into five parts. I first set the stage by describing some of the challenges facing ecology and conservation. Some of these challenges stem from the changing world, others from how science and scientific thinking progress. "Change," however, can take many forms, so I also consider the multiple ways ecological systems can change. In Part II, I discuss several forces behind the changes that make ecology and conservation so challenging, so interesting, and so important. These changes create a variety of conundrums for conservation, which I address in Part III. These "wicked problems" must be tackled if conservation is to achieve its goals; continuing with "business as usual" will not meet the challenges. In Part IV, I comment on some aspects of how ecology and conservation are done: methods, communication, dealing with differences of opinion, and debates. I also comment on some philosophical and ethical undercurrents that run through these debates and the historiographies. I conclude in Part V by attempting to salvage some hope out of what may seem to have been a cataloging of daunting and dismal challenges.

To accustom you to the interspersion of essays with the chapters and to provide background for what follows, I begin with three essays. The first (Essay 1, *How did I get here*? (page 1)) gives some perspective on where I've come from—what might have led me to think and write the way I do? The second (Essay 2, *Found*! *The survivor in the swamps (2005)* (page 4)) provides a parable of sorts for conservation, embodying many of the themes that drive conservation, but especially hope. The third (Essay 3, *In defense of footnotes (2014)* (page 7)) explains why I use footnotes throughout the book.

Acknowledgments

A simple listing of people would not do justice to the many sources, in many places, at many times, that have nurtured my thinking and shaped my perspectives. These are what make this book a personal tour through some topics that have interested me rather than a comprehensive review of the state of the sciences. Teachers, students, colleagues, adversaries—all have contributed to what's in this book, even though they may not realize it. However, several people have had a special role in helping me mold a confusing mélange of thoughts, experiences, paragraphs, and sentence fragments into something coherent, and they deserve credit.

First, Alan Crowden. Alan is a long-time friend who, in his role as Editor of the *Bulletin of the British Ecological Society*, first encouraged me to try my hand at writing an essay now and then. "Now and then" became more regular, leading to most of the essays included in this book. Alan carelessly suggested that I write on whatever struck my fancy, so the responsibility for the eclectic nature of the collection rests on his shoulders—I simply took the bait.

Whatever grace and style the essays have is due to my daughter, Kyra, who served as my in-house editor for most of the essays. After laboring to get just the right phrasing, I'd send Kyra what I imagined to be a final version, just needing a tweak here and there. Kyra always—*always*—found a better way to say things, a more logical flow, a way to shorten my sentences to crystallize the meaning. She left little for Alan to do except pass an essay on to the printer.

Two colleagues have been sounding boards for my thoughts over the years. Mike Scott and I have critiqued each other's ideas and collaborated for nearly 50 years, and Richard Hobbs became a kindred spirit when my attention shifted to landscape ecology, and then to conservation and restoration ecology. Richard also orchestrated an appointment for me as a Winthrop Research Professor at the University of Western Australia, which has provided contemplative as well as physical space, recharged my Aussiephile batteries, and enabled me to access their library holdings from my desk in Oregon.

Others have read bits and pieces of the writings to validate (or, more often, correct) my thinking. I especially thank Jerry Franklin, Dale Goble, Richard Hobbs, Bob Lackey, Ariel Lugo, Michael Nelson, Dick Norgaard, Mike Scott, and Tom Spies. Pete Warzybok and Jamie Jahncke of Point Blue Conservation Science provided graphics and updates on the saga of Cassin's auklets on the Farallon Islands, a great example of all sorts of things (not least the value of long-term monitoring). Cliff Dahm, Peter Goodwin, the staff of the Delta Science Program, and my colleagues on the Delta Independent Science Board helped me begin to understand the Sacramento–San Joaquin Delta in California, where all the challenges and conundrums explored in this book seem to converge in a single place.

That this book emphasizes conservation so much is a consequence of my move from the sheltered halls of academia to the world of conservation NGOs. Steve McCormick took the risk of hiring an ivory-tower professor to be a Lead Scientist at The Nature Conservancy, for which I am forever grateful. Together with Peter Kareiva and M. Sanjayan, the other Lead Scientists, I experienced the challenges of incorporating science into the real-world practice of conservation, and the delight when it happened. Ellie Cohen and the enthusiastic scientists at PRBO Conservation Science (now Point Blue Conservation Science) provided the opportunity to learn about conservation issues in that peculiar state called California.

I've sometimes wondered why authors always acknowledge the support of their family. Is this just something that's politically correct, or is it really that important? Now I know: it's absolutely essential! There's putting up with the aberrant behavior of someone experiencing writer's block,[1] or abruptly leaving the dinner table to write furiously when the blockage suddenly breaks. But there's also the deeper understanding and belief in the value of what you're doing. Everyone in my family—Bea, Kyra, Taryn, Ann, Mike, Dave, Susan—is a scientist, artist, or writer, so their understanding extended well beyond what I was trying to do to what I was actually saying. A very special kind of support, for which I'm forever grateful.

[1] A condition, according to Wikipedia, in which "an author loses the ability to produce new work or experiences a creative slowdown." I know it well.

ESSAY 1

How did I get here?

Essays and writings are by their nature personal, reflecting the personality and perspective of their writers. Writers carry the baggage of their past—their childhood, their family, their teachers, their friends and colleagues, their jobs, their experiences—which affects how they view the world and how they write about it. It's therefore important, I think, that you know something about what's in my bags—how I came to write about the things I do in the ways I do. There are several threads to weave together.

One thread is natural history. I grew up in central Oklahoma. It was a time when kids could roam freely, and I did, riding my bike out of town to spend the days exploring nature. For reasons I can't quite fathom, I almost always headed west, to the prairies, rather than to the woodlands east of town. This was when I developed a fascination for birds. I became a boy birdwatcher, saved my allowance to buy binoculars and a field guide, and joined the Cleveland County Bird Club. As I grew older, my interest in birds was nurtured by several mentors—George Miksch Sutton, Charles Carpenter, Margaret Morse Nice—who took me into the field with them and slowly but surely revealed how birdwatching could be a science as well as a passion. When I decided to major in Zoology as an undergraduate, Sutton hired me as a research assistant to collect birds for the university museum. Such collecting has now fallen out of favor as taxonomy and systematics have become increasingly molecular, but then, in the early 1960s, it was still the foundation of museum work. So I spent my free time (and some time I should have been in classes) roaming the state with David Ligon, observing and collecting. I learned more about ecology, behavior, and natural history in those days in the field than I did in the classroom.

When I went to Wisconsin for graduate work, I joined John Emlen's large group of students. Everyone else was studying behavior and ecology, so I joined in and left museum collecting and systematics behind. I was on the pathway to becoming a behavioral ecologist.

However, I didn't start out to be a scientist. My parents were university professors in the humanities, and my boyhood friends were the sons and daughters of academics, none of them scientists. Little wonder, then, that I enjoyed things academic—reading, writing, and debating. I was much more attracted to the currency of ideas than the cataloging of facts, to asking "why" (or "why not") questions more than "what" or "how." In high school, my favorite classes were

Ecological Challenges and Conservation Conundrums: Essays and Reflections for a Changing World, First Edition. John A. Wiens.

in English literature, where Francis Dunham alerted me to the nuances of writing and the joys of critical reading. When I started college, my majors (several, at various times) were in the humanities rather than science. Courses in philosophy encouraged challenging of conventional modes of thought, and critical dissections of writing styles and composition structure were the bulwark of my classes in English literature, which I continued to take even after changing to major in Zoology. How many Zoology undergraduates would have time now to take a graduate-level course in Shakespeare from a renowned Milton scholar, as I was able to do?

This background in the humanities fostered my interests in writing, but also my gravitation toward concepts and ideas, along with an inherent skepticism. These traits were slow to surface, however. When I began my graduate studies (that is, when I began to do real research rather than just read about it), ecology was dominated by concepts of equilibrium and order in nature. Perhaps because it appealed to my love of ideas (or simply because it was the prevailing paradigm of the time), I embraced this thinking. My PhD dissertation examined how habitat was partitioned in an assemblage of grassland birds, and I had no reservations in concluding that the community was exquisitely (but subtlety) structured so as to circumvent competition among the species. Only later, after I had expanded my studies into a broader range of grasslands and shrubsteppe over western North America, did I begin to realize that my difficulties in getting my observations to match theory might have less to do with the observations than with the theory itself. Competition theory did not explain everything, everywhere. I became an iconoclast (and, perhaps as I've grown older, a curmudgeon).

What led me to expand the scope of my studies beyond the 100-acre grassland in Wisconsin? Perhaps it was simply an extension of the roaming I had done to the prairies west of town in my childhood, or traversing the state during my museum days in Oklahoma, or perhaps because I had moved to the forested landscapes of western Oregon and yearned for the open spaces of grasslands and shrubsteppe. Whatever the impetus, the broadened perspective alerted me to the importance of scale, which became another thread in the fabric of my thinking. I soon saw that the patterns I had so carefully documented at a single site in my PhD studies began to erode as I expanded the scale of analysis. The habitat associations of species, for example, could change completely with a shift from a local to a regional scale. As my students and I conducted repeated surveys over several years, the patterns we documented changed depending on whether we analyzed one or several years. Scales in space and time affected what we saw and how we interpreted our results. This is now almost a trivial observation, but at that time it was a new insight, at least for me.

All of these studies fell under the rubric of basic science: gathering observations to test theories, accumulating knowledge for knowledge's sake, and pursuing questions because they were interesting. At some point, however, I began to move toward addressing more practical problems. It wasn't because I was suddenly motivated to do something relevant, however. Instead, two of my graduate students who were interested in seabirds saw an opportunity to fund their studies in a program to assess oil development off Alaska. We applied, got the funding, and spent several years doing field studies and modeling. Years later, this work led to a quarter-century involvement in assessing the effects of the *Exxon Valdez* oil spill on marine birds, deepening my involvement in real-world issues.

As an ecologist, I had long had an interest in conservation. This latent interest was activated when I was asked to lead a review of science in The Nature Conservancy (TNC). As we conducted the review and I learned more about conservation in practice, I became increasingly intrigued. One thing led to another, and I ended up leaving the sheltered halls of academia to join TNC as a Lead Scientist. After several years, I moved to smaller, west-coast organization, PRBO Conservation Science, to see how the lessons I had learned could be applied at a different scale. This brought me back, probably not coincidentally, to birds. And then I decided to retire so I could write this book.

So, that's my story. Careers in science often do not develop following the lines of some carefully scripted plan; at least mine hasn't. The threads of thinking that I bring to these essays and writings—a perspective grounded in natural history, the humanities, and an appreciation for writing; a fondness for concepts combined with a skepticism about their veracity; a joy in the complexities of spatial and temporal variation and scale; and a meshing of science driven by what interests me at the moment with science that is pragmatic and applied—have developed as a sinuously branching braided river, directed by the currents of what came before and wandering this way and that, only then to merge again into something new and different.

ESSAY 2

Found! The survivor in the swamps (2005)*

This essay provides somewhat of a parable, illustrating several of the themes that characterize conservation. The story begins in the Big Woods of eastern Arkansas in early 2004, when a lone kayaker caught a glimpse of an extraordinarily large woodpecker. Its field marks matched those of an ivory-billed woodpecker, thought for a half-century to be extinct. A fuzzy video recorded two months later appeared to support the identification. Subsequent searches yielded tantalizing hints that the species really was there, and in April 2005 the findings were released to the public.[1] Excitement spread. That's when I wrote this essay. For several years, teams scoured the Big Woods and other likely areas seeking definitive proof that the bird still existed. There were sightings in other areas, but no irrefutable proof. Hope has given way to disappointment.

What does this say about conservation? Like many species that attract conservation attention, the ivory-bill had iconic status: it was large, visually striking, hung out in nearly impenetrable swamps, and seemed to be remarkably persistent (it had been thought extinct and then rediscovered twice before). I doubt that a less charismatic species would have garnered the same attention. The shreds of evidence that emerged from time to time gave rise to hope—the conservationist's mainstay—that the species might still exist in the depths of the few remaining southern swamps. Like the vast majority of imperiled species, ivory-bills suffered from human actions, primarily logging and the fragmentation and disappearance of its forest habitat.

The responses to the announcement of the sightings in 2005 are also symptomatic of conservation. Non-governmental conservation organizations (NGOs) such as The Nature Conservancy set about quietly buying land to supplement the habitat already protected in wildlife refuges. Government agencies allocated funds to conservation efforts, and the U.S. Fish and Wildlife Service prepared a recovery plan under the auspices of the Endangered Species Act, just to be ready in case tangible evidence of more than a single individual emerged. These actions were driven by the excitement of the discovery rather than a thoughtful and deliberate prioritization process, and they inevitably entailed tradeoffs; the funds allocated to ivory-bill conservation were taken from programs to support other endangered species.[2]

These themes—the importance of charisma, the effects of habitat loss and fragmentation, the role of NGOs and government actions, the lack of prioritization, and especially the driving force of unrelenting hope—will emerge time and again in the writings and essays that follow.

*Wiens, J. A. 2005. Found! The survivor in the swamps. *Watershed Journal (Brown University)*. 1(3):11. Reproduced with permission of the Watershed Journal.

[1] Fitzpatrick et al. (2005).

[2] Additional information about ivory-billed woodpeckers and their conservation may be found in Tanner (1942) and Hoose (2004).

Ecological Challenges and Conservation Conundrums: Essays and Reflections for a Changing World, First Edition. John A. Wiens.

Once upon a time, virgin forests of mixed hardwoods and softwoods blanketed much of the South. The Civil War and industrialization changed all that. Mature timber was cleared from the uplands and the old forests were increasingly confined to the swampy bottomland. Eventually, most of the bottomland forests also disappeared, drained for agriculture or harvested for their huge trees. Such is the all-too-familiar nature of progress.

These disappearing forests were the province of the ivory-billed woodpecker, the largest and most spectacular woodpecker in North America. Nearly two feet long, with a wingspan of two and a half feet and a flaming red crest, the woodpecker inspired awe from early naturalists and backwoods settlers alike. Individual birds covered immense territories in search of the huge beetle larvae on which they fed, larvae that occurred only in the largest dead and dying trees. With the disappearance of large expanses of forest and the fragmentation of much of what remained into small patches, the woodpecker ran out of habitat and their numbers dwindled. By the 1930s only a few remained, and by the 1940s the species was widely presumed to be extinct. Even so, occasional sightings in the few remaining, remote bottomland swamp forests in scattered places kept the hopes of diehard believers alive.

But now, with the recent sightings and recordings of the Ivory-bill in the Big Woods of eastern Arkansas, we no longer need use the past tense. There *are* Ivory-bills, lingering still in the depths of backwater swamps and forests.

I remember as a boy birder in the early 1950s buying a tattered copy of James Tanner's monograph on Ivory-bills, being enthralled with the magnificence and mystique of the bird, and dreaming that someday someone would find it again. There was something about the allure of this bird and the eerie places it haunted to give one hope.

All of us who care about nature and conservation are heartened by the very idea that this noble bird could persist for so long, beyond our notice, and it gives us renewed confidence that our efforts to protect endangered species and places such as the Big Woods are, indeed, worthwhile.

However, once the euphoria of the moment has faded, what then? Already some scientists are questioning the reliability of the sightings and sound recordings, and while the government rushed to pledge $10 million to protect the woodpecker's habitat, critics note that this cash infusion has come while budgets for the recovery of other endangered species have been slashed. More to the point, even the best evidence confirms the existence of only one, or perhaps two, woodpeckers. What if this is the last lone bird, its cries for companions echoing silently through the bayous? Will that bring our optimism and hope crashing down? Does that mean that our efforts in protecting the Big Woods or other places have been for naught?

Surely not! The message of this bird is in its startling affirmation of the core philosophy of conservation, of the value of building over many years a place where Ivory-bills *can* live long and, if all goes well, prosper. Without places of sufficient variety and size, the loss of species like the Ivory-bill will be an inevitable consequence of the inexorable march of progress. But "progress" is not incompatible with conservation. People have been hunting and fishing in the Big Woods for more than a century, and the woodpeckers don't seem to have minded much.

Perhaps these will turn out to be the last sightings of the last of a species. We hope not. But even if it is, this bird has shown us that not all lost causes are really lost, that by protecting such places we keep alive the hope. Somewhere, perhaps, another budding birder may be dreaming of the day when someone, somewhere, hears the "tin-horn" call of an Ivory-bill and is as startled by the vision of a majestic black bird with expansive white wing patches fleeing into the depths of a swampy forest.

ESSAY 3

In defense of footnotes (2014)*

I believe that this essay speaks for itself. It is my justification for using footnotes throughout this book.

I should note, however, that footnotes, by definition, are placed at the bottom of the page containing the reference. When this and other essays were published in the Bulletin, *however, the footnotes were gathered together at the end. Instead of footnotes, this made them endnotes, which I disparaged. Here I've put them back at the bottom of the page.*[1]

I like footnotes. I think it goes back to my childhood. Both of my parents were professors in the humanities. Most of the books that surrounded me were about art, literature, philosophy, history, and language. They were full of footnotes. Aside from the science fiction that I bought, I don't recall there being a single science book in the house.

When I went to college, I focused on English literature, continuing to read heavily footnoted texts. It was only in my third year of studies that my earlier interest in birds re-emerged. I switched to major in Zoology and started down the road to becoming a scientist, only to discover that those discursive footnotes I had learned to love for their interesting digressions, commentaries, and speculations were nowhere to be seen.

Such footnotes are generally frowned upon in scientific journals and books. They disrupt the text, rather like speed bumps or detours in the flow of scientific prose. Curious readers will wonder what's behind those little numbers and pause to look, while others will ignore them and miss any nuggets they might contain.

*Wiens J.A. 2014. In defense of footnotes. *Bulletin of the British Ecological Society* 45(3): 41–42. Reproduced with permission of the British Ecological Society.

[1] Where they belong; that's why they're called *foot*notes.

Ecological Challenges and Conservation Conundrums: Essays and Reflections for a Changing World, First Edition. John A. Wiens.

If they are lengthy,[2] they clutter the page with small type.[3] They may also express opinions, which, as we all know, are dangerous and by definition unsubstantiated. They have no place in scientific writing.

I would argue instead that by banishing footnotes and leaving no room for personal viewpoints, scientific writing is itself diminished. Although footnotes in literature, history, and the arts were initially used to demonstrate the thoroughness of scholarship,[4,5] they soon took on the additional role of providing commentary, digressions, or opinions, rather like the *sotto voce* asides that figure so prominently in many of Shakespeare's plays.[6] This is what gives footnotes their particular value and makes them more than just another way of citing sources. The really interesting stuff is often in the footnotes. One cannot fully understand the progression of Sir Karl Popper's thinking in *The Logic of Scientific Discovery*,[7] for example, without reading his footnotes. Legal arguments, which rely on precedents to a greater extent than most other areas of scholarship, often contain copious footnote references to previous case law. But here, also, the asides and digressions in footnotes can be extraordinarily important, leading some to suggest that "the most eagerly studied parts of Supreme Court opinions are the footnotes."[8]

Such "wandering footnotes"[9] are largely absent from scientific writing. Here, however, the prose of Stephen Jay Gould provides a refreshing exception. In his massive synthesis of evolutionary theory,[10] Gould used standard scientific notation (author, year) for reference citations and parenthetical clauses (or sentences,

---

[2]As some tend to be. The prize for the longest footnote on record apparently belongs to the 165-page entry that John Hodgson, a 19th century British vicar and antiquarian, wrote in the *History of Northumberland*; if you're really interested, see Creighton (1891), or look it up on Wikipedia. [Bowing to convention, I've assembled full citations at the end of this book rather than including them in the footnotes.] One has to wonder, even with a footnote much less lengthy, whether it is possible to go back to pick up whatever flow of thought was interrupted—as you're probably wondering right now.

[3]Worse yet are endnotes, which force the reader to flip to the end of the paper, chapter, or entire book to find the information, which (to make things worse) may list citations in the order in which they appear in the text rather than alphabetically, creating even more of a challenge. See *Science* or *Nature* for examples.

[4]Grafton (1997).

[5]Scientists do this too, by citing obscure or foreign references or, more often, the works of their close colleagues (or potential reviewers).

[6]See Hirsh (2003).

[7]Popper (1959).

[8]Balkin (1989). Balkin devoted most of a 45-page paper originally published in the *Northwestern University Law Review* to the lasting impact of a single footnote (known in legal circles as "The Footnote") written by Justice Harlan Fiske Stone in 1938 to an opinion in United States v. Carolene Products, which had to do with interstate shipment of a product containing skimmed milk.

[9]Horowitz (2011).

[10]Gould (2002). At 1,433 pages and 5.0 lbs, it is scarcely light reading, despite its readability (Gould wrote in the first person throughout).

or entire paragraphs) to embellish a point. His infrequent footnotes, however, provide delightful and astute asides. In a footnote dealing with controversies over the unit of selection in Darwinian theory[11] (p. 598), he observes, "I don't think that mere personal stupidity underlies my puzzlement—or rather, if so, the mental limitations must be largely collective, because other participants share the same struggle and express the same frustrations." He then goes on to wonder if this reflects an underlying wiring of the human brain to deal in dichotomies. Other footnotes refer to his grandparents (p. 684) or to his graduate (p. 1231) or undergraduate (p. 1290) experiences.

In fact, whether and how footnotes are used is one of the clearest demarcations between writing in the sciences and in the humanities—scientists avoid footnotes, humanists embrace them. At one level, this distinction may simply reflect differences in the cultures of science and the humanities, reinforced by the conventions of publication in scholarly outlets.[12] We become habituated to the mode of referencing in our respective disciplines—scientists, for example, are more likely to be jolted by the intrusions of footnotes into a text than are those in the humanities, who scarcely notice the skipping back and forth. More deeply, however, there may be fundamental differences in the way scientists and humanists think (or are trained to think). Scientists tend to think linearly, from cause to effect, theory to hypothesis to test. Footnotes, especially digressive ones, represent a shift in thought. They seem symptomatic of disorderly thinking, something to be avoided in scientific writing. Perhaps humanists think differently, pursuing a thought and then thinking of other related things, in a process more closely resembling a fisherman's net than a taut line. This is the stuff of footnotes. By shunning footnotes, scientists are deprived of an outlet for their interesting thoughts, opinions, and asides.

I realize now that, by my (over)use of footnotes, I may have had an effect opposite to what I intended, distracting you from my main point. Footnotes, by providing a way to separate opinions, speculations, and digressions from the mainstream of a scientific text, allow those opinions, speculations, and digressions to come forth. These "tangentia" come from thinking about the science, adding a twist or a novel interpretation, or pointing out a relationship that may be the seed of innovation. They should not be lost or suppressed.

There is, of course, some risk in allowing such tangentia to intrude into scientific writing. When we read a paper in a scientific journal or a chapter in a science book, we expect that what we read has an empirical or theoretical foundation and the work has followed appropriate scientific methods. This is what peer review is supposed to assure. But now, if opinions and speculations are allowed to creep into

---

[11] Interestingly, Darwin did not use footnotes in *On the Origin of Species* (published in 1859), although he did in other works, such as *Journal of Researches into the Geology and Natural History of the Various Countries Visited by H.M.S. Beagle, under the Command of Captain Fitzroy, R.N., from 1832 to 1836* (published in 1839) or *The Descent of Man and Selection in Relation to Sex* (published in 1871).

[12] A clear signpost of Snow's *Two Cultures* (1963).

footnotes, can their inclusion in the main text be far behind? Would such footnotes call into question the credibility of the science in the main text? Would the role of science as an objective arbiter of policy disputes be jeopardized?[13] Would the use of footnotes further blur the line between science and advocacy?

These are not easy questions. They are not confined to footnotes. The avenues for communicating science are rapidly diversifying through online journals, blogs, podcasts, TED talks, Twitter, and the like.

But I still like footnotes.

[13]Leaving aside the question of whether science really does play this role.

PART I

Change, the challenge

The title of this book gives equal billing to ecology and conservation. My emphasis, however, is on the conservation side of things. Yet ecology and conservation are closely intertwined, so it is hard to separate the two. Ecology provides much of the scientific foundation for conservation, but conservation is setting a good share of the 21st century agenda for ecology. And both must deal with change. Changes in the environment—climate change and sea-level rise, land-use change, and natural and human-induced shifts in species' distributions chief among them—are altering the setting for conservation and ecology and creating new challenges. However, there are also changes taking place in how ecology and conservation are done. Both also seem to have a built-in inertia that may limit how nimble they can be in responding to rapid environmental changes.

In this part, I develop a context for thinking about change. I'll begin by describing the multiple perceptions of what "conservation" means and offer a summary of thoughts about how the practice of conservation may need to adjust to a changing world. I'll then consider how a particular aspect of science—the development and dominance of paradigms—may contribute to inertia and resistance to change. Following that, I'll elaborate a bit on how "change" has been viewed in ecology and conservation. I'll end the part with a brief discourse about "disturbance" and its consequences.

Ecological Challenges and Conservation Conundrums: Essays and Reflections for a Changing World, First Edition. John A. Wiens.

CHAPTER 1

Conservation and change

In 1862, early in the Civil War and 1 month before signing the Emancipation Proclamation, Abraham Lincoln sent his annual message to the Congress. His concluding words encapsulate the challenge facing conservation and the essential theme of this book:

> *The dogmas of the quiet past, are inadequate to the stormy present. The occasion is piled high with difficulty, and we must rise – with the occasion. As our case is new, so we must think anew, and act anew.*

Conservation is at a crossroads. Change is everywhere. Habitat is being lost to crop production, draining of wetlands, and development at an alarming rate and fragmentation is leaving isolated remnants of habitat scattered like broken glass across a kitchen floor. The combination of climate change and sea-level rise threatens to overwhelm all other sources of environmental change. Protecting the natural world and conserving the richness of biological diversity require that we rethink what has worked in the past and consider fresh approaches.

The context in which conservation is conducted is also changing. How people use lands and waters is undergoing transformation as the tentacles of urbanization reach farther into the rural countryside and population and economic growth increase the demands for goods and services. Economic globalization has created a web of interdependencies, so what happens in one place immediately sends waves across the globe. Changing societal attitudes about the environment, the natural world, and conservation are intertwined with changing political and cultural forces.

How should conservation, and its ecological underpinnings, adjust to this interwoven mélange of change? How can ecological science be applied to advance the conservation of biological diversity—species, ecosystems, habitats, landscapes—in short, "nature"? Answering such questions requires that we consider the ambiguity about what "conservation" really means. For many people, the word conjures up images of pandas, tigers, polar bears, gorillas, and the like—the charismatic animals favored as icons by large conservation organizations. Others equate it with solar energy, clean water, or anything green. "Conservation" means different things to different people.

Faced with ambiguity, one can always turn to the Oxford English Dictionary. The online dictionary offers two definitions: "preservation, protection, or restoration

---

Ecological Challenges and Conservation Conundrums: Essays and Reflections for a Changing World, First Edition. John A. Wiens.

of the natural environment, natural ecosystems, vegetation, and wildlife," and "prevention of excessive or wasteful use of a resource."[1] These definitions mirror two distinctly different perspectives on conservation that have both historical and philosophical roots.

The first definition guides the work of most conservation biologists, conservation organizations, and environmentalists. It is grounded in the natural philosophies of Ralph Waldo Emerson and Henry David Thoreau, the activism of John Muir, and, later, in the land ethic of Aldo Leopold.[2] To these writers, nature has standing in and of itself—what environmental ethicists term *intrinsic* value. Consequently, people have a moral imperative to preserve and protect nature and wilderness.[3] This preservationist philosophy underlies the work of many nongovernmental conservation organizations. The view that nature has intrinsic value also motivates environmental activists, who use lobbying, litigation, and education to defend the environment.

The second Oxford definition reflects a quite different perspective, following the utilitarian philosophies of John Stuart Mill and others. Nature is something to be used by people, and "conservation" means wise use and development of natural resources, not setting them aside in parks or wilderness. Nature's values are *instrumental*, determined by their value to people. Only people have intrinsic value, giving them moral primacy over everything else. This view was perhaps most forcefully advanced in conservation by Gifford Pinchot in the first part of the 20th century. It was put into practice in the 1930s, when the ravages of the drought that led to the Dust Bowl[4] prompted the formation of the Soil Erosion Service (now the Natural Resources Conservation Service). Other federal and state agencies in the United States have a similar responsibility to protect and manage resources for productive uses by people.

Both perspectives—the preservationist/idealist and the utilitarian/pragmatist—are embodied in laws and regulations that govern how or whether species or other natural resources are to be protected, managed, or used. Some laws and regulations in the United States, such as the Endangered Species Act (passed in 1973), give priority to protection (and are viewed by developers as impeding progress or putting fish above people). Others, such as the General Mining Act (1872) or the Taylor Grazing Act (1934), give priority to the extraction or use of resources (and are an anathema to conservationists, who think that regulations encourage degradation of the environment). Policy and regulations are inseparable from politics, of course, and therefore tend to vary depending on whether (politically) the environment is accorded intrinsic or instrumental value (although politicians do not normally think in such philosophical terms).

[1] http://www.oxforddictionaries.com/us/definition/american_english/conservation.
[2] Leopold (1949) and Callicott (1990).
[3] Rolston (2012).
[4] Chronicled by Paul Sears in *Deserts on the March* (1935) and John Steinbeck in *The Grapes of Wrath* (1939).

Science also plays a role in supporting both perspectives. Conservation science uses the tools of science—study design, sampling, statistics, modeling, hypothesis formulation, and the like—to protect and preserve species and places *and* to manage natural resources so that they can be used by people. Although it draws largely from the biological sciences, especially ecology, more broadly it includes elements of economics, sociology, political science, and other applied disciplines.

The different perceptions of "conservation" lead people to be concerned about different aspects of change. To those who think of conservation as wise use, changes in the economics of supply and demand may be most important. To environmentalists, changing policies and societal attitudes that affect their advocacy positions may hold the greatest sway. To those concerned about the persistence of biological diversity, changes in land uses may be paramount. These different perspectives can lead to different attitudes about how conservation should adjust to changing circumstances. However, the dimensions of change intersect and blend together, as do the preservationist/idealist and utilitarian/pragmatist perspectives. To be effective, conservation needs to consider all of them.

A central thesis of this book is that traditional approaches to conservation are inadequate for dealing with the challenges of a changing world—the "stormy present" or the even stormier future. Although conservation and its ecological foundations are rooted in the past, they are not locked into the past; both *have* been changing, of course. The past emphasis of conservation on the protection of pretty places, preservation of charismatic species, and restoration of damaged ecosystems has expanded. Conservation research and applications have been transformed by new technologies; benefits to people—ecosystem services—are receiving increasing attention; and climate change has been part of the discussion for some time.[5] Ecology has also changed, moving from assumptions about equilibrium, community assembly, and orderly succession to consider nonequilibrium concepts, neutral models of community dynamics, and ideas about ecosystem resilience.[6]

Beneath these changes, however, there remains a resistant core. Conservation of threatened species or habitats is still dominated by the "protected areas paradigm," the expectation that protecting critical places will return disproportionate conservation benefits. Organizations such as Conservation International, for example, have emphasized protecting "hotspots" of global biodiversity, a limited number of areas that contain a substantial share of the earth's species.[7] Many restoration projects still have the goal of returning a degraded place to a former "natural"

---

[5]For example, Mace et al. (1998), Lovejoy and Hannah (2005), Brodie et al. (2013).

[6]Rohde (2005), Walker and Salt (2006).

[7]The hotspots concept was initially proposed by Myers (1988). Conservation International has identified 34 global hotspots of biodiversity whose intact remnants cover only 2.3% of the earth's land surface (Mittermeier et al. 2005). A counter-argument, that conservation should also focus on unique but low-biodiversity environments, is offered by Kareiva and Marvier (2003).

state. Notions of feedback control and broad-scale stability still hold sway over a good deal of ecological thinking. I'll have more to say about this in Chapter 3.

If conservation needs to adapt to changes in the environment and societal context, how should it change? I'll reflect on several emerging (or emerged) avenues of change in the essays and chapters that follow, but I can summarize them here.[8] To move forward, conservation needs to:

- Recognize that nature is dynamic; stability is a myth, and extreme events may have unexpected consequences.
- Deal with places in their broader landscape context; protected areas are not isolated from their surroundings.
- Consider the effects of scale; the structure and dynamics of ecological systems differ depending on the scale in time or space, so conservation is also scale-dependent—what works at one scale may not work at another.
- Embrace uncertainty; change begets uncertainty, which only increases as we peer into the future.
- Move beyond a focus on "nature" untrammeled by people to include people and the places they live and work; people as part of nature, and nature as part of people.
- Re-evaluate the targets of conservation; are they genetically distinct populations, species, metapopulations, community structure, ecosystem function, something else, or all of these?
- Include ecosystem services as part of what is being protected and used, but avoid thinking that they substitute for conserving biodiversity for its own sake.
- Recognize that the "nature" of the future may be quite different from that of the present or the past; change will create novel ecosystems, which may require novel conservation and management approaches.
- Incorporate considerations of system resiliency and thresholds; ecological systems have evolved in changing environments and may be resilient to change, but once a threshold is passed the dynamics of the system may be fundamentally altered, and so also must be the conservation approaches.
- Understand that most threatened species, places, or ecosystems are "conservation reliant" and will require long-term, continuing management.
- Recognize that history provides guidance but not targets for conservation and restoration; change means that the targets are moving, and returning to past conditions is generally not feasible (or even desirable).
- Prioritize conservation efforts; resources are insufficient to protect and preserve everything that merits attention, and the queue will only grow longer as the environment continues to change; decisions must be made, including decisions about what *not* to conserve (i.e., triage).

[8] I apologize for using a bulleted list, something I found useful in my dealings with the corporate and political worlds. The staccato nature of bullets, however, does not make for relaxing, readable, or interesting prose. I'll avoid them hereafter.

- Don't neglect what environmental ethics has to tell us; environmental ethicists are giving fresh thought to the place of humans in nature, which should help guide how we think about conservation actions and priorities.
- Consider how ecologists and conservationists should advocate for environmental policies, if at all; be alert for biases that may creep into the science along with advocacy.

These are some components of an emerging view of conservation, what some might call a "new conservation paradigm." But what does it mean to talk of a "new conservation paradigm"? Is it really useful to think of the shifts in focus I've described in terms of paradigms, or does it confuse things with philosophical gobbledygook? Time, then, for a digression on paradigms, expanding on a theme I considered in Essay 4, *The power of paradigms (2014)* (page 23).

CHAPTER 2

A digression on paradigms

The notion of scientific paradigms developed by Thomas Kuhn in *The Structure of Scientific Revolutions*[1] has generated debate among philosophers and historians of science, as well as introspection among scientists, for decades.[2] Kuhn posited that sciences go through phases in their development, progressing in fits and starts in a not entirely rational way. As it matures, a discipline begins to coalesce about a set of theoretical beliefs, methods, assumptions, modes of explanation, standards, values, and knowledge—a worldview, really—that is broadly shared by a community of scientists. Kuhn called such research traditions *paradigms*.

By providing the conceptual framework, tools, and models of exemplary science to address the outstanding questions in a discipline, a paradigm circumscribes a field and sets the stage for what Kuhn called *normal science*. In this phase, scientific progress is seemingly rapid, as practitioners delve into solving the puzzles raised by the paradigm in ever greater detail. Because the research follows well-established methods of puzzle solving to address questions deemed to be important by the community of scientists, the findings largely confirm and enrich the paradigm.[3] Normal science continues.

Once established, a paradigm does more than create puzzles and provide the tools needed to solve them. Adherents to a paradigm form a tight, self-reinforcing community of scientists. Students are trained within the framework of the

[1]Kuhn (1970).
[2]See Lakatos and Musgrave (1970) and Nickles (2003).
[3]Here there is a stark contrast between Kuhn's view of how science operates and that espoused by Karl Popper in *The Logic of Scientific Discovery* (1959). Popper proposed that science advances by testing falsifiable hypotheses generated by a theory; falsification of a hypothesis invalidates (or at least brings into question) the theory itself. Progress is measured by the accumulation of plausible hypotheses that have been falsified and by the few hypotheses that withstand falsification and, consequently, are more likely to be "true." Kuhn suggested that scientists conducting normal science would instead frame hypotheses so that rejection of the hypothesis would confirm the theory, and thus strengthen the paradigm. The difference is between framing hypotheses to reject or cast doubt on a theory versus testing hypotheses to save the theory. This difference was (philosophically at least) the basis for a debate about concepts of community assembly and the use of null models in ecology that pitted Dan Simberloff and Donald Strong against Jared Diamond and Michael Gilpin (see Chapter 23).

Ecological Challenges and Conservation Conundrums: Essays and Reflections for a Changing World, First Edition. John A. Wiens.

paradigm, and, as illustrated in Essay 4, *The power of paradigms (2014)* (page 23), textbooks simplify and reinforce the teachings of the paradigm.[4] Papers and grant proposals that offer solutions to puzzles within the domain of the paradigm are viewed favorably by reviewers (who, operating under the umbrella of the paradigm, are like-minded). Papers or proposals that challenge the paradigm face much tougher sledding, and contrarian scientists may be ostracized and their views suppressed.

Kuhn proposed that such comfortable stability in a science may not last. Inevitably, observations emerge that cannot be reconciled with the paradigm—the puzzles defy simple solutions. Initially, such anomalies can be ignored or discounted as exceptions, or the paradigm may be tweaked to accommodate them. At some point, however, enough troublesome anomalies accumulate to challenge the sanctity of the paradigm, creating what Kuhn termed a *crisis*. Uncertainty grows and festers. The door is opened for better ideas, new approaches, and explanations that encompass the anomalies as well as many of the puzzles that have already been solved. A "Kuhn cycle" has been set in motion (Figure 2.1).[5] If a new paradigm emerges that does a better job of explaining troublesome anomalies, there can be a rapid paradigm shift—a *scientific revolution*.

With the new paradigm in place, an exciting array of new puzzles is at hand, fodder for a new phase of normal science guided by the theoretical beliefs, methods, assumptions, modes of explanation, standards, values, and knowledge of the new paradigm. Scientists now see things in an entirely different way. Rather than viewing science as progressing in an orderly, rational way in which new theories and understanding build progressively on earlier thinking (a steady march toward truth), Kuhn envisioned a process of abrupt, nonlinear changes in the trajectory of thinking and practice. To those who viewed science as the epitome of rationality, this was unsettling.

In 1959, while his thinking in *The Structure of Scientific Revolutions* was still evolving, Kuhn published an essay[6] in which he suggested that, despite this seeming irrationality, the cycling between periods of normal science and the paradigm shifts of scientific revolutions could actually promote scientific progress. He drew attention to an *essential tension* between two ways of thinking among scientists. "Divergent thinking" or innovation—rejecting existing ideas and striking out in new directions—was (and is) often claimed to be the hallmark of science and the source

---

[4]In his essay *The Essential Tension* (1959), Kuhn observed that textbooks serve to consolidate thinking around a consensus view to a much greater extent in the sciences than in other disciplines, especially the humanities.

[5]Not to be confused with the Gus Kuhn cycle. Gus Kuhn was involved in motorcycle racing between the First and Second World Wars. He founded Gus Kuhn Motors in London in 1932 to manufacture and sell motorcycles. Gus Kuhn died in 1966 but the business thrived until 1989, when it was sold. A bit of trivia that turned up in a search for "Kuhn cycle" on the internet.

[6]Kuhn (1959).

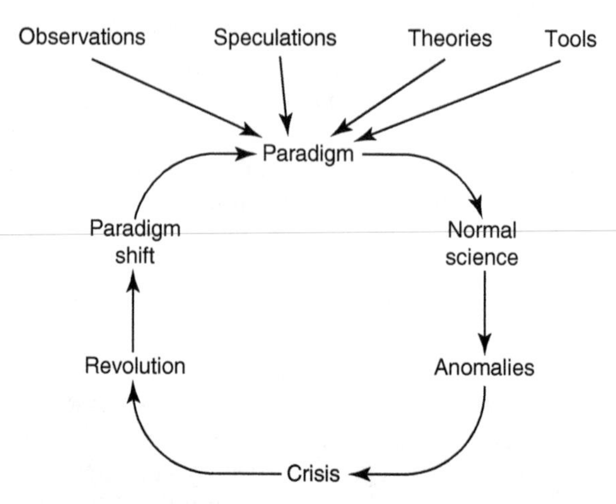

Figure 2.1 The "Kuhn cycle." Initially, observations, speculations, theories, tools and methods, and other features of a developing area of activity within a science coalesce into a cohesive worldview—a paradigm—which guides research to explore the problems and puzzles posed by the paradigm—normal science. As anomalous observations and results that do not mesh comfortably with the paradigm accumulate, a crisis and scientific revolution may follow, producing a shift to a replacement paradigm that does a better job of explaining things, and the cycle continues.

of scientific progress. Kuhn suggested that "convergent thinking"—following the consensus or traditions in vogue at the time—could be every bit as important to advancing science. It is the baseline provided by convergent thinking and normal science that allows anomalies and unsolvable puzzles to become obvious, fueling the innovative solutions that may underlie a paradigm shift. Such "thinking outside the box"[7] is stimulated when something goes wrong, when the results can't be explained by the prevailing theories. Scientific progress, according to Kuhn, involves an interplay between convergent and divergent thinking.

Kuhn supported his ideas about paradigms and scientific revolutions with examples from physics, chemistry, and evolutionary biology—disciplines with long histories and dramatic shifts in which one worldview completely overturned a previous one. Some question whether ecology and conservation have had or have real Kuhnian paradigms,[8] although several examples seem to come close.

The "balance of nature" paradigm and the associated concept of ecological communities as tightly integrated systems dominated ecological thinking from

[7] See Essay 19, *Moving outside the box (2009)*, page 164.
[8] See Wiens (1983a) and Schoener (1983) for a small taste of the debate; Cuddington and Beisner (2005) provide multiple perspectives on paradigms in ecology.

early in the last century into the 1950s and 1960s. This view was expressed in Frederic Clements' "supraorganismic" view of plant communities and the equilibrium-based theories of G. Evelyn Hutchinson and Robert MacArthur, each of which has been labeled a paradigm.[9] Each of these conceptual frameworks prompted a burst of studies exploring its central themes,[10] something Kuhn would have recognized as normal science. I consider one example in Essay 5, *Patterns, paradigms, and preconceptions (2013)* (page 29).

But anomalies accumulated. When it became apparent that plant communities were much less orderly than portrayed by Clements, for example, ecologists resurrected the "individualistic" concept that had been advanced by Henry Gleason in the 1920s, reinforced by new approaches to gradient analysis and the vegetation continuum concept.[11] There was something approximating a paradigm shift. In the 1970s, as the effects of environmental variability became more widely recognized and the role of interspecific competition in structuring communities was challenged, the Hutchinson–MacArthur paradigm lost some of its allure. In the absence of an alternative, however, ecology simply became more pluralistic—more of a paradigm dissolution than a shift. More broadly, however, the balance of nature paradigm gave way to a nonequilibrium, "nature in flux" paradigm.[12] One aspect of this flux, temporal variations in resource levels in relation to competition among species, is the focus of Essay 6, *Fat times, lean times and competition among predators (1993)* (page 33).

Other ecological paradigms are still going strong. Arthur Tansley's concept of the ecosystem, which was first advanced in the 1930s, gave rise to a full-fledged paradigm beginning in the 1950s and 1960s, bolstered by developments in systems ecology and biogeochemistry and the "big science" of the International Biological Program.[13] More recently, the "paradigm" label has been applied rather liberally to almost any ecological concept that has gained some level of popularity, in ways that would probably dismay Kuhn.

What about conservation? Given the multiple perceptions of what conservation is, it is not surprising that the discipline has not developed a single, unifying paradigm. Instead, the number of "mini-paradigms" has proliferated. Some, such as the metapopulation paradigm, the island biogeography paradigm, or the population regulation paradigm, developed in ecology and have been

---

[9]See, for example, Kricher (2009) and Wiens (1989a). I'm not sure what it says about my academic upbringing, but the Clementsian paradigm was the conceptual cornerstone of the textbook in the ecology class that I took as an undergraduate in 1960 (Allee et al. 1949).

[10]See, for example, Deevey (1972) and Cody and Diamond (1975).

[11]Gleason (1926), Whittaker (1967), and McIntosh (1993). I consider the tensions between Clements and Gleason in Chapter 23.

[12]Wu and Loucks (1995).

[13]Tansley's development of the ecosystem concept was at least in part a reaction to extreme versions of the Clementsian community paradigm. Although E.P. Odum was instrumental in promoting ecosystem thinking beginning in the 1950s, his views also incorporated elements of Clements' holism. Golley (1996) reviews this history.

applied to conservation.[14] Others, such as the protected area paradigm, the fragmentation paradigm, or Graeme Caughley's declining-population paradigm and small-population paradigm, encapsulate particular approaches to conservation problems.[15] These are scarcely complete paradigms in the Kuhnian sense; rather, they are useful concepts for organizing and guiding conservation thinking and practice.

Kuhn suggested that paradigms are a sign of maturity in a science, so the lack of an overarching paradigm in conservation may reflect its relative youth. It may also be related to the nature of conservation as an applied science. In *The Essential Tension,* Kuhn commented that because applied scientists aim to address specific, practical problems rather than to solve intellectually interesting puzzles, they are less likely to be engaged in normal science and to be guided by a paradigm.[16] In addition, the emergence of the internet, global networking, and the rise of online journals has promoted a broad exchange of information, quickening the pace of change in all areas of science. Consequently, exceptions and anomalous observations proliferate and spread before a proper paradigm can become firmly entrenched. The power that paradigms once held over a science has dissipated.

Where does this leave us? Thinking in ecology is shifting to consider many of the points listed in Chapter 1, and conservation is following suit. Labeling something a "paradigm" is always a bit touchy, since it's easy to equate "paradigm" with "dogma." Whether conservation has a paradigm in the strict Kuhnian sense, however, is not really the point. The benefit of considering a paradigm is that it draws attention to the historical flow of concepts and how they can coalesce disparate elements into a coherent body of knowledge, approaches, and questions. In view of the varied perceptions about what "conservation" is, the discipline could perhaps stand a bit more cohesion. On the other hand, a paradigm can so dominate a discipline that it stifles the innovative thinking needed to adapt to change. Either way, thinking about paradigms can help us understand the historical and philosophical currents that can simultaneously enhance or impede progress in dealing with a changing world.

[14] Ale and Howe (2010).
[15] Caughley (1994) and Lindenmayer and Fischer (2006).
[16] Kuhn (1959, 238).

ESSAY 4

The power of paradigms (2014)*

For some reason, I've always been attracted to Thomas Kuhn's ideas about paradigms and scientific revolutions. Perhaps it's because his book, The Structure of Scientific Revolutions,[1] *appeared during my impressionable years as a graduate student just beginning to get a feel for science. Or it might be because, a few years later, I became more aware of how a paradigm can grip the hearts and minds of scientists, and I experienced firsthand the obstacles that confront one who questions a paradigm. Whatever the reason, I came to see many of the elements of Kuhn's thesis expressed in ecology as theories, concepts, and approaches came and (sometimes) went. Kuhn's thesis itself became somewhat of a paradigm for how ecology progresses.*

In this essay, I describe how a paradigm can dominate a discipline, particularly by influencing how ideas and information are presented in the textbooks that introduce eager young minds to thinking and practice in a science.

I should also note that I probably would never have written this essay had we not built a new house (see Essay 37, Being green isn't easy (2010) (page 295)), forcing me to unpack and sort through years of accumulated papers (as well as assorted specimens, my childhood rock collection, 35-mm slides from long-forgotten talks, and well-worn field notebooks—the sorts of things ecologists hoard away).

Some time ago, going through a long-forgotten box of papers, I discovered some old correspondence. My exchange of letters[2] with Robert Whittaker in the mid-1970s prompted me to think about paradigms[3] and to re-read Thomas Kuhn's book, *The Structure of Scientific Revolutions* (1962, 1970). In the very first paragraph of his book, Kuhn emphasized the pivotal role that textbooks play in reinforcing and promoting a paradigm.

*Wiens, J.A. 2014. The power of paradigms. *Bulletin of the British Ecological Society* 45(2): 48–49. Reproduced with permission of the British Ecological Society.

[1]Kuhn (1962).

[2]A mode of communication used in an earlier era, but scarcely practiced today.

[3]Kuhn thought of a paradigm as a coherent research tradition in a science, a set of methods, theories, modes of explanation, standards, and knowledge that is widely accepted by the community of scientists. Cuddington and Beisner (2005) present multiple perspectives on paradigms in ecology.

Ecological Challenges and Conservation Conundrums: Essays and Reflections for a Changing World, First Edition. John A. Wiens.

On first encountering the subject matter of a discipline, Kuhn argued, a student learns the concepts, methods, and problem solutions through textbooks and compelling (usually simplified) examples that align with the established paradigm. In this way, the paradigm "exerts a deep hold on the scientific mind" (1970, p. 5). It's rather like a process of imprinting, in which, having been exposed to the worldview of the paradigm at a sensitive stage of development, the budding scientist finds it almost impossible to view the world in a different way. This makes it possible to rationalize ignoring anomalous or contrary observations, strengthening the belief that the paradigm explains more than it actually does.

Whittaker was working on the revision of his textbook, *Communities and Ecosystems* (1975) first published in 1970) and had read my paper on grassland bird communities (Wiens 1973). He wrote to ask for data he could use to illustrate niche displacement among coexisting species—in this case, how species of different sizes partitioned the sizes of prey in their diets. He even drew a sketch of what he had in mind (Figure 1). Such niche partitioning was a central tenet of the competition paradigm that dominated animal community ecology at the time. Delighted that one of the leaders in American ecology would ask, I immediately sent the data. A few weeks later he replied (Figure 2). After carefully plotting my data (Figure 3), he concluded that it really wasn't what he was looking for after all, so he would find a better example. And he did (Figure 4). My data presented an anomaly, an outlier observation that had no place in an introductory textbook. Other, more confirmatory examples were available.

But my story doesn't end there. Some years later, I was given a copy of the lecture notes Whittaker had used in his graduate-level community ecology class in 1976. I recently looked at the notes to see how Whittaker presented niche displacement. The neatly typewritten notes followed the treatment in his textbook, but he had penciled in a marginal note, under "qualifications": "Wiens—summer grassland birds differ less in food-size, probably winter-limited in the tropics." So he had found a way to reconcile my observations with the paradigm; the observations were just made at the wrong time and place, when conditions weren't limiting and the assumptions of the paradigm didn't apply. The competition paradigm remained intact.

I don't intend this story to disparage Whittaker in any way—far from it! He was a brilliant, gifted, and gracefully articulate ecologist who made wonderful contributions to plant ecology and died much too young, at the age of 59 in 1980; but he was writing a textbook. As Kuhn observed, textbooks "expound the body of accepted theory, illustrate many or all of its successful applications, and compare these applications with exemplary observations and experiments" (1970, p. 10). That's what Whittaker did. Anomalous observations, such as mine, would only cloud the picture and confuse the student. They were not appropriate.

So what's the message here? Paradigms can constrain thought and the exploration of novelties. At their worst, they suppress contrary observations that threaten to subvert the paradigm. At the same time, a well-established paradigm

COLLEGE OF ARTS & SCIENCES

NEW YORK STATE
COLLEGE OF AGRICULTURE AND LIFE SCIENCES
A Statutory College of the State University

CORNELL UNIVERSITY
DIVISION OF BIOLOGICAL SCIENCES
ITHACA, N. Y. 14850

SECTION OF ECOLOGY & SYSTEMATICS

February 5, 1974

Dr. John A. Wiens
Oregon State

Dear John:

I have been reading your inpressive monograph on grassland bird communities with interest. I wondered from it if you might have data of a kind I am looking for. I would like, in the revision of "Communities and Ecosystems," to use a figure showing a niche size-sequence of animals together with the means and frequency distributions of their food sizes. Would you have this and be willing to share it? (I am looking at Table 13). It could involve, say, horned lark, grasshopper sparrow, and meadowlark at Cottonwood, or dombining stations.

Yours sincerely

R. H. Whittaker

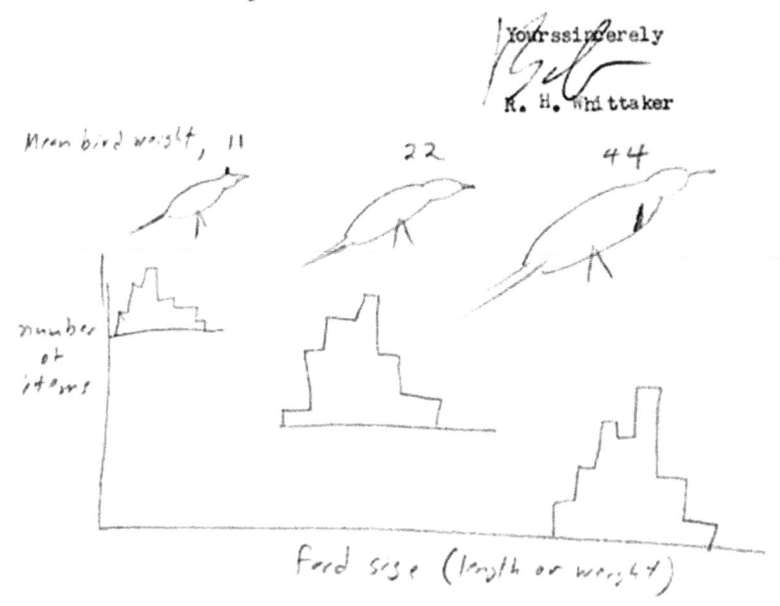

Figure 1 The initial letter I received from Robert Whittaker asking for material for his textbook and kindly suggesting what a figure might look like.

COLLEGE OF ARTS & SCIENCES

NEW YORK STATE
COLLEGE OF AGRICULTURE AND LIFE SCIENCES
A Statutory College of the State University

CORNELL UNIVERSITY
Division of Biological Sciences
Ithaca, N.Y. 14850

Section of Ecology & Systematics

March 25, 1974

Dear John:

Thanks very much for your data and the manuscript copy.

I plotted up your data for insects. There is reason to use either length cubed, as proportional to weight, or a log plot (or even a log plot of length cubed), of course, and I used the familiar octave log plot as simplest. The results are attached, and the curves overlap broadly. I am afraid they support not my purpose but your article.

I'll try other data, but appreciate having had these.

Yours sincerely

R. H. Whittaker

(Let me know if you would like those data sheets back)

Figure 2 Whittaker's reply after looking over the material I sent him.

can unleash a flurry of research that converges about the pregnant questions and problems highlighted by a paradigm. The excitement is palpable.

That is how it was in the 1960s and 1970s, when Whittaker was writing his textbook and the paradigm of equilibrium communities structured by interspecific competition held sway. However, that paradigm eventually lost its grip, eroded by a growing body of evidence that did not fit comfortably into its framework. Equilibrium views expanded to admit nonequilibrium conditions. Single-factor explanations morphed into multiple-factor explanations. Rather than experiencing a Kuhnian shift from one paradigm to another, community ecology became more pluralistic. Perhaps the ascendant ecological paradigm was "it all depends."

Paradigms are not restricted to ecology, or to the physical and biological sciences. Economics is famous for its paradigms—Keynesian General Theory versus supply-side economics, for example. There are programming paradigms

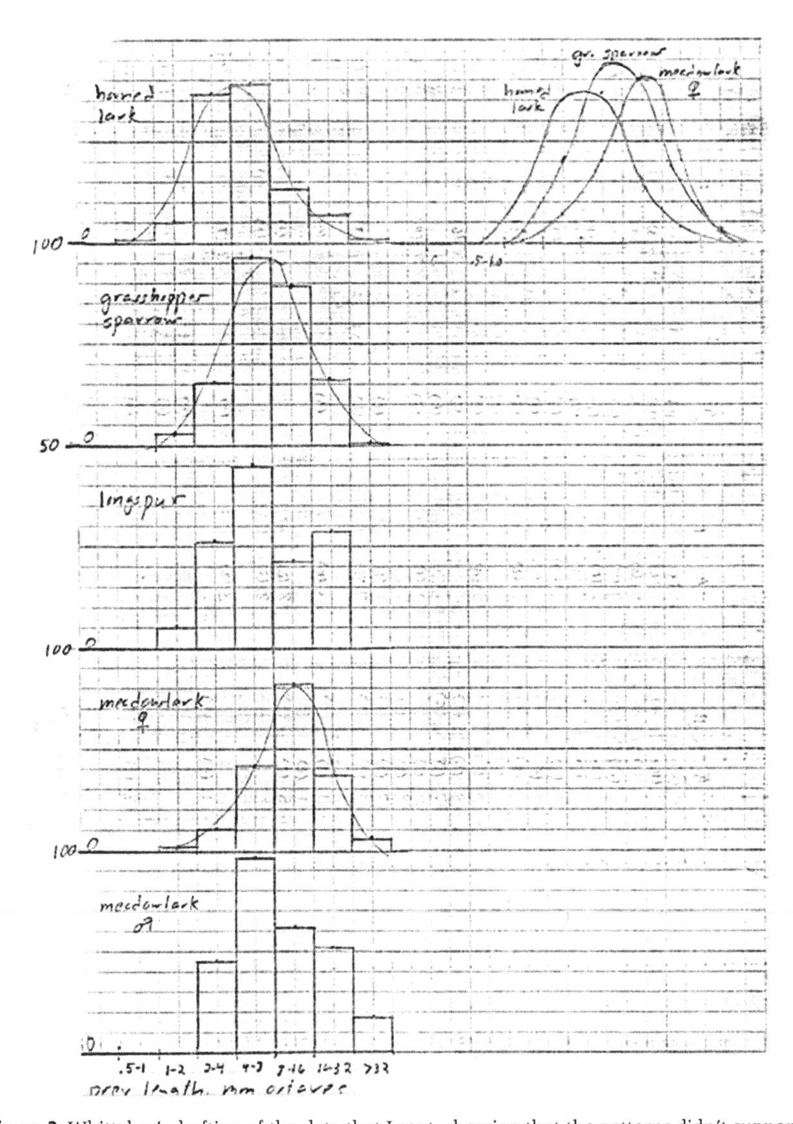

Figure 3 Whittaker's drafting of the data that I sent, showing that the patterns didn't support what he was looking for.

in computer science. There are paradigm salad dressings, paradigm nightclubs, paradigm plastic bottles, and paradigm wines.[4] Paradigms are pervasive. Why, just this morning I heard a report on the shift away from the paradigm that has dominated the health sciences for decades, that saturated fats in the diet are invariably bad, toward a view favoring a more diversified diet. In this case, the

[4]I am indebted to Richard Hobbs for pointing this out.

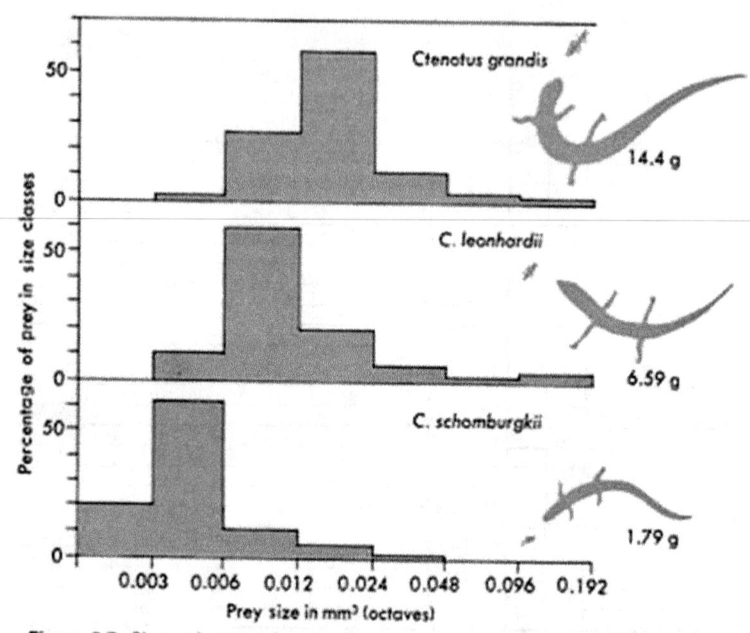

Figure 3.7. Sizes of prey taken by three species of a lizard uild. Per cents of prey are plotted for prey size classes on a logarithmic scale (by octaves or doubling units) for three species of the genus *Ctenotus* feeding mainly on termites in an Australian desert. Mean weights of the lizard species are given on the right. [Data of Pianka, *Ecology* **50**:1012 (1969).]

Figure 4 The figure that appeared in Whittaker's revision of *Communities and Ecosystems*, showing differences in prey sizes among coexisting species of *Ctenotus* lizards. Source: Whittaker (1975).

"fats are bad" paradigm had such a grip that studies finding contrary evidence had difficulty being published.

Why should this matter? My story, I think, should lead us to think carefully about what is written and widely accepted, not just in textbooks but also in mainstream scientific papers. Paradigms, even old ones, may still "exert a deep hold on the scientific mind." Some contemporary ecology textbooks still include examples of niche displacement not unlike the *Ctenotus* lizards in Whittaker's figure. Although paradigms should not be abandoned willy-nilly on the basis of a few anomalies, neither should they shroud our thinking from contrary evidence and alternative explanations. There is no substitute in science for the open mind (Oppenheimer 1955).

ESSAY 5

Patterns, paradigms, and preconceptions (2013)*

Science is all about detecting and explaining patterns. The patterns of nature, however, are often cloaked by an array of confounding factors—unmeasured or irrelevant variables, unanticipated or undetected changes in the study system, seemingly random environmental variation, and so on. The goal of science is to pull aside the cloak to see what patterns lie beneath. Paradigms help us do this by directing our attention to the interesting patterns that may be mired in a confusing mass of observations. And the interesting patterns, the ones that pop out at us, are more likely than not to be the ones that accord with our expectations. The challenge is to recognize the patterns we don't expect, those anomalous surprises that create a feeling of unease or distrust about a paradigm.

The story I recount here illustrated to me that there's more to science than the facts. Jared Diamond and I looked at the same mass of data and saw different things. Each of us was both helped and constrained by our preconceptions about what to look for—what was interesting.

And we all have preconceptions, whether we recognize them or not. So it's little wonder that scientists sometimes disagree. The art, as well as the difficulty, of science is in looking beyond the preconceptions to see the unexpected as well as the expected patterns.

"To do science is to search for repeated patterns, not simply to accumulate facts" (Robert MacArthur, in *Geographical Ecology*, 1972, p. 1)

"All we want are the facts, ma'am" (Sgt. Joe Friday in the television series *Dragnet*, in the early 1950s)

Science, as MacArthur observed, is about detecting patterns and relationships and deriving causal explanations. Like many ecologists, MacArthur interpreted patterns through their alignment with theory—a deductive approach. Sgt. Friday's approach (like Sherlock Holmes' before him) was more inductive—accumulate facts and then, through inspired guesswork, connect the dots to see the pattern and solve the crime. MacArthur's approach was analytical, Friday's more intuitive.

*Wiens, J.A. 2013. Patterns, paradigms, and preconceptions. *Bulletin of the British Ecological Society* 44(3): 39–40. Reproduced with permission of the British Ecological Society.

Ecological Challenges and Conservation Conundrums: Essays and Reflections for a Changing World, First Edition. John A. Wiens.

While we may think of MacArthur's approach as more scientific than Friday's (who was, after all, not a scientist but a fictional detective), both approaches may be important as we are confronted by a growing deluge of data ("facts") about all sorts of things. Separating the signal (an interesting pattern) from the noise of seemingly irrelevant data has become a major challenge, spawning increasingly complex and sophisticated statistics, data mining and knowledge discovery procedures, and the pattern-recognition software used in DNA screening or the detection of E-mail spam. We ask computers to do the work of detecting patterns and telling us whether they are real (by which we mean statistically significant) or not. Against this backdrop, is there still a role for the practiced eye of a keen observer to see a pattern in a mass of data?

Back when I was a brash young ecologist in the 1970s (I'm no longer young, and arguably not so brash), ecology was roiled by intense debates about whether interspecific competition determined community structure. Part of the debate revolved around how one marshaled evidence. In one camp, proponents such as Jared Diamond, Martin Cody, or Tom Schoener looked for examples of niche-displacement patterns between co-occurring species that agreed with the predictions of competition theory. In the other camp, opponents such as Dan Simberloff and Don Strong (and I was among them) challenged the ubiquity of theory-based explanations by assessing niche overlap among many co-occurring species, noting the many patterns that did not match predictions and asking whether the occurrence of confirmatory patterns was any different from what one would expect by chance.

The distinction between the approaches was brought home to me one spring when Diamond visited my lab group. Jared is a gentle and gracious man, and the acrimony that had fueled some of the past debates was left behind. At one point, I spread out on the lab bench a large table showing multiple bird species and niche variables. I said something like, "See, there's no consistency. This is why I have difficulty accepting competition as the driving force." Jared looked over the matrix of numbers. Without hesitation he pointed to several cells. "You're looking at it the wrong way. Here's a perfect example of niche differentiation, and here's another. You're letting all those other numbers confuse you." I had missed the patterns that he saw, and he was unbothered by the nonconforming observations that so interested me.

I don't think that either of us convinced the other that day. There's a more important message here, however, than who was right or wrong. Both of us looked at the data—the facts—through different lenses, colored by our preconceptions of how natural communities are put together. Jared's perspective was shaped by his experiences in New Guinea and by competition and niche theory, which dominated community ecology at that time. Mine reflected my experiences in highly variable grasslands and shrub deserts and a growing view that environmental variation can erode the patterns expected from equilibrium-based theory. The data set contained something for both of us.

All of this resembles what Thomas Kuhn described in his ideas about paradigms and scientific revolutions (Kuhn, 1970). Paradigms—a widely accepted body of theories, methods, and examples that embodies a view of how the world works—influence the questions researchers ask and what they look for (and are prepared to accept) in their results. Nonconforming patterns are considered anomalies that can be ignored or explained away without jeopardizing the power of the paradigm. Under the sway of a paradigm, the search for patterns is not an aimless wandering or mindless data dredging, but a focused effort in which the expected patterns seem to stand out in bold relief against a background of irrelevant detail. Jared was able to see patterns in my data because the prevailing competition paradigm told him what to look for.

But surely statistics provides a check against the urge to see only confirmatory patterns? Well, not necessarily. Of course, scientific hypotheses are supposed to be appropriately null and statistical tests designed to place the burden on rejection of the null hypothesis so that one has confidence in the (statistical) veracity of an alternative hypothesis (e.g., that niche differences among similar species are greater than one would expect by chance alone). But it is not at all difficult to structure hypotheses and tests to favor the expected patterns (and their interpretations). After all, Mark Twain popularized the phrase that "There are lies, damned lies, and statistics"[1] and Darrell Huff's 1954 book, *How to Lie with Statistics*, has sold well over a million copies. Automated pattern-detection algorithms may seem less vulnerable to conscious or unconscious bias, but someone needs to write the code to tell the computer what to look for, leaving ample room for paradigm-derived preconceptions to influence the search.

More to the point, a rigid adherence to statistical analyses or computerized searches for patterns may blind us to the outlier patterns that occur at the fringes of data sets—the patterns that Jared saw in my data. These may indeed be examples that confirm the paradigm, or they may be chance occurrences that are inevitable in a large sample or collection of data. They may also be those unexpected but real patterns that lead us to doubt the ubiquity of a paradigm, the anomalous observations that are the stuff of Kuhn's scientific revolutions. The trick is in knowing when such patterns deserve attention and when they don't; preconceptions can make them too easily ignored.

All of this becomes more relevant in the context of climate change. Paradigms hold true when nature's rules—cause–effect relationships—remain relatively steady. Science thrives by filling in the details. The anomalous observations arise when nature bends or changes the rules. Climate change will certainly alter, if not actually change, the rules and how they are expressed in ecological systems. No-analog species assemblages and novel ecosystems will change interaction webs and ecological processes in unexpected ways. Extreme events will push systems beyond thresholds into unknown territory. Things will no longer be as they ought to be, at least according to the paradigms that have guided our thinking in the past.

---

[1] "Chapters from My Autobiography". http://www.gutenberg.org/files/19987/19987.txt.

We shouldn't abandon the statistics and modeling that have contributed so much to the rigor of ecology. However, we must also be more attentive to the new, novel patterns that emerge, for these hold the keys to adapting our concepts, theories, and methodologies—our paradigms—to the new realities of a nature out of balance.

ESSAY 6

Fat times, lean times, and competition among predators (1993)*

> *When I first started doing ecology in the 1960s, the prevailing view was that populations were usually limited by resources (particularly food), which would lead to competition among species that used resources in similar ways. This, in turn, would lead to divergence in resource use—niche partitioning—that would permit the species to coexist, resulting in ecological communities at or close to an equilibrium determined by resource availability and niche overlap ("limiting similarity").*
>
> *By the early 1990s, when this essay was written, this paradigm no longer held a grip on the thinking of ecologists. It had been challenged by a mounting body of observations that failed to match expectations and by the realization that niche overlap might change as resource levels varied between abundant ("fat times") and scarce ("lean times"). The underlying assumption of equilibrium was unlikely to hold in a dynamic, changing environment.*
>
> *In this essay I reviewed studies of an assemblage of predators in Chile that provided a particularly clear documentation of the intricacies and inconsistencies of resource use and niche overlap in a varying environment. This led to the unsurprising recognition that "fat times" and "lean times" must be considered relative to the level at which resources are actually limiting. A given change from "fat" to "lean" resource levels can lead to quite different results depending on how that change relates to a resource-limitation threshold. In other words, the one true law of ecology prevails: it all depends.*[1]

Classical competition theory predicts that coexisting species that share limiting resources should compete. For coexistence to continue, the species should diverge in resource use, reducing niche overlap[1,2]. This view of nature, which was prevalent in ecology during the 1960s and early 1970s, has been questioned on several counts[3,4]. The effects of environmental variation on resource levels and competition have received particular attention. If environmental conditions

*Wiens, J.A. 1993. Fat times, lean times and competition among predators. *Trends in Ecology and Evolution* 8: 348–349. Reproduced with permission of Elsevier.

[1]To be consistent with the original publication of this essay, I've cited references in their order of appearance in the text.

Ecological Challenges and Conservation Conundrums: Essays and Reflections for a Changing World, First Edition. John A. Wiens.

vary (as they do), then resources such as food may be abundant at some times ("fat" times) and scarce at others ("lean" times). Accordingly, we might expect niche overlap among species to be greater during the fat times than the lean times, when competition is presumably more severe[2,3,5].

In the early 1980s, Schoener[3] tested this expectation by surveying some 30 studies in which changes in niche overlap among coexisting species had been documented between seasons or years. In most cases, overlap was indeed less during the relatively lean period, usually the winter.

Does this mean that the species are competing during the lean times but not during the fat times? Not necessarily. Competition is related to resource limitation, and without direct information about resource levels, terms such as "fat" or "lean" tell one little about whether or not a resource-limitation threshold has been passed (Figure 1a). Also, niche overlap may not change monotonically with a reduction in resource supplies. Coexisting species may respond opportunistically to certain resources when they are superabundant, specialize as resources become more limiting, but then converge again as resources become extremely scarce, leading to a high–low–high sequence of niche overlaps (Figure 1b).

A recent report by Jaksić et al.[6] at least partially addresses these problems, and shows that the "fat–lean" scenario may be an oversimplification. Jaksić et al. studied the food habits and guild structure of predatory vertebrates over a 4-year period in semi-arid scrub desert north of Santiago, Chile. These predators (four owl species, four falconiform hawk species and two foxes) are well-suited to an investigation of niche dynamics. They are large and conspicuous and habitually use sites such as roosts or dens where their pellets or feces may be collected, and remains of prey in these castings can be identified with some precision[7]. This is important, for the determination of the trophic guild structure of a community is sensitive to the level to which prey items can be resolved taxonomically[8]. In this study, prey remains in the castings were identified to species for vertebrate prey and to orders for invertebrate prey—not ideal resolution, but better than that in most such studies.

In this system, bird and arthropod prey populations showed the expected pattern of an increase in abundance associated with the flush of vegetation growth at the onset of the breeding period and a decline at the end of the season—breeding and nonbreeding seasons were relatively "fat" and "lean" for consumers of these prey. Small mammal populations, on the other hand, showed no such changes. Instead, they irrupted early in the study, apparently in response to unusually high rainfall and vegetation production[9]. Numbers declined dramatically throughout the remainder of the study, and estimated densities at the lowest level were only 7% of those recorded at the abundance peak. For predators on small mammals, then, "fat" and "lean" were expressed on a scale of years, not seasons.

Under these conditions, one might anticipate that omnivorous predators should shift diets seasonally, reducing overlap during the nonbreeding season, whereas predators on mammals might emigrate or shift to other prey as the mammal populations crashed, increasing overlap with the omnivores. However, despite the fact

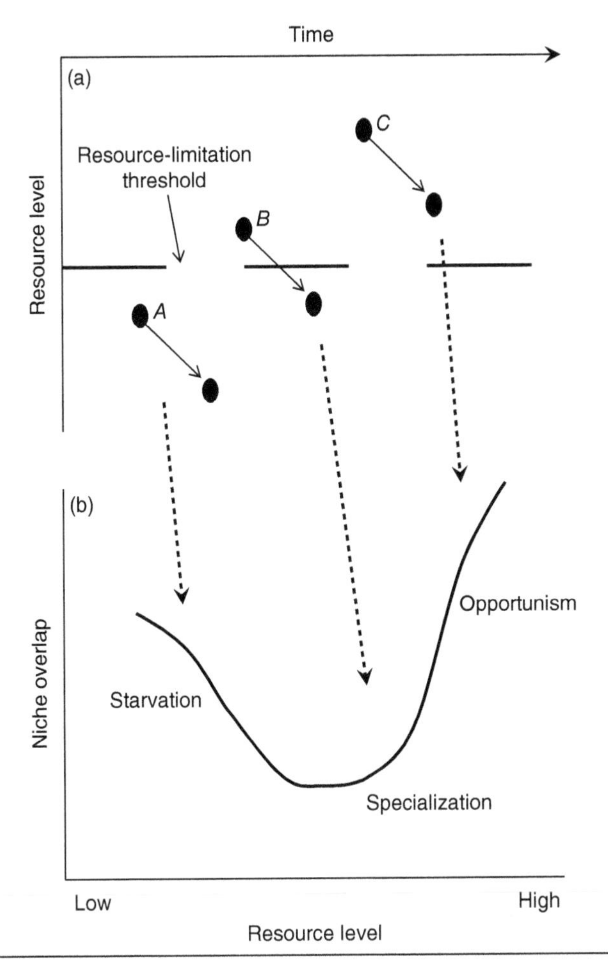

Figure 1 (a) The consequences of changes in resource levels (relatively "fat" to "lean" times) depend on the relation of resource levels to a resource-limitation threshold. In *A*, resources are always limiting, in *C* they are always superabundant; only in *B* do conditions change from abundant to limiting, as envisioned in the "fat–lean" scenario. These changes in resource levels relate to the pattern of niche overlap expected among potentially competing species. (b) When resources are superabundant, the species may opportunistically use the same abundant resources, and overlap is high. As resources become limiting, the species may specialize following species-specific resource preferences or as a result of interspecific competition. Resource overlap decreases. When resources are extremely limited, the species may all be forced to use the few remaining resources, and overlap increases. Source: Wiens, J.A. 1993. Fat times, lean times and competition among predators. *Trends in Ecology and Evolution* 8: 348–349. Reproduced with permission from Elsevier.

that nearly every predator species displayed a different response to the seasonal and yearly changes in food abundance, the basic guild structure of the assemblage was stable. One mammal-eating guild, containing the owls *Tyto* and *Bubo*, was always present, and these species exhibited high dietary overlap. During the few survey periods when *Tyto* was not recorded, *Bubo* remained isolated in this guild. Neither species showed any dietary shifts in response to the decline in small mammal abundance.

A second group (two other owls and the two fox species) formed a tight omnivorous feeding guild based on arthropods and a broad range of vertebrate prey. Only one of these predators showed any response to the seasonal pattern of "leanness" of these prey. The four falconiform predators shifted seasonally from occupying unique guilds to affiliating with either the omnivorous or the mammalivorous guild, with little apparent relation to levels of food abundance. Generally, guild clustering was tighter during the nonbreeding season, indicating, if anything, greater rather than reduced dietary overlap during the relatively "lean" period.

Why was the classic "fat–lean" pattern of niche overlap not evident in this assemblage? Jaksić et al. offer several possibilities. One of these may have to do with the nature of the resources used by these predators. Rather than differentiating prey on a continuous scale such as size, as planktivorous fish or granivorous birds do, these predators feed on distinct types of prey that may require distinctive foraging techniques. This discreteness of prey may impose constraints on prey switching by predators, enhancing guild stability and niche overlap. A second possibility is biogeographic. The region in Chile where Jaksić et al. conducted their studies is enclosed by ocean to the west and south, by the Andes to the east, and by the Atacama desert to the north[10]. Perhaps as a result, true migrants are absent from the regional predator assemblage. Such species might respond opportunistically to prey abundance during fat times and thereby increase guild membership and niche overlap, as they do on other continents[7].

The third possibility returns us to the scenarios pictured in Figure 1. Although Jaksić et al. estimated the density of small-mammal prey populations, they lacked the information to relate these estimates to resource-limitation thresholds for the consumer populations. Despite the enormous decrease in the abundance of mammalian prey during the study, these resources may never have become truly limiting, and the competitive conditions that would produce the expected changes in niche overlap may not have arisen. Conditions may simply have gone from "fat" to "less fat".

This last possibility indicates something of the quandary one faces in testing the fat–lean hypothesis. The mechanism underlying the predicted niche-overlap changes is competition, which requires resource limitation. This is difficult to document in field settings. One can attempt to measure or estimate resource abundance, as Jaksić et al. did, but limitation is related to the relation between resource *availability* and consumer demands, not simply resource abundance.

Changes in niche overlap may also be related to variations in consumer population densities[11]. If densities are low during relatively lean times, intraspecific competition may be lessened and niche breadths reduced[12], diminishing niche overlap among species whether or not interspecific competition is important.

Testing the hypothesis, of course, also requires that the critical resources have been identified correctly and that the community has been properly delineated. Perhaps because it is obvious that energy must at some point be limiting to populations, food has become the ecologist's favorite niche dimension. But there are other niche dimensions ("space" and "habitat", for example), and a failure to record the expected niche-overlap dynamics during presumed fat and lean times may simply mean that the niche dimension in which competition occurs was not considered. If the designation of community membership is incomplete or overly inclusive, the patterns may also be disrupted. The Chilean study area of Jaksić et al., for example, contained snake and lizard populations that use the same prey as the other vertebrate predators in the system. Because these reptiles are endothermic and have low prey consumption rates per individual and limited seasonal activity, Jaksić et al. felt justified in excluding them from the community. Their analyses probably did include all of the major predators of small mammals in the system, although other consumers of arthropods (such as spiders) were not considered.

Despite these shortcomings (which plague nearly all field studies of competition), the work of Jaksić et al. has some important implications. Clearly, the generality of the notion that niche overlap is reduced during "lean" times must be questioned[2,10], especially in systems where the dynamics of resources are expressed on different time scales. In such situations, the "snapshot" documentations of feeding relationships that are the foundation of a good deal of food-web theory[13,14] may produce misleading patterns. Detecting multiple scales in resource dynamics requires a long-term perspective. This study adds one more reason to the growing list of justifications for conducting long-term studies in ecology[15].

References

1 Keddy, P.A. (1989) *Competition*, Chapman and Hall.
2 Wiens, J.A. (1989) *The Ecology of Bird Communities (Vol. 2): Processes and Variations*. Cambridge.
3 Schoener, T.W. (1982) *Amer. Sci.* 70, 586–595.
4 Chesson, P.L. and Case, T.J. (1986) in *Community Ecology* Diamond, J. and Case, T.J., eds, pp. 229–239, Harper & Row.
5 Wiens, J.A. (1977) *Amer. Sci.* 65, 590–597.
6 Jaksić, F.M., Feinsinger, P. and Jiménez, J.E. (1993) *Oikos* 67, 87–96.
7 Marti, C.D., Korpimäki, E. and Jaksić, F.M. (1993) in *Current Ornithology, Vol 10* Power, D.M., ed., pp. 47–137, Plenum.
8 Greene, H.W., and Jaksić, F.M. (1983) *Oikos* 40, 151–154.
9 Jiménez, J.E., Feinsinger, P. and Jaksić, F.M. (1992) *J. Mammal.* 73, 356–364.
10 Jaksić, F.M., Jiménez, J.E. and Feinsinger P. (1990) *Proc. XX Int. Ornithol. Comgr.*, 1480–1488.

11 Llewellyn, J.B. and Jenkins, S.H. (1987) *Amer. Nat.* 129, 365–381.
12 Svärdson, G. (1949) *Oikos* 1, 157–174.
13 Pimm, S.L. (1982) *Food Webs*, Chapman and Hall.
14 Cohen, J.E., Briand, F. and Newman, C.M. (1990) *Community Food Webs: Data and Theory*, Springer-Verlag.
15 Likens, G.E. (ed.) (1989) *Long-term Studies in Ecology: Approaches and Alternatives*, Springer-Verlag.

CHAPTER 3

Equilibrium, stability, and change

Thinking about environmental change is nothing new. For millennia, people have had difficulty dealing with change and uncertainty and have sought to find order and stability in their world. The early humans who survived were those who could detect order and use it to their advantage. This is the psychological (and perhaps evolutionary) foundation of a balance-of-nature worldview. The intellectual framework for thinking in terms of order and balance was developed by Greek philosophers, and this perspective carried forward as an almost uninterrupted undercurrent into natural philosophy and the rise of modern science. So the belief in balance, stability, equilibrium—whatever label one applies—is old, and its hold on our psyche has been remarkably strong. It had all the trappings of a paradigm before there were such things.

This belief in order and stability[1] was deeply embedded in ecology as it grew from its roots in natural history to become a recognized science in the first decades of the 20th century.[2] The mathematical expressions of population dynamics developed by Alfred Lotka, Vito Volterra, Raymond Pearl, and others; Frederic Clements' concept of ecological succession and climax communities; the perspectives on ecological niches advanced by Joseph Grinnell and Charles Elton; Arthur Tansley's notion of the ecosystem; the link between energy flow and food webs developed by Raymond Lindeman and G. Evelyn Hutchinson—all were based on the premise that natural systems tend toward equilibrium and stability, and all became part of the foundations of modern ecology. The approaches to community ecology of Robert MacArthur, Jared Diamond, and others; to ecosystem ecology of Gene Odum, Gene Likens, and Herbert Bormann; or of systems ecology of Kenneth Watt, H.T. Odum, Bernard Patten, and others deepened and strengthened the conceptual and theoretical basis of equilibrium thinking and feedback controls in ecology.

The expectation of balance and stability had a profound influence on how ecology and conservation were (and to some degree still are) conducted. If systems are in equilibrium, they will promptly return to a stable configuration following any perturbations. Any effects of past history will quickly be erased. Consequently, 2 or 3 years of study (the normal duration of a graduate research project

[1]The belief, as Bronowski (1977: 12) put it, that lurking "under the colored chaos there rules a more profound unity."
[2]Kingsland (1985) provides a nice perspective on the development of this thinking.

Ecological Challenges and Conservation Conundrums: Essays and Reflections for a Changing World, First Edition. John A. Wiens.

or a research grant) should be sufficient to detect important patterns and relationships and encompass any relevant dynamics. Periodic seasonal variations can be great, of course, and disturbances may temporarily disrupt populations and communities. By restricting studies to only part of the year[3] and avoiding obviously disturbed areas, however, such complications can be avoided. Besides, systems should return to equilibrium quickly, so annual variation about a stable long-term average should be small.

At least that was the thinking. Short-term studies became the norm. Spatial variation could also confound the expected patterns. Any effects of spatial variation, however, could be minimized by using small, homogeneous study areas.[4] Study results were extrapolated to other times and places with little hesitation.

The presumption of stability also encouraged short-term conservation and management actions. Once a place was protected or restored, one could expect it to persist with little additional attention, and the tools and methods that worked at one time and place would be likely to work at others. Ecology and conservation flourished, fostered by the myopia of the present—the inclination to think that the past and future were and will be like the present.[5]

But the belief in equilibrium and stability was illusory, fueled by a desire to see things as we wish them to be rather than as they are. To be sure, there were occasional challenges to the prevailing equilibrium views in ecology. Frederic Clements' notions of orderly succession leading to a well-integrated stable community in which each species had its place were challenged by Henry Gleason in the 1920s.[6] In the 1950s, the logistic growth model spawned concepts of density-dependent population regulation that were staunchly defended by David Lack against the contrary views of H.G. Andrewartha and Charles Birch.[7] The proposal that food-web or ecosystem complexity enhances system stability was championed by Gene Odum and Charles Elton in the 1950s.[8] In most instances,

---

[3]An inordinate proportion of the studies that contributed to the development of community theory during the 1950s–1970s were conducted on birds during the breeding season.

[4]Kareiva and Andersen (1988) demonstrated how ecological field studies were biased toward small spatial scales.

[5]Captured, for example, in the doxology of many Christian traditions as "... as it was in the beginning, is now, and ever shall be ..."

[6]Drury (1998) provides a thoughtful discussion of the development of succession thinking. Beyond this, his book constitutes an eloquent argument in support of the view that ecological systems are normally in a state of "comfortable disorder."

[7]Lack (1954) and Andrewartha and Birch (1954); see also Chapter 23. Later on, Wynne-Edwards (1962) added to this debate by expounding at length (652 pages) his proposition that animal populations are regulated by social behavior, in which "epideictic" displays provide the feedback control mechanisms. His thesis entailed evolution by group selection, a proposition that was contentious at the time but which he subsequently (1986) elaborated, to even greater contention. One of my earliest papers, and the first one of any intellectual substance (Wiens 1966), addressed the dependence of Wynne-Edwards' behavioral ideas on his arguments favoring group selection.

[8]Odum (1953) and Elton (1958).

a concept based on feedbacks and stability was challenged by an alternative, laissez-faire argument that communities or populations were scarcely regulated, if at all, and were subject to constant change. Yet the stability views prevailed, at least for a while.[9] The environments that people studied and in which they lived also affected their perspectives on environmental stability, however; Richard Hobbs and I explored this possibility in Essay 7, *From our southern correspondent(s) (2011)* (page 44).

That was then. By the 1970s, evidence was beginning to accumulate that ecological systems might not be as well-balanced, stable, and tightly organized as people had thought. The ubiquity of competition as a force leading to stable and optimally structured communities was questioned on the basis of both field observations and models; I described one example in Essay 6, *Fat times, lean times, and competition among predators (1993)* (page 33). Ecologists no longer used hushed tones when talking of nonequilibrium as an alternative to the equilibrium thinking that had prevailed for so long. Definitions of "stability" proliferated, drawing from advances in chaos theory, systems theory, and the dynamics of self-organizing systems. Ecologists began to view variation as a topic worthy of investigation in its own right rather than something to be ignored or swept aside by statistics. Skepticism flourished. Change, and talk about change, was in the air.

"Change", however, comes in many forms. Put simply, "change" is to become something that is not the same. However, that's far too simple to be useful. Some forms of change may have much more profound effects on ecology and conservation than others, so it's important to be clear about what we're talking about.

Change encompasses a spectrum of dynamics. At one end of the spectrum, conceptually, is the theoretical ideal of a steady-state equilibrium (the traditional meaning of a balance of nature), in which variation about a constant condition is miniscule. At the other end lie thresholds, "tipping points" where a system shifts to an alternative state with qualitatively different attributes and dynamics (Figure 3.1).[10] Between these extremes variation may be periodic, as in seasonal change, or erratic but constrained within an envelope of variation—a dynamic equilibrium. When variation is even greater, there may be occasional or frequent extremes, rendering the arithmetic mean ecologically meaningless. This is especially so in systems that undergo extreme "boom and bust" dynamics, such as the Australian Outback that Richard Hobbs and I contemplated in Essay 8, *Boom and bust: lessons from the Australian outback (2014)* (page 51). A trend becomes evident when average conditions change directionally and systematically, and a threshold occurs when a trend becomes an abrupt and massive change.[11]

[9]I consider these debates and their philosophical underpinnings in Chapters 23 and 24.

[10]I discuss these dynamics, along with an actual, real-world example, in Essay 17, *Tipping points in the balance of nature (2010)* (page 144).

[11]In *Panarchy* (2002), Gunderson, Holling, and their colleagues extend this spectrum of dynamics even further, suggesting that systems may undergo cyclic transitions between phases. The concept requires some deep thinking, but the book is well worth a read, and the first two chapters are especially relevant to the points I make here.

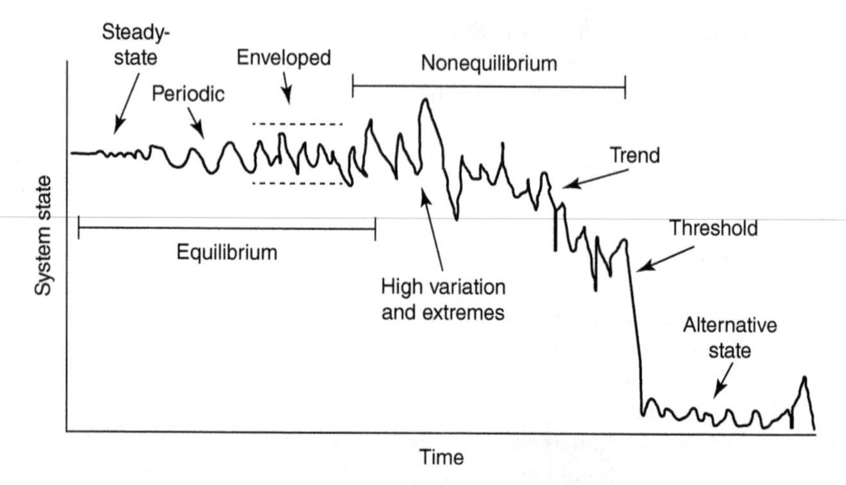

Figure 3.1 A conceptualization of the changes that may occur in ecological systems. The graph is not intended to illustrate a particular time sequence, but a spectrum of dynamics from steady-state equilibrium to thresholds and alternative system states. The horizontal bars illustrating equilibrium and nonequilibrium dynamics show the sorts of variations that have been considered by ecologists who have espoused one or the other view of ecological systems. Although the diagram illustrates different sorts of temporal dynamics, it may also apply to variations in space. "Equilibrium" and "nonequilibrium" are equivalent to "homogeneous" and "heterogeneous," respectively. "Trends" may indicate gradients or ecotones, while "thresholds" may be the counterpart of sharp boundaries or discontinuities between places, habitats, or cover types in a landscape.

Whether one sees stability or seemingly chaotic variation is affected by scale. A small sliver of any of the trajectories in Figure 3.1, such as might be documented in a short-term or spatially restricted study, can easily give the impression of stability. Looking only at "boom" or "bust" years, as might occur in a 2–3-year study in central Australia, would provide a warped view of the system. When the scale is expanded, however, the scope of variation and change will emerge. For example, considered over very broad scales of time and space, the earth's climate and its ecological systems have always been in a state of flux. Paleoecological studies document the discordant distributional shifts of species and entire communities that occurred during and following glacial advances and retreats over past millennia.[12] Human actions have further transformed the earth's physical and natural environments.[13] These changes have been so profound since the onset of the industrial revolution that some have suggested that we are in a new geological epoch, the Anthropocene, in which the effects of human activities dominate the earth's

[12]See, for example, Jackson (2012).
[13]Thomas (1956) and Turner et al. (1990).

atmosphere and ecosystems.[14] However, humans were altering the environment for millennia before then, by using fire to manage habitats, overharvesting timber and other resources, developing agriculture, intentionally or inadvertently moving species about, and forming towns and cities.[15]

So change has been ceaseless. Now, however, the pace and magnitude of environmental change are increasing, largely as consequences of human actions—the dynamics are shifting toward the right side of Figure 3.1. The current and coming changes may be so great that the challenges facing ecology and conservation will be qualitatively different, requiring that we "think anew, and act anew." If this is the case, we shouldn't be debating about the stability or equilibrium of ecological systems, but should instead focus on the form and magnitude of variation and change. We need to shift from a paradigm of balance of nature to one of variability and change. This requires some understanding of the forces creating variability and change, which I consider in Part II. However, because any change can be thought of as disturbing the status quo of a system (especially if one is thinking in equilibrial terms), a brief digression on disturbance is in order.

[14]Crutzen and Stoermer (2000). Lewis and Maslin (2015) propose that the Anthropocene began in 1610.

[15]Diamond (2005) and Delcourt and Delcourt (2008). Mulvaney and Kamminga (1999) and Gammage (2011) recount how Australian aborigines have been altering landscapes for some 50,000 years, perhaps much longer.

ESSAY 7

From our southern correspondent(s) (2011)*

A while back my Australian colleague Richard Hobbs and I got to talking about how much fun it would be to get together more often. It might even be productive, although we had no delusions about that. One thing led to another, and somehow Richard convinced the University of Western Australia to appoint me to a Winthrop Research Professorship. This enabled me to travel to Perth every year or two to spend time with Richard and his group. We also found the opportunity to visit a local brewery in Fremantle, appropriately called Little Creatures,[1] *sampling their wares. Like me, Richard had also been writing essays for the Bulletin of the British Ecological Society for some time, so we decided to cast our lot together and meet the looming deadline with a joint essay.*

As we cast about for ideas, we began contemplating how the differences between Australian and North American environments might have contributed to differences in how ecologists think about and conduct ecological studies. The ecosystems of the continents are different, to be sure, so what is at hand to study also differs. But have the differences influenced how ecologists see things, so they end up asking different questions or using different approaches? It turned out to be an interesting conversation (we thought so, at least), so we tried to convey some of that by framing the essay as a conversation. Regardless, the beers were very good.[2]

Richard Hobbs and John Wiens are sitting enjoying a beer at Little Creatures in Fremantle, Western Australia, on a warm November evening. Wiens is in town visiting Hobbs to develop collaborative projects, but this evening they've decided to concoct a plan to pull together a joint offering for the BES ***Bulletin****, rather than each having to write one. They*

*Wiens, J.A. and Hobbs, R.J. 2011. From our southern correspondent(s). *Bulletin of the British Ecological Society* 42(1): 49–51. Reproduced with permission of the British Ecological Society.

[1]Their web site, https://littlecreatures.com.au, describes their history and their beers. I learned, for example, that the inspiration for starting the brewery came from tasting a pale ale from the Pacific Northwest of the U.S., where I live. An interesting, but totally irrelevant, coincidence.

[2]I should explain the reference to the mysterious "Alan" in our conversation. Alan Crowden edits the *Bulletin*, in which this and other essays appeared. Both Richard Hobbs and I are more-or-less regular contributors (Alan would prefer "more"), and Alan regularly badgers us about deadlines and such. When he's with us we make sure he buys the beers, promising of course that we'll meet the next deadline if he does. Not bloody likely.

Ecological Challenges and Conservation Conundrums: Essays and Reflections for a Changing World, First Edition. John A. Wiens.

Figure 1 John Wiens and Richard Hobbs collaborating on an essay at Little Creatures, Fremantle, WA, Australia. Source: Photograph by John Wiens.

know that this will both save some individual effort and send the ***Bulletin*** *Editor into temporary apoplexy. They take a prolonged swig of pale ale and start discussing what the essay might be about (Figure* 1).

John: Well, we really ought to have some sort of integrating theme to this essay. What can we write about that's interesting and topical, and might also have an important message for ***Bulletin*** readers?

Richard: When I start writing these pieces for the ***Bulletin***, I'm never actually sure where they are going to end up until I get started. It's actually one of the rare occasions that I get to write in a relatively unconstrained way without having to worry about what journal editors and reviewers will have to say. Don't tell Alan, but I actually really enjoy that opportunity.

John: I think he suspects. Of course I could claim that I have a much more reasoned and deliberative approach, but who would believe that? Spontaneity is my muse. And this is a great time and place to be spontaneous.

They both take another lengthy swig of beer and stare into the distance for a while.

John: So, old chap, if we're doing this jointly, shouldn't we have some sort of idea about what our main message is going to be ... ?

Richard: Well' I'll be going to the Ecological Society of Australia conference in a couple of weeks, and this year is the 50th anniversary of the founding of the society. How about we look at the history of ecology in Australia and think about whether its development here is different from that in the Northern Hemisphere and, if so, why?

John: Ah, yes, that has possibilities. Coming here from the north, one is certainly struck by how different things are. Big herbivores hop instead of run, one looks long and hard for anything resembling a proper mountain, and people speak a strange form of English. So why wouldn't Australian ecology be different?

Richard tries to say something but John continues, warming to the theme and rising to erudition.

John: We could argue, in fact, that much like the systems we study, ecology and ecologists bear the imprints of their past history. As ecology has developed around separate geographic nodes over the past century, it has taken on a regional flavor that reflects those geographies. This is what the American ecologist Paul Sears referred to as the "ecology of ecologists," the notion that their environment has shaped the thinking of those who set ecology on its multiple trajectories to the present. In North America, for example, the development of Clements' view of communities as tightly integrated units (his "supraorganisms") grew from his studies of grasslands in the Great Plains, where sets of plant species closely mirrored discontinuous soil conditions. The philosophically opposing individualistic view of communities espoused by Gleason reflected his experiences in forested landscapes in the Midwest, where vegetation composition changed more or less continuously across environmental gradients.

Later on, the flowering of American ecology in the 1950s and 1960s followed two conceptually distinct threads—one emphasized population and community dynamics that was heavily based on mathematical theory (the Hutchinson–MacArthur theme) and another focused on ecosystem dynamics and biogeochemistry that was grounded in systems thinking (the Odum(s)–Bormann/ Likens theme). Despite their differences, both drew inspiration from analogues in the industrial world (e.g., the emphasis on competition, feedback regulation, and systems engineering), and both developed in the setting of eastern North American forests. It strikes me that the strong seasonality of these forests, their well-developed structure, the seemingly orderly patterns of succession, and especially their seeming consistency from year to year drew ecologists to emphasize such things as niche partitioning, density-dependence, competition, and well-defined pathways of nutrient and energy flows. They saw nature in balance, particularly over the 2 to 3-year duration of a typical study.

Both Hobbs and Wiens need time to reflect, not to mention a loo break, after this unexpected torrent of insight from Wiens, On returning, they order another pale ale and continue.

Richard: OK, so you're considering all the founding fathers of ecology in that light. MacArthur, the Odums and others—and in Britain people like Tansley, Lack, and Elton. All products of their time, place, and environmental setting? So, what are the counterparts in Australia? Andrewartha and Birch and the whole density independent debate they engendered? Caughley and others observing a fauna that apparently didn't follow Northern Hemisphere rules? Is it not just that Australian ecosystems are different (the title of an ESA conference some decades ago), but Australian ecologists, being influenced by those systems, are also different in how they think? Even back then, folks knew that North-temperate ecologists and Australian ecologists saw nature differently—they just didn't ask why. But the "why" has a lot to do with the nature of the continents, perhaps especially the highly variable climate and impoverished soils of much of Australia. This, and the lengthy isolation of Australia, moved adaptation of plants and animals along different pathways, so when ecologists got around to studying them they had a different set of building blocks to play with. Also, I wonder how much is shaped by the glacial versus nonglacial histories of the continents, the flatness and aridness of Australia, nutrient-rich versus nutrient-poor soils, and so on? We're certainly thinking that way here in the southwest, where the antiquity of the landscape plays a big part in how things work.

John: But didn't the Australian ecologists mostly come from somewhere else—predominantly in the Northern Hemisphere, and especially the UK? So were they more adaptable than their counterparts who stayed at home, or what? Did they start thinking differently as soon as they experienced the new environment, or did it take a while for the "Australian-ness" of the ecology to sink in?

Richard: Interesting point. Certainly, a lot of the environmental pickles we are in now arose because the management schemes that developed in the Northern Hemisphere did not translate easily to the Australian environment: agriculture is the prime example. But we need to remember that Britain had the Empire and a long tradition of exploration, study and, of course, exploitation of the biology in far-flung places.

John: Ah, yes—the Empire. Of course, you're an import, exile, or whatever from the UK too. Do you think Australia has made you think differently about things?

Richard: Certainly! The ecosystems here could hardly be more different than in my native Scotland. Everything seemed entirely weird when I first got here—the plants, the animals, how the ecosystems worked, what the key environmental variables were. I guess what I've been trying to figure out ever since is whether they are really entirely different, or whether the ecosystems operate in the same way with the same sorts of mechanisms but with a different cast of players.

John: Right-o. So if the worldview of ecologists is shaped by the systems they study, they may believe that what they see is what is "normal" and things elsewhere are "unusual" or even "weird". There's a real temptation to think that the concepts and theories about how nature works that are derived from such "normal" situations should hold elsewhere. I think I had pretty much the same experience as you. When I came to Australia years back to study bird communities in shrub deserts, I fully expected that they'd align nicely with the explanations that I'd developed from a decade of work in "normal" shrub deserts in North America. But they didn't. It took a long time for me to realize that, by focusing on the similarities in vegetation structure but ignoring the differences in environmental variability, I was trying to fit a square peg into a round hole. My faith in normality and my belief in the generality of ecological theory were shattered. An appreciation for the variability of nature has infected my thinking ever since.

This realization that normality is illusory causes both Wiens and Hobbs to stare into their near-empty beer glasses for a while. Hobbs suddenly resumes.

Richard: Of course, things have changed a lot since the early days. During the 1950s and 1960s ecologists didn't travel much and most meetings and journals were regional, so people with similar perspectives and experiences talked mostly to one another. The relative isolation of ecologists in different parts of the world may have strengthened the impact of the environmental setting on the development and divergence of ecological thinking. We were both fortunate to be able to move about early in our careers, which may have caused us to be just a shade contrarian, suspicious of the "normal" and skeptical of general ecological theory.

John: Speak for yourself, mate! But there's no question that ecology has become a global discipline. Conferences are held in far-flung places, journals pride themselves on being international, and the internet connects everyone, everywhere, all the time. Colleagues have inter-hemispheric confabs to think great (or not so great) thoughts and drink beers together. Students travel across the globe to conduct research and studies. Why, you probably have students from dozens of countries in your group right now.

Richard: 8, actually.

John: OK, I was rounding up.

Richard: But I'll buy your point (if you'll buy the next round). The globalization of ecology has made "normal" an increasingly irrelevant concept. More and more ecologists delight in a variety of systems histories, contexts, and dynamics—*vive la différence* is the new banner of ecology. However, on the other hand, there may be a downside to all this globalization. It may also promote an increasing homogenization of ecology—perpetrated especially by journals and the intense pressure to publish stories that don't rock the boat, since a single negative review is often enough to doom a submission. Hence, stuff that doesn't fit the "normal" story may be increasingly hard to publish. Maybe Kuhn's "normal science" is running rampant in ecology?

John: That's too many meanings of "normal" for this time of night. It seems to me that Australian ecologists don't have much problem getting their voices heard on the international scene. But if there are different stories to be told here and elsewhere (I'm sure New Zealanders, South Africans, and South Americans could chime into this discussion as well), what does this say about the "holy grail" of ecology—general theories that hold generally? Do we always need to add, parenthetically, "except in Australia" (and where else?)?

Richard: Ah, GUTs!

John: I beg your pardon?

Richard: Grand unified theories! What the physicists always go on about. Can we find these in ecology? Or are we dealing with systems that are too complex and idiosyncratic? Or are ecologists simply too complex and idiosyncratic?

John: You men, "Do ecologists in different parts of the world see things differently?"

Richard: Wasn't this where we came in?

An attractive wait-person approaches, having been listening in the background for a while.

Attractive wait-person: Listen you two—isn't it pretty obvious that people living in different environments are going to perceive differences in the systems they study and mold their scientific worldview in relation to these? Australian ecology will always have a different flavor to ecology from the Northern Hemisphere, and a jolly good

> thing too, I say. Sure, you can mix people around and make sure people experience different systems and different ways of thinking—but if all the systems get homogenized (as they are likely to with globalization, invasive species, and whatever) and the thinking also gets homogenized, there would be no point in going different places, visiting different labs or anything. Another pale ale?

Stunned by this succinct summation of what they had been trying to enunciate all evening, Wiens and Hobbs both nod and wait in silence until two more beers are delivered to the table. They spend the rest of the evening contemplating where they might have gone wrong and why they didn't ask the attractive wait-person in the first place.

ESSAY 8

Boom and bust: lessons from the outback (2014)*

September 2014 found me once again in Australia, sharing thoughts and beers with Richard Hobbs. Because we were in Alice Springs, in the center of the Red Centre of Australia, our thoughts turned to environmental variation—boom and bust as it is known there. Also, I had just come across a delightful little book[1] that captured my affection for both birds and arid places. So it seemed fitting to ponder extreme variation in rainfall in places like Australia and California (then in the midst of a multiyear drought that might seem short by Australian standards). Such variations are likely to become more frequent and more severe in these places as climate change unfurls.

A side note: we mentioned the Todd River, which runs through Alice Springs and is normally bone dry. In early January 2015 a deep tropical low-pressure cell met humid monsoon air from the north, dropping up to 1 cm of rain in some parts of the river catchment. It was enough to cause the river to flow knee-deep, making the national news. With characteristic understatement, a meteorologist noted that the circumstances leading to the heavy rain were "quite uncommon." Of course, where I live in western Oregon we receive that amount of rain during a normal drizzly winter day. "Booms" and "busts" are relative.

"I love a sunburnt country/A land of sweeping plains/Of ragged mountain ranges/Of droughts and flooding rains. . . ." John was in a particularly poetic mood, reciting Dorothea Mackellar's My Country (1908) freely and accurately, while Richard brought another round of beers.

The two of us were in Alice Springs for the annual meeting of the Ecological Society of Australia. We were pondering booms and busts over beers (as ecologists are wont to do) while we watched the afternoon sun highlighting the river red gums bordering the dry, sandy bed of the Todd River in Alice Springs. It's the site of the famous Henley-on-Todd Regatta, in which the Australians (ever creative) race down the dry riverbed, running barefoot in bottomless boats. Only once in

*Wiens, J.A. and and R.J. Hobbs. 2014. Boom and bust: lessons from the outback. *Bulletin of the British Ecological Society* 45(4): 40–41. Reproduced with permission of the British Ecological Society.

[1]Robin et al. (2009).

Ecological Challenges and Conservation Conundrums: Essays and Reflections for a Changing World, First Edition. John A. Wiens.

Figure 1 The Todd River as it normally is (bust), Alice Springs, NT, Australia. Photograph by Steve Morton.

more than 50 years has the race been cancelled because there was actually water in the river. It's the epitome of boom (the rare floods) and bust (the rest of the time).

Central Australia (the "Outback" of legends and tourist brochures) is a vast arid land, one of the largest relatively intact environments on Earth. It's not only a sunburnt country but, most of the time, a parched country. But now and then, with little warning, the flooding rains come. It's a country where "the creeks run dry or ten feet high."[2] A boom and bust country (Figures 1–3).

The ESA meeting was replete with talks focusing on different aspects of this boom and bust environment. Flora and fauna, terrestrial and aquatic, adapted in a myriad of ways to coping with long periods without water and capitalizing on the sudden, brief periods where water is abundant. Birds, mammals, and insects covering large distances to reach isolated water sources or adapting behaviorally and physiologically to long-term scarcity. Complex interactions developing when water suddenly becomes abundant, plants can grow, and animals of all sorts can thrive and reproduce. Truly amazing human understanding and use of the landscape that allowed aboriginal people to persist through boom and bust over millennia. And gradual development of pastoral practices that are more in tune with the country's dynamics.

Boom and bust dynamics are more evident in arid Australia than in many other places, but such extremes characterize most arid and semiarid parts of the world.

[2] Friedel et al. (1990).

Figure 2 The Todd River running full (boom), Alice Springs, NT, Australia. Photograph by Margaret Friedel.

Figure 3 Another view of the Todd River running. Photograph by Ashley Sparrow.

And they're all about water—too much of it or far too little. Past civilizations have risen on the booms and fallen on the busts. For example, a severe, prolonged drought in the 12th century that followed an unusually wet period may have led to the decline and disappearance of the Anasazi culture in the American southwest. The Hohokam culture in what is now Arizona lasted longer, into the 15th century, perhaps because they had built elaborate irrigation systems to support agriculture with water from rivers in what was otherwise a desert. But then they too suddenly disappeared.

Boom and bust episodes are not confined to the sparsely populated Australian Outback or past civilizations, however. California, which is at the opposite extreme in population and agricultural production, also has boom and bust water dynamics. As we write, California is in the third year of a drought that is nearly unprecedented in historical times. In 2013 and 2014, a vast region of persistently high atmospheric pressure over the northeastern Pacific Ocean—the "Ridiculously Resilient Ridge"—prevented typical winter storms from reaching California, bringing record-low precipitation and record-high temperatures.[3] Many reservoirs that store water for the dry spells are at less than 15% of their operating capacity. Domestic water use has been curtailed, wells in towns have run dry, and some farmers are receiving no irrigation water at all this year. Californians are even being asked not to wash their cars or water their lawns!

The biota are also suffering. The wetlands that were once traditional wintering habitats of millions of migratory waterbirds are long since gone, replaced by agriculture. The birds have come to depend instead on flooded rice fields, but this winter many of those fields are dry and barren. Salmon, Delta smelt, and several other fish are legally protected under the US Endangered Species Act, so water must be released from the reservoirs to meet the needs of the fish. That's water the farmers don't get, sparking debates about water for "stupid little fish" that's allowed to flow out to sea, and therefore "wasted." As the drought continues, the debates get angrier.[4]

That's the bust. But California also has booms. In the midst of the current drought, southern California was deluged by the remnants of a tropical storm that flooded neighborhoods and submerged cars, although it did little to alleviate the effects of the drought. In 1862, the city of Sacramento was virtually destroyed by the "Great Flood" resulting from rains of Biblical scale (actually, 45 days). Every few years the "pineapple express" (the technical term is atmospheric rivers) brings pulses of moisture-laden Pacific air to the state, dumping more water than the streams and rivers can handle. But not lately.

Southern California, where much of the water normally goes, used to be a desert. Unlike the Outback, however, it had sources of water from elsewhere—northern California, the Colorado River—that could be tapped, albeit

[3]Herring et al. (2014).

[4]John has written about this in a previous essay (Essay 13, *Wildlife, people, and water: who wins?* (2012) (page 100)).

over long distances and at great cost. Over the past century, Californians have adapted to the boom and bust dynamics, not by adjusting their agricultural practices or water use, but by building a vast infrastructure of dams, channels, pumps, and aqueducts that stores water and then conveys it into the desert, creating an agricultural breadbasket and fueling the explosive growth of the Los Angeles Basin.[5] The current drought has brought calls for more dams, built higher, and ever greater technology to soften the boom and bust water dynamics. In the meantime, farmers have responded to global demands by planting high-income crops such as almonds, which require more and more irrigation water and are therefore increasingly vulnerable to busts.

How will this end? Eventually the current drought in California will be broken and the rains and floods will return.[6] The boom may erase memories of the bust, but the adaptation by engineering will continue apace. Yet there are limits (recent history notwithstanding). Climate-change models uniformly predict that the frequency and magnitude of extreme events—the booms and busts—will increase. In past millennia, California has experienced droughts that lasted 30, 50, or even 500 years.

California may seem far removed from the Australian Outback. Certainly, the long-term aridity of the Outback, the extended drys, the sparse population, and the lack of free-flowing rivers fed by snowfall in distant mountains have precluded the development of the infrastructure that has (so far) enabled Californians to cope with much lesser drys. It seems unlikely that central Australia will ever see the sort of development going on in California. On the other hand, it is also unlikely to run into the problems now being faced by Californians. The same is not true for the less arid fringes of Australia, though, where California-type issues are being faced for very similar reasons.

"Wot's all this about?" interjected Alan Crowden as he appeared, looking for a beer. "Not likely to be very interesting to BES members in the UK, is it? Rains all the time there, doesn't it?" We wondered whether to try to ignore Alan as usual, but thought he had a fair point. Wrong, but fair. The boom and bust cycles of interior Australia may seem a long way removed from a rainy United Kingdom, but it may be that even the United Kingdom could yet experience such dynamics, albeit less dramatic and on a shorter timeframe. Droughts do occur in the United Kingdom and there is evidence to suggest that these may become more frequent and intense.[7] Because of high population densities and low storage capacities, droughts can get serious quickly in the United Kingdom—busts, if you like. On the flip side, storms and flood events also seem to be making news more often in the United Kingdom—in line with many other parts of the world.

[5] Chronicled in detail by Marc Reisner in *Desert Cadillac* (1993).
[6] Update December 2015: the California drought intensified during 2015. A massive El Nino developed late in the year, threatening to bring floods and landslides to the parched landscapes. Whether El Nino will break the drought is yet unknown.
[7] Rahiz and New (2013).

So, perhaps instead of being a weird outlier, the boom and bust interior of Australia could serve as a useful bell-weather for changing dynamics in many other places, even those traditionally seen as having more equable and predictable climates. Anyway, that seemed like a reasonable premise as we watched the sun finally set behind the red rocks of the range bordering Alice Springs. "This could even be a good topic for a BES Bulletin essay" quipped John. "When's the deadline again?" asked Richard. "Typical" grumbled Alan. And the attractive wait-person brought another round of beers.

CHAPTER 4

A digression on disturbance

According to most dictionaries, "disturbance" is a departure from normal or average conditions. The Oxford English Dictionary puts it more quaintly, as an "interruption of a settled and peaceful condition."

Such definitions assume that there is in fact a "normal," against which a departure can be gauged. Without saying as much, this notion of a disturbance embodies an equilibrium or balance-of-nature worldview. It begs the question of how far something must depart from "normal" to be considered a disturbance. In a dynamic environment, variation is the norm and the "peaceful condition" of nature is always under attack. "Normal" is constantly changing, and average conditions exist only as numerical artifacts.

The definition of disturbance offered by Peter White and Steward Pickett in *The Ecology of Natural Disturbance and Patch Dynamics*[1] is ecologically more sensible. A disturbance is a "relatively discrete event in time that disrupts ecosystem, community, or population structure and changes resources, substrate availability, or the physical environment." By these measures, any factor that changes the condition of a system, whether the system is in equilibrium or far from it, is a disturbance.

Ecologists have long thought and written about disturbance.[2] In practice, the disturbances that most concern conservationists and managers are those that stem from events external to the system and that have significant ecological consequences. Sometimes the disturbances are unintentional results of human activities. Large oil spills, for example, are clearly unintentional—no oil company wants to lose that much oil, suffer so much bad publicity, fight so many lawsuits, or cause so much environmental damage. The *Exxon Valdez* oil spill resulted in the deaths of several hundred thousand seabirds, fouled over 2,000 km of shoreline in Prince William Sound, Alaska, and led to a quarter-century of legal action and scientific debate. The environmental and social consequences of the *Deepwater Horizon*

[1] White and Pickett (1985: 7).

[2] In fact, some of the first ecologists to think about disturbance did so in the context of a strongly deterministic and equilibrial worldview. With the emergence of a nonequilibrium perspective, ecologists have turned more to considering the nuances of variation rather than lumping everything under the rubric of "disturbance."

Ecological Challenges and Conservation Conundrums: Essays and Reflections for a Changing World, First Edition. John A. Wiens.

disaster in the Gulf of Mexico are still unfolding.[3] I comment on these two spills in Essay 9, *Oil, oil, everywhere ... (2010)* (page 60).

Usually, however, the disturbances are the results of how land or waters are used or managed. Clearing of a forest or plowing of a prairie to make way for agricultural crops or replacing agriculture with suburbia creates disturbances that often have more profound and far-reaching ecological effects than do large natural disturbances such as earthquakes, floods, landslides, or volcanoes. Such anthropogenic disturbances are intentional, although they may have unintended consequences. The suppression of fire in U.S. forests for much of the 20th century, for example, was a product of well-intentioned (at the time) policies. The result, however, was the growth of dense forest stands and understory brush that fueled widespread fires that have altered landscapes, especially in the West. Outbreaks of bark beetles over vast areas of drought-stressed forests—perhaps associated with climate change—have further increased vulnerability to fires. These are disturbances on a grand scale.[4]

The ecological effects of a disturbance depend not only on its magnitude and spatial extent, but also on its duration. Ecologists distinguish between *pulse* disturbances—sudden, one-time disruptions of a system—and *press* disturbances, in which the alteration is sustained and continuing. An earthquake or an oil spill is a pulse disturbance. Chronic, small releases of oil into a marine environment from recreational and fishing vessels are a press disturbance.[5] When the concept and terms were introduced in the 1980s, equilibrium thinking was still prevalent, and pulse perturbations were defined in part by a return to a previous equilibrium state.[6] We now realize that "recovery" from a pulse disturbance can follow multiple trajectories, depending on the nature and magnitude of the disturbance, on what else happens in the following years, and a host of other factors. If the disturbance is sufficient to move a system beyond its zone of resilience and across a threshold to another state (as envisioned in state-and-transition models), a return to the previous condition may be unlikely.[7] A fire in sagebrush areas in the American West, for example, may result in a conversion to grasslands that may

---

[3]See National Research Council (2013) and Wiens (2013a).
[4]I comment further on such disturbances in the context of their effects on "ecosystem health" in Essay 28, *A metaphor meets an abstraction: The issue of "healthy ecosystem" (2015)* (page 230).
[5]Modeling of oil-spill effects on seabirds by Ford et al. (1982) suggested that such chronic effects may have greater impacts than a large, one-time oil spill; see also Wiese and Robertson (2004).
[6]This view remains embodied in the Natural Resource Damage Assessment (NRDA) regulations that guide the determination of damages from oil spills, in which "recovery" is defined as the return of an injured natural resource (e.g., a species) to the condition that would have existed had the oil spill not occurred—generally determined by a comparison with conditions existing before the spill (i.e., equilibrium). See Wiens (2013a: 24–26).
[7]Bestelmeyer (2006).

persist for many decades, particularly if the burned area is invaded by cheatgrass, which fosters recurrent fires. The conditions created by a press disturbance, on the other hand, may persist as a new quasi-equilibrium, but only so long as the disturbance continues. I discuss such dynamics in Chapter 13 and Essay 17, *Tipping points in the balance of nature (2010)* (page 144).

ESSAY 9

Oil, oil, everywhere ... (2010)*

A large marine oil spill is the epitome of an environmental disaster. The oil can't be contained, and sooner or later everything—beaches, birds, boats—is coated with a sticky, gooey mess. Oil is black, and thus inherently evil. Large spills are associated with "big oil" and a public perception (not always justified) of corporate greed. Spills are a consequence of modern society's thirst for oil, addiction to fossil fuels, and reluctance to live on less energy.

The Exxon Valdez *and* Deepwater Horizon *oil spills (Figures 1 and 2), which I compared in this essay, occurred in particularly sensitive marine environments—in the cold, stormy waters of Prince William Sound, Alaska, and in the warm and (usually) calmer Gulf of Mexico. The environmental and legal issues of the* Exxon Valdez *spill have now been resolved, and affected species and communities have recovered, but it took over a quarter-century to reach this point.*[1] *The* Deepwater Horizon *blowout and spill happened in 2010, so the story is yet unfolding and may do so for some time. Already, however, there has been a shift in how impacts are assessed. The emphasis following the* Exxon Valdez *spill was on injury and subsequent recovery of natural resources, primarily bird, mammal, and fish species, some of economic or cultural importance and others not. While there have been similar concerns about effects on species from the* Deepwater Horizon *spill, a new focus has emerged on how benefits to people from the Gulf and its resources—ecosystem services—have been affected.*[2] *This represents something of a new paradigm, one that is more closely tied to the utilitarian than the preservationist views that I discussed in Chapter 1.*

This essay concluded with a plea for a more thoughtful assessment of our dependence on fossil fuels, particularly oil. Although the cast has changed, the story remains much the same. Vast quantities of oil are now being produced by hydraulic fracturing ("fracking"). Despite concerns about groundwater contamination and increased seismic risks, fracking continues unabated. Currently, the largest oil fields in North America are in the Bakken Formation in North Dakota, Montana, Saskatchewan, and Manitoba. The amount of oil being produced substantially exceeds the capacity of existing pipelines to transport it, so much of the oil is being shipped by mile-long trains to coastal terminals and refineries. In addition to the attendant risks of oil spills, the Bakken oil is especially volatile, and tanker explosions in densely populated areas are heightening public concerns. But the rush to pull oil from the ground continues, driven by increasing domestic demands, fears of political instability in oil-producing areas elsewhere in the world, and emerging markets for oil in China and elsewhere.

Although there are cautionary lessons from the Exxon Valdez, Deepwater Horizon, *and other oil spills, we seem slow to learn them.*

*Wiens, J.A. 2010. Oil, oil, everywhere ... *Bulletin of the British Ecological Society* 41(3): 25–28. Reproduced with permission of the British Ecological Society.

[1]The consequences of the *Exxon Valdez* oil spill and what was learned about conducting scientific studies in a difficult and contentious setting are summarized in Wiens (2013a).

[2]A report from the National Research Council of the National Academies (National Research Council 2013) articulates this approach in detail.

Ecological Challenges and Conservation Conundrums: Essays and Reflections for a Changing World, First Edition. John A. Wiens.

Figure 1 The *Exxon Valdez* off the coast of Alaska, with booms deployed in a futile attempt to limit the spread of oil. Source: Photograph by NOAA.

Figure 2 The Deepwater Horizon oil rig ablaze on April 21, 2010. Source: United States Coast Guard.

Figure 3 Knight Island in Prince William Sound, Alaska, 2 years after the shorelines were heavily oiled by the *Exxon Valdez* oil spill. Photograph by John Wiens.

When I heard the news of the *Deepwater Horizon* disaster, I had that *déjà vu* feeling all over again. Once again, an oil spill has fouled our oceans and coasts. Once again, we see heart-wrenching images of seabirds struggling helplessly to escape the heavy mantle of black goo. Once again, we ask how, in this age of advanced technology, such things can happen. Then, if the past is any guide, the images will fade from our memories, politicians will dither, and stock prices for oil companies will rebound. Our insatiable thirst for fossil fuels will drive us to seek oil in more extreme and riskier environments. This time it must be different.

The explosion and blowout of the BP-leased drilling platform *Deepwater Horizon* and the Macondo-1 exploratory well brought back memories of the *Exxon Valdez* oil spill. More than two decades ago, the tanker *Exxon Valdez* ran aground on Bligh Reef, spilling some 260,000 barrels of Alaska North Slope crude oil into the waters of Prince William Sound, Alaska (Figures 3 and 4). Within a month, I began documenting the spill's impacts on marine birds—and I'm still at it. What I've learned from that experience may provide some perspective on the *Deepwater Horizon* spill and how scientists should assess its ecological effects.[3]

Before April, the *Exxon Valdez* spill held the record for the largest oil spill in North American waters. Estimates of the size of the *Deepwater Horizon* spill change

[3]For updated information on the *Deepwater Horizon* spill, see http://response.restoration.noaa.gov/deepwaterhorizon and http://www.deepwaterhorizonresponse.com/go/doctype/2931/55963.

(a)

(b)

Figure 4 Two photographs of a beach on Green Island, Prince William Sound, Alaska taken soon after the spill (a) and 2 years later after human and natural cleanup (b). Source: © Exxon Mobil Corporation. Reproduced with permission. Exxon Mobil Corporation.

as the oil continues to flow. By mid-June, perhaps 2.7 million barrels had been released, far surpassing the *Exxon Valdez* spill. It's little solace that spills elsewhere in the world have been much larger. The largest marine oil spill (so far) was the intentional release of 11 million barrels of oil by retreating Iraqi military during the first Gulf War. The Ixtoc I oil platform blowout off the coast of Mexico in 1979 continued to release oil for 9 months, eventually spilling 3.5 million barrels. In 1978, the *Amoco Cadiz* wreck spilled 1.6 million barrels on the coastline of Brittany. In 1967, the breakup of the supertanker *Torrey Canyon* on the coast of Cornwall released some 860,000 barrels. The *Exxon Valdez* spill ranks "only" 35th worldwide. The amount of oil spilled by the *Exxon Valdez* grounding is spilled *every year* during "normal" operations in Nigeria. Some records are not made to be broken.

Initial responses to the *Exxon Valdez* oil spill and the *Deepwater Horizon* blowout were similar, some disturbingly so. There was a strong emotional response from the public, amplified in the *Deepwater Horizon* explosion by the loss of human lives. There were immediate concerns about damages to marine life and human livelihoods. There was outrage at "big oil" and a rush to assign blame. Enough lawsuits were filed to keep lawyers busy for years. There were difficulties in mobilizing a rapid response to protect coastal areas, exacerbated by disputes among government agencies and corporate interests about who should do what, where, and when. Pundits pontificated. The media flocked to the scene, along with advocates eager to advance their agendas. And scientists were there too, including some who were all too willing to oblige the media with premature prognostications about long-term environmental damages.

The two oil spills differ in important ways. The *Exxon Valdez* spill released a large amount of oil all at once onto the surface of icy cold Alaskan waters in an area studded with bays and islands, and an early spring storm pushed the spreading oil onto shorelines. In contrast, oil from the *Deepwater Horizon* continues to spew from deep in the ocean, creating an underwater plume of oil and methane. As oil reached the warm surface waters it spread slowly, reaching land only weeks later (Figure 5). Oil from the *Exxon* Valdez was deposited on steep, rocky coastlines, and cleanup efforts were helped by the winter storms that buffet Prince William Sound. The shoreline of the Gulf of Mexico is dissected by countless bayous, marshes, and wetlands (Figure 6), the beaches are shallow and sandy, and (aside from hurricanes) the weather is more benign. Even with a massive effort, cleanup will be difficult.

The *Exxon Valdez* spill also occurred in an area of incredibly abundant wildlife, resulting in the mortality of hundreds of thousands of marine birds and mammals and countless fish. Within weeks, oiled carcasses of over 35,000 birds and several hundred sea otters had been retrieved; overall, perhaps 250,000 birds and as many as 1,000 sea otters may have died. It is much too early to project the direct effects of the *Deepwater Horizon* spill on wildlife, but as I write this fully two months after the blowout, oiled carcasses of 934 birds, 380 sea turtles, and 46 marine mammals have been retrieved. Oil from the *Deepwater Horizon* has yet to have the significant impacts on coastal breeding colonies of pelicans, terns, and other species that many expect, but it is doubtful that wildlife mortality will approach that from the

Figure 5 Oil on the surface of the water in the Gulf of Mexico, 12 May 2010. Photograph by NOAA.

Figure 6 Saint Mark's National Wildlife Refuge in North West Florida is at risk of oil impact. If oil comes ashore among the low-lying, marshy wetland, cleanup will prove an even more difficult challenge than that presented by the steep rocky coast of Alaska. Photograph by U.S. Fish & Wildlife Service.

Exxon Valdez spill. On the other hand, coastal marshes and estuaries and marine life may be heavily impacted, and if the underwater plume of oil causes a crash in phytoplankton production, entire food webs may be imperiled.

The role of scientists in assessing the ecological consequences of environmental accidents such as the *Exxon Valdez* or *Deepwater Horizon* is to provide reasoned, rational, and objective answers to two questions: "Are there ecological effects?" and "Are the effects we see due to oil, or to something else?" Framing the questions is important. It's all too easy to ignore the first question (aren't dead birds evidence enough?) and ask instead "How bad was it?" This leads one to look only for negative impacts, which may so narrow the focus of investigations that we fail to understand the overall effects. In our studies of the *Exxon Valdez* spill, for example, we found that marine bird species differed in their responses to the spill in ways related to their ecology and behavior, with some reoccupying oiled habitats within a month or two.

Addressing these questions requires careful attention to research design. Counts of oiled carcasses can be extrapolated to estimates of overall mortality, providing a gauge of immediate ecological effects. But what are the effects on population dynamics, abundance, or the use of habitats? Does the immediate mortality translate into long-term consequences, or are there effects on individuals other than mortality? Do populations recover, and if so, when? Answering these questions for the *Exxon Valdez* spill has proven to be difficult and contentious (which is why I'm still at it).

Ideally, information from before the spill is available to serve as a baseline for comparison. Oil spills rarely happen in well-studied areas, however, so scientists instead must compare areas that were affected by the spill with other, unaffected areas some distance away. Such comparisons assume *ceteris paribus* ("all else being equal"): the only thing that differs (or the only difference that matters) between the areas is the exposure to oil, so any differences in population dynamics or distributions among habitats must be due to oil.

However, *ceteris paribus* rarely holds in nature. In the years before the *Exxon Valdez* spill, for example, there were major changes in oceanographic conditions and the prey available to seabirds, and populations of some species had been declining regionally for some time. Within the spill area, shoreline habitats varied from place to place, and the oil was not randomly distributed among habitats. Some of the areas used as unoiled reference sites for oiled areas were close to glaciated fjords or in areas uplifted by the 1964 Alaska earthquake, so the habitats differed in many ways other than oiling. Such environmental variations make it difficult to pinpoint whether the observed differences in wildlife populations are due to oil or something else. There are ways to design studies and analyze results to disentangle the effects of multiple factors, but they must be part of the plan from the outset rather than being afterthoughts.

Ceteris paribus also presumes a balance of nature, that within some narrow window of variation, things will stay the same. A world at equilibrium. As time passes after a spill, however, other things happen that destroy any semblance of equilibrium and may set systems onto different trajectories of change. The *Exxon Valdez* spill was followed by large *El Niño* events, and some species continued to decline regionally while others increased. Distinguishing any remaining effects of a spill from the accumulating "other things" becomes increasingly difficult.

Attention also shifts to evaluating species' recovery. In many previous oil spills, wildlife populations appeared to recover within a few years, leading to the expectation that ecological systems may be resilient to such affronts. Although our studies in Alaska indicated rapid reoccupancy of oiled habitats by several seabird species, others took nearly a decade. Other studies have suggested that some species may not yet have fully recovered from the spill.

It is not easy to define "recovery," however. Regulations used to assess responsibility for damages to natural resources in the United States stipulate that recovery occurs when a species returns to the levels that would have obtained had the spill not occurred. How do we determine what "would have obtained" in environments as variable as those in Prince William Sound or the Gulf of Mexico? And what effects constitute continuing impacts of a spill? Is it sufficient that there are no longer statistically significant differences in population levels or habitat occupancy that can be attributed to the spill, or must there be no indications of *any* lingering effects in individual physiology or genetics, whether or not they result in population-level consequences? There have already been pronouncements that the biological systems in the Gulf of Mexico "will never be the same." Determining whether or not this is true will require an operational definition of "the same" (i.e., recovery) that is linked to clear biological mechanisms and ecologically significant consequences, that recognizes the reality of natural environmental variation, and that acknowledges that a failure to be "the same" may be due to causes other than oil.

Oil spills have impacts that extend well beyond wildlife and the environment, of course. The *Exxon Valdez* spill changed the way of life for many people in coastal Alaska. The social and economic impacts of the *Deepwater Horizon* blowout in Gulf coastal states are already profound and are likely to worsen before they get better. Public attitudes are also affected. The blowout of a drilling platform off of Santa Barbara, California, in 1969, for example, mobilized support for important environmental legislation and led to a Congressional moratorium on offshore oil leasing in many areas that lasted for 27 years. Whether one believes that offshore drilling is a recipe for more disasters or an essential ingredient for economic growth, the legacies of such events complicate efforts to develop a comprehensive energy policy, one that can foster the development of alternative "green" energy sources. The mantra of "drill, baby, drill" no longer resonates with as many people.

The *Deepwater Horizon* blowout was not anticipated. Such "rare events," however, are becoming less rare. Eight months before the *Deepwater Horizon* explosion, the *Montara* well platform in the Timor Sea blew out, creating Australia's largest oil spill (another record!). In its exploration plan for the *Deepwater Horizon*, BP assured itself, its shareholders, politicians, government regulators (who, it turns out, were not regulating), and the public that the risks of deepwater drilling were too small to be of concern. The risks may indeed have been minimal, but they obviously weren't zero, and we, and the environment, are suffering the consequences.

Who is to blame for the *Deepwater Horizon* blowout? In a sense, all of us are. Until we find alternative ways to quench our thirst for energy or learn how to live on less energy, the risks of accidents associated with the exploration, extraction, and transport of fossil fuels will grow and more records will be broken. The economic, social, and environmental costs of accidents will escalate as well. The overriding message of *Deepwater Horizon*, and before that of the *Exxon Valdez* and a litany of other oil spills, is that we must wean ourselves from our dependence on oil sooner rather than later. Until that happens, however, we need to develop comprehensive ways of assessing and balancing *all* the benefits, costs, and risks of our oil dependence. The *Deepwater Horizon* disaster tells us that, no matter how small, no risk should be ignored.

PART II
The forces of change

This is a book about change. Change is all around us. It always has been, but we are more painfully aware of it now than even a few years ago. These changes—in the environment, in scientific knowledge and tools, and in economics and society—affect how ecologists and conservationists think about the environment and how they go about their work. In Part III, I'll explore some of the challenges and conundrums facing ecologists and conservationists that are created by the changing world. In this part, I'll delve a bit more deeply into the forces of change and how they are affecting the environment and biodiversity that ecologists study and conservationists strive to preserve.

I highlight five forces in the following chapters: climate change and sea-level rise, land-use change, distributional changes and invasive species, sociopolitical change, and population growth. Many books have been written on each of these. My comments are intended to highlight some points that seem particularly relevant to where ecology and conservation are headed. I'll begin with climate change.

Ecological Challenges and Conservation Conundrums: Essays and Reflections for a Changing World, First Edition. John A. Wiens.

PART II

The future of change

This is a book about change. Change is all around us. It always has been [illegible] more rapidly [illegible] of [illegible] [illegible] in scientific [illegible] and society affect [illegible] ecologists and conservationists [illegible] and how they go about their work. In Part II, they explore some of the challenges and [illegible] communities [illegible] ecologists and conservationists that are [illegible] this part [illegible] more deeply into the [illegible] plants [illegible] and conservationists [illegible]

[illegible] the following chapters, climate change, [illegible] land use change, distributional changes, and invasive species [illegible] have [illegible] and conservation are needed. [illegible] begin with climate change.

[illegible] First Edition. [illegible]
© 2015 by John Wiley & Sons, Ltd.

CHAPTER 5

Climate change and sea-level rise

Climate is the blanket that envelops environments, habitats, and life itself. If climate changes, everything changes.[1] So changes in climate are of paramount importance to anyone interested in ecology or conservation.

Several decades ago, however, few people concerned about the environment raised the issue of climate change. A quick perusal of two books on my shelf from the 1960s prognosticating about the future[2] uncovered no mention of climate change, although "weathermaking"—using technology to change weather to our liking, as by cloud seeding—was briefly addressed. The grip of equilibrium thinking at that time is nowhere clearer than in this passage:

> *The greatest concern, however, is whether the weathermakers might unpremeditatedly and innocently set off a disaster that was nonetheless irretrievable and irreparable. Several things hinder them, and reassure the fearful. First, the "balance of nature" opposes the weathermaker as it does, say, the jungle clearer. The natural climate is the balance among forces whose immensity boggles human schemes, Hence, even those man-made perturbations involving the energy of atomic explosions are soon damped out.*[3]

It's a sign of how far we have come that climate change, and the attendant rise in sea levels, is no longer the elephant in the room, an obvious problem that no one recognizes or wants to discuss.

Well, almost no one. Nearly all climate scientists[4] agree that climate change is real, it is with us now, and it is largely a consequence of human activities. In 2014, the scientists on the Intergovernmental Panel on Climate Change (IPCC) considered the increased atmospheric concentrations of CO_2, methane, and nitrous oxide

[1]Even bedrock, although the rate of change is glacial.
[2]Fraser Darling and John Milton's *Future Environments of North America* (1966) and Stewart Udall's *1976: Agenda for Tomorrow* (1968).
[3]Paul Waggoner in Darling and Milton (1966: 89–90).
[4]The oft-cited number of 97% is based on the number of published scientific papers drawing this conclusion (Cook et al. 2013). Because climate change has become somewhat of a paradigm among climate scientists, there may be a bit of confirmation bias in this finding. A broader survey of scientists in multiple disciplines drawn from the membership of the American Association for the Advancement of Science (Pew Research Center 2015) found that 86% agreed that human activity is driving global warming, a remarkable degree of agreement, given the contrarian attitudes of many scientists.

Ecological Challenges and Conservation Conundrums: Essays and Reflections for a Changing World, First Edition. John A. Wiens.

(which are unprecedented in the past 8,00,000 years) as "extremely likely" to be the dominant cause of the observed warming since the mid-20th century.[5] Despite this extraordinary degree of scientific consensus, over a third of the American public think that scientists disagree about whether the earth is getting warmer due to human activities.[6] And, although many politicians in the United States now accept the reality of climate change, some still deny any connection with human causes, expressing concerns about scientific uncertainties and the economic costs of dealing with the problem and a general distrust of science.[7] Among the American public, a recent Gallup poll (2014) indicates that, although two-thirds agree that the earth has been getting warmer, only one-third consider climate change to be a serious threat to their way of life.[8] Distressingly (to a scientist), respondents are evenly split between those trusting what scientists tell them about climate change and those believing that scientists manipulate their findings for political reasons.

Beliefs and public opinion aside, there is little doubt that the earth's climate is changing rapidly, in ways that are either consistent with or exceed model-based projections of future climate changes.[9] Rising temperatures garner perhaps the greatest attention (after all, climate change is also called global warming). Temperature records are now broken with monotonous regularity. Global temperatures have continued to rise, particularly since the 1960s. 2015 was the hottest year ever recorded globally. The average temperature in California in 2014 shattered all records since measurements began in 1895 (Figure 5.1). Extreme events such as this or heat waves, droughts, or torrential rains are becoming more frequent and last longer.

Meteorologists distinguish between weather—what happens to the atmosphere at local scales over minutes to months—and climate—what happens over broader areas over longer time periods. People experience weather in their daily lives. Climate doesn't have the same immediacy; it's what scientists study and model. It's not surprising, then, that scientists and the public have different perceptions of what is happening—they are operating at different scales of time and space. Because temperature and precipitation vary considerably among days, weeks, seasons, or years (e.g., Figure 5.1), the trends in climate that have become so overpowering are not evident to people attuned to weather. Consequently, a few bitterly cold winter storms may assuage those who doubt that climate change is real. Disappearing glaciers, however, are harder to deny. As one of many examples, the

---

[5]IPCC (2013).

[6]Pew Research Center (2015).

[7]I am reminded of the image of an ostrich with its head in the sand. Ostriches do not actually do that, of course, but the image is nonetheless apt.

[8]The percentages of respondents believing that global warming is happening now or will happen during their lifetimes (65–75%) and that it will pose a serious threat to their way of life (31–40%) scarcely changed between 2001 and 2014; http://www.gallup.com/poll/167879/not-global-warming-serious-threat.aspx.

[9]IPCC (2014), AAAS (2014), and National Academy of Sciences and The Royal Society (2014).

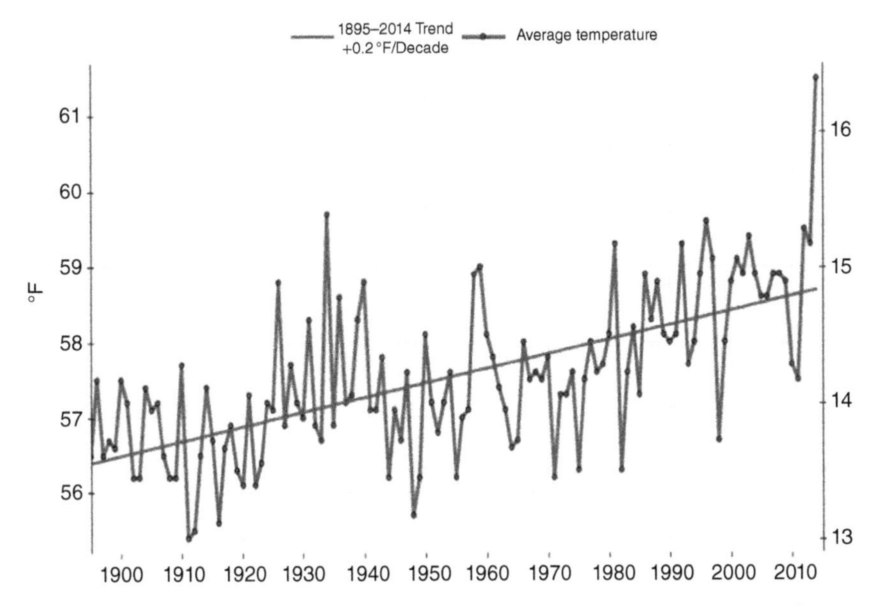

Figure 5.1 Yearly average temperature for California since records began in 1895. Note the high annual variation, the long-term trend, and the extreme value for 2014, which surpassed previous records by 1.8 °F. This graph provides an empirical example of some of the forms of change depicted conceptually in Figure 3.1. Source: NOAA.

Columbia glacier, which flows out of the Chugach Mountains into Prince William Sound in southeast Alaska, changed little between the time it was first surveyed in 1794 and 1980; since then, it has retreated some 20 km and has thinned by several hundred meters.

The oceans are another matter. Because of their vastness, yearly variations are diluted. Nonetheless, long-term records document the inexorable trends. Somewhat lost in all the talk about temperature increases on land (which is where people live and notice such things) is the accelerating warming of the world's oceans (Figure 5.2). More than 90% of the human-generated warming of the earth goes into the oceans—a gigantic heat storage. As ocean circulation transfers this accumulated heat to the water's surface, it will unleash accelerated melting of sea ice and affect atmospheric circulation, increasing the likelihood of extreme weather events. Thermal expansion of the ocean's water, along with melting of the huge ice caps of Greenland and Antarctica, will exacerbate sea-level rise.[10] Globally, sea levels have risen some 20 cm over the past century; the rate of increase has doubled over the last two decades. Oceans are changing in other ways as well.

[10] In Greenland, southeast Alaska, and elsewhere, land is rebounding and rising as the weight of coastal glaciers melts away, leading to a lowering rather than a rising of sea level in those places.

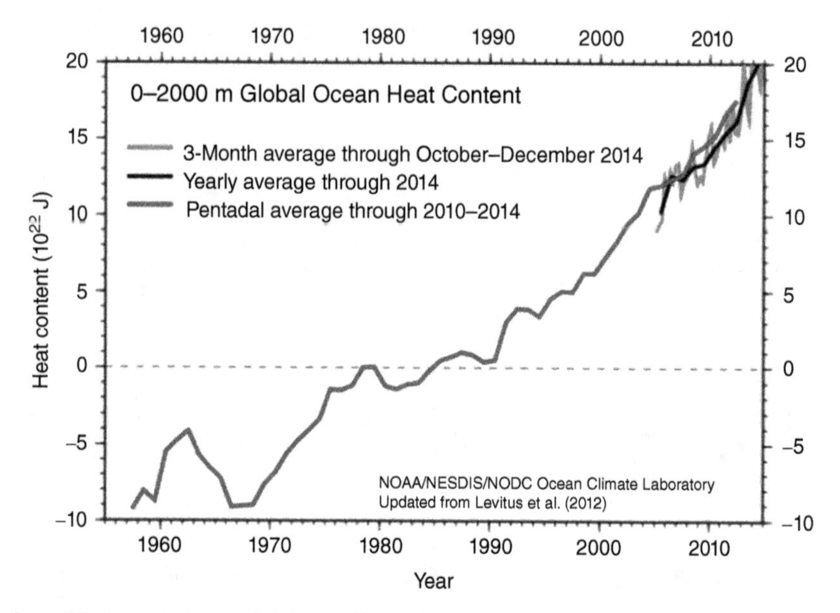

Figure 5.2 Annual average global ocean heat content, 1955–2014. Source: NOAA National Oceanographic Data Center (http://www.nodc.noaa.gov/OC5/3M_HEAT_CONTENT/).

Because they absorb CO_2, oceans are now 26% more acidic than they were at the beginning of the industrial era.[11] Estuaries are becoming more saline as tides intrude farther inland.

As Arctic sea ice has been melting, new sea lanes have opened to shipping (and potential transport of invasive species) and shorelines have been increasingly exposed to storm surges and erosion. People living in low-lying coastal areas in Alaska don't need data or models to tell them what is happening; some coastal communities have been forced to relocate inland (where they encounter melting permafrost). Polar bears may face a more daunting challenge, as they are almost entirely dependent on sea ice for hunting for their seal prey (see Essay 10, *Polar bears, golden toads, and conservation futures (2008)* (page 76)).

This litany of changes in climate, sea levels, and oceans will be familiar to anyone who has gotten this far in this book. The changes that are now underway will only become greater over the coming decades. The driving force—increased emissions of CO_2 and other greenhouse gasses—shows little sign of abating. The levels of CO_2 atop Mauna Loa in Hawai'i (Figure 5.3), which first drew widespread attention to escalating greenhouse gases, passed 400 parts per million (ppm) on April 29, 2013, which is well beyond the upper boundary

[11] IPCC (2014).

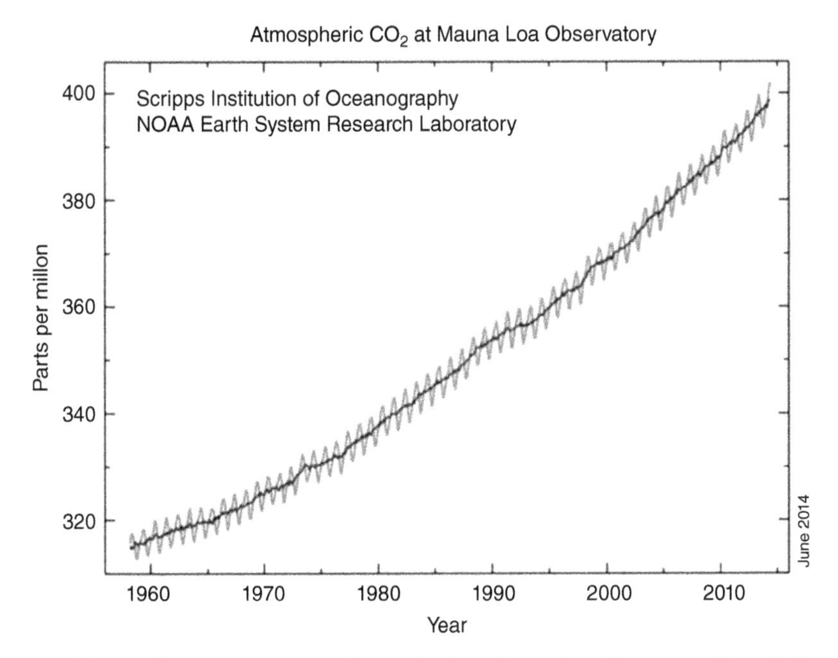

Figure 5.3 Monthly mean atmospheric levels of CO_2 at Mauna Loa Observatory, Hawai'i. The trend line represents the seasonally corrected data. Data are reported as a dry mole fraction defined as the number of molecules of carbon dioxide divided by the number of molecules of dry air multiplied by 1 million (ppm). Source: NOAA National Oceanographic Data Center (http://www.esrl.noaa.gov/gmd/ccgg/trends/).

of 350 ppm advocated by many climate scientists.[12] Even if effective emission controls were to be implemented now, climate change and its effects would continue to be felt for centuries. Attention is shifting from how to mitigate these effects, which seems to be politically intractable, to how people, agriculture, and infrastructure can adapt to the coming changes. These changes in people and their activities will have profound effects on the environmental as well as social context of conservation, effects that have scarcely been acknowledged.[13] Most conservation and management approaches have been designed for the climate of the past 50–100 years, not the one we will have in the future. Lincoln's quote, once again, is apropos.

I like metaphors, so here's one. The train (climate change) left the station long ago. We only recently realized that there even was a train, much less where it was going. We'd like to bring it to a stop before there's a wreck. We have some ideas how, but it seems too hard and it will take time. So we remain on the station platform, waiting. For what?

[12] Hansen et al. (2008).
[13] Watson (2014).

ESSAY 10

Polar bears, golden toads, and conservation futures (2008)*

I wrote this essay shortly after the polar bear was listed as a threatened species under the U.S. Endangered Species Act. The listing attracted considerable attention because the main threat to the bears—melting of Arctic sea ice—was directly linked to global warming (even though, to be politically acceptable, the listing specifically precluded identifying global warming as a threat). Since 2008, loss of polar bear habitat has accelerated as sea ice continues to disappear. Whether this has yet affected bear abundance is uncertain. People are seeing more bears more often near coastal villages in the Arctic. This has led some to conclude that their numbers are increasing; others counter that it is more likely that bears are encountered more often because they are forced into coastal areas by the lack of sea ice. Polar bears live in remote places, so it is nearly impossible to estimate overall population size accurately. Consequently, controversies have developed over how many bears there actually are and whether numbers are decreasing or increasing. There seems little doubt, however, that sea ice will continue to disappear from the Arctic. Whether polar bears will adapt to use terrestrial habitats (as some hope) or will be unable to adjust and will disappear along with the sea ice (as many fear) remains to be seen.

*The golden toad story has also continued to evolve, not because the status of the species has changed (it's still extinct) but because the explanations of the extinction have deepened. Although a relationship to climate change and a shifting of the cloud-forest habitat that I mentioned remains possible, it seems that the extinction was more likely driven by several more immediate factors. The Monteverde forest was extraordinarily dry and hot following a severe El Niño event in 1986–1987. Although over 1,500 adults were seen at breeding pools in early 1987, few tadpoles were produced from the drying ponds. Only 11 adults were found the following year, and only one (the last one seen) in 1989. A pathogenic fungus (*Batrachochytrium dendrobatidis*), which has been implicated as a cause of amphibian extinctions throughout the world,[1] may also have been involved. There is no shortage of hypotheses, but the subject of the hypotheses, the golden toad, is no longer around.*

I ended this essay with a call for "conservation futures"—designing conservation approaches with an eye toward what lies ahead rather than what brought us to where we are now. It's too late for the golden toad; I hope we still have polar bears—in nature, not just in zoos—50 or 100 years from now.

*Wiens, J.A. 2008. Polar bears, golden toads, and conservation futures. *Bulletin of the British Ecological Society* 39(3): 31–32. Reproduced with permission of the British Ecological Society.

[1] Anchukaitis and Evans (2010).

Ecological Challenges and Conservation Conundrums: Essays and Reflections for a Changing World, First Edition. John A. Wiens.

The polar bear (*Ursus maritimus*) is an icon of the Arctic. It figures prominently (as Nanook) in Inuit legends and art and, more recently, as Lyra's strong protector in Phillip Pullman's *His Dark Materials* trilogy. Polar bears are charismatic species in what is, quite literally, a coldspot of global biodiversity.

They are also the subject of mounting conservation concern. So much so that, after considerable foot-dragging and administrative dithering, in May of this year the United States government listed the polar bear as a "threatened species" under the Endangered Species Act. This means that the species is likely to become endangered (i.e., "in danger of extinction throughout all or a significant portion of its range") in the foreseeable future unless actions are taken now. This decision is noteworthy because, unlike other species that have been afforded federal protection under this act, the threat to the long-term viability of polar bears as a species stems directly from the effects of global climate change. Polar bears are dependent on sea ice for hunting ringed seals, their primary prey. And as the rate of climate warming in the Arctic accelerates even beyond the projections of the recent Intergovernmental Panel on Climate Change (IPCC) report, the threat of a complete disappearance of this essential habitat within the next few decades becomes ever more real. Based on even moderate climate-model projections, scientists with the United States Geological Survey have predicted that two-thirds of the world's polar bears are likely to disappear by 2050.

The listing of the polar bear under the Endangered Species Act was overdue (the International Union for the Conservation of Nature—IUCN—classified the species as 'vulnerable' in 2006), but this "victory" for conservation has a hollow ring. The listing has been opposed by native groups in Canada and by the State of Alaska over concerns about constraints on hunting rights and economic development, and the listing itself contains an exclusionary clause that precludes restrictions on oil and gas development that might result from the listing. Other groups have challenged the science behind the decision, arguing that it is based on computer models and laced with uncertainties. And despite the clear identification of the loss of sea ice as the primary threat to the persistence of the species, the listing explicitly excludes global warming from the threats that must be considered when protecting the bear's habitat.

The listing of the polar bear also highlights a conservation conundrum. When a species is listed under the Endangered Species Act, a process is initiated to prepare, and then implement, a recovery plan for abating the threats, arresting a population's decline, and restoring the distribution and abundance of the species to the point where its long-term survival in the wild is ensured. Recovery plans usually entail population management, captive breeding and release, control of predators or competitors, habitat protection, or similar actions. But the success rate over the 30+ years of the Act's existence is not encouraging, and a recent unpublished analysis by Mike Scott and his colleagues suggests that as many as 80% of the currently listed species may be "conservation reliant" (Scott et al. 2005), requiring continuing, species-specific conservation management to ensure their persistence.

So here's the conundrum: the decision to list a species often comes too late; most of the currently imperiled species are conservation reliant and will require long-term management actions to forestall their extinction; the threat to polar bears (either the loss of sea ice or the global warming that drives this loss) is beyond the scope of feasible management actions; many more species will be threatened with extinction as the environmental changes already instigated by climate change are played out; as these extinctions occur, and as species distributions shift with climate changes, new community complexes will be created. The resulting management challenges will far exceed our wildest dreams (or nightmares).

I suggest that it's time to rethink our approach to biodiversity conservation. We are dealing with global processes and threats that have local consequences, and we are likely to see nature rearranged in ways that are beyond our previous experience (and science). As the queue of species heading toward extinction becomes longer and more crowded and unruly, we cannot continue to manage them all individually, especially over a long haul. We need to revisit the original intent of the Endangered Species Act, "to provide a means whereby the ecosystem upon which endangered species and threatened species depend may be conserved." We need to think about conservation of the broad functionality of Nature.

This is where ecologists can help—by putting flesh on the bones of phrases such as "ecological integrity" or "ecosystem health"; by conducting the research that explores how community and ecosystem processes are likely to change with accelerating turnover in species assemblages that create new complexes and new pathways of species interactions; by devising robust ways of grouping species that will respond to management actions in similar ways; and by assessing where and how species and ecological systems are likely to be most vulnerable to the cascading consequences of climate change.

So what about golden toads? Golden toads inhabited a small area of cloud-forest habitat near Monteverde, Costa Rica. They are now extinct, most likely as a consequence of climate changes that shifted the distribution and occurrence of the cloud zone on the mountain that was critical to their survival. They were victims of forces acting at a global scale driving regional and local changes that occurred rapidly and were beyond the reach of conservation management. Polar bears are wide-ranging apex predators in an entirely different sort of ecosystem, but the same forces may imperil their survival. And the list will grow. Conservation of the future cannot rely on approaches of the past.

CHAPTER 6

Land-use change

To many people, climate change is still an abstraction based on uncertain science and models. The reality of land-use change, however, is not something that can be argued or denied: cut a forest, plow a prairie, or build houses in farmland and the results are all too obvious. Changes in how people use land and waters affect all aspects of ecological systems (see Essay 11, *Will land-use change erode our conservation gains? (2007)*, page 85).

As land has been cleared for timber, agriculture, or living space, habitat has been lost and fragmented and connections among the remnants have been fractured. John Curtis[1] provided one of the earliest and most compelling documentations of such changes, depicting the reduction in forest cover in Cadiz Township, Wisconsin, as land was cleared for agriculture following European settlement in the 1830s (Figure 6.1). By 1954, less than 4% of the original forest cover remained, and much of that was grazed by cattle; Curtis estimated that less than 1% was still in a semi-natural state.

Examples of such changes in forest cover are now legion. Amazonian forests in Brazil, for example, have been cleared and converted to pasture or agriculture at alarming rates—some 27,000 km^2 in 2004. Although deforestation rates have declined, clearing of Amazonian forests continues at an unsustainable rate. In Indonesia, oil palm plantations expanded from 6,000 km^2 in 1985 to over 60,000 km^2 in 2007,[2] mostly at the expense of tropical forests. Deforestation is not confined to the tropics; between 1996 and 2011, for example, nearly 43,000 km^2 of forest cover in the coastal areas of the United States were lost, much of it to development.[3]

I can relate another example from personal experience. In 1991, Brandon Bestelmeyer and I initiated a study of the effects of land use on biodiversity

[1] Curtis (1956). Curtis also discussed other direct and indirect effects of the changes in forest and grassland cover in the North American Midwest.
[2] http://wwf.panda.org/what_we_do/where_we_work/borneo_forests/borneo_deforestation/.
[3] NOAA's Coastal Change Analysis Program Land Cover Atlas (http://oceanservice.noaa.gov/news/sep14/land-cover.html).

Ecological Challenges and Conservation Conundrums: Essays and Reflections for a Changing World, First Edition. John A. Wiens.

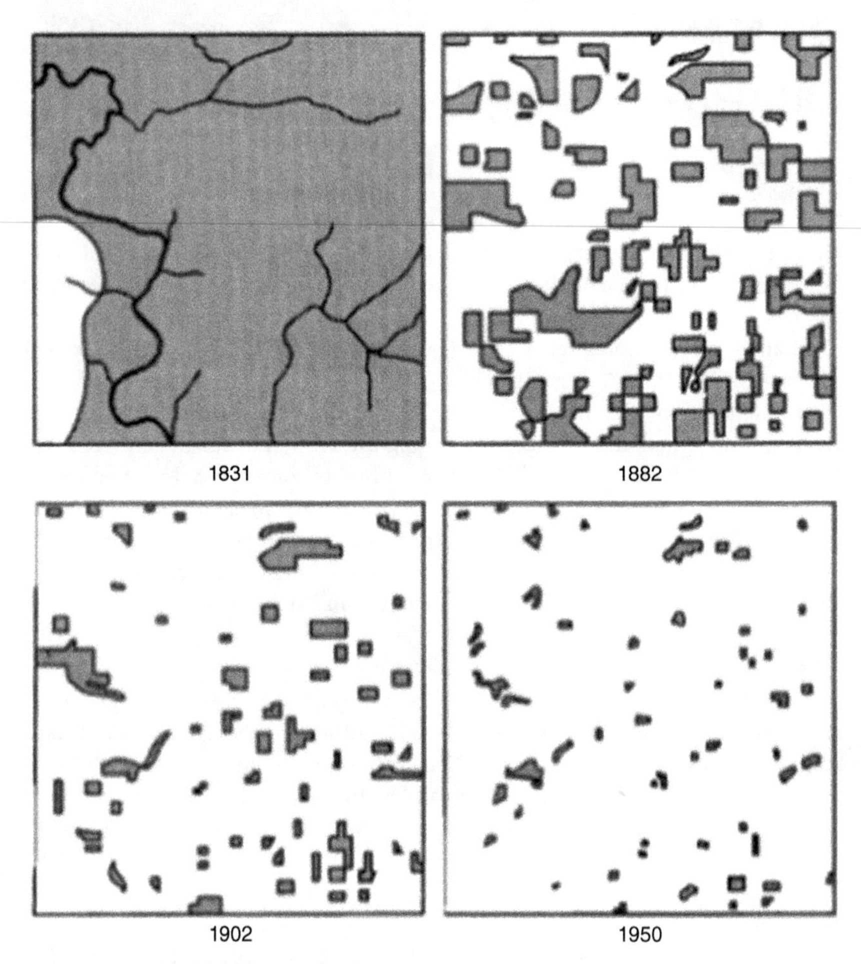

Figure 6.1 Changes in forest cover in Cadiz Township, Wisconsin, from the time of European settlement in the 1830s until 1950. .Source: Curtis (1956). Reproduced with permission of the University of Chicago Press

(as measured by diversity of ground-foraging ant species[4]) in the Chaco woodlands of northern Argentina. We sampled a gradient of land use, from intensely grazed areas close to *puestos* (the dwellings of local campesinos and the water source), to more lightly grazed areas farther from the water, to areas in which grazing was closely managed and some semblance of historical Chaco vegetation

[4]Why ants? We actually intended to survey birds and small mammals, but birds were too difficult to monitor in the dense underbrush and small-mammal densities were abysmally low, at which point we looked down and noticed all the ants running about on the ground.

structure had recovered—a classic disturbance gradient. Based on the ecological theory of the time, we expected ant diversity to peak where grazing pressure was intermediate and to be reduced at either end of the gradient—the "intermediate disturbance hypothesis." We found quite the opposite. Land use clearly affected ant diversity, but not in the way that theory predicted. Being good ecologists, we came up with a nifty explanation, and even made some management and conservation recommendations.[5]

That land use was ecologically important, however, is only part of the story. Brandon has been keeping tabs on what's been happening to our site in the Chaco. In recent years, conversion to cropland and pasture has eroded the forested area around our study site.[6] These local changes mirror regional changes in the Gran Chaco of Argentina, Bolivia, and Paraguay, where deforestation was faster between 2000 and 2012 than anywhere in the world.

Such changes have been especially great in places most suitable for productive agriculture. Almost all of the tallgrass prairie that covered nearly 1 million km^2 in North America at the time of westward expansion in the 19th century was long ago plowed under and its wetlands drained, replaced by ever-larger expanses of pastures, corn, wheat, and other crops. With advances in technologies such as center-pivot irrigation and shifts in commodity prices, the conversion of grasslands has moved west into increasingly arid areas, where crops are more susceptible to droughts.[7] In the wheatlands of Western Australia, less than 10% of the original shrubland and eucalypt woodland vegetation remains, as land has been cleared for wheat and sheep production over the last two centuries. The removal of the woody vegetation has caused the water table to rise, increasing soil salinity, rendering land unsuitable for agriculture and leading to irreversible changes in wetlands.[8] Similar conversions to agriculture have affected river floodplains and deltas, where recurrent flooding over centuries enhanced soil productivity by depositing sediments over large areas. The vast expanses of tidal wetlands encountered by early European explorers in the delta of the Sacramento and San Joaquin rivers in California, for example, have largely disappeared as the marshes were drained and diked to create some of the most productive farmlands in the United States.[9]

Land-use changes do not always lead to the blanket replacement of one cover type by another. David Theobald and others have drawn attention to *exurban* development—the extension of low-density residential areas into rural or forested areas beyond the urban fringe.[10] As people have sought to avoid urban sprawl, first

[5]Described in Bestelmeyer and Wiens (1996).
[6]Bestelmeyer (2014).
[7]Wright and Wimberly (2013).
[8]Cramer and Hobbs (2002).
[9]Whipple et al. (2012) have recounted the historical ecology of the Delta in fascinating detail.
[10]Theobald (2005) defines exurban residential development as 0.68–16.18 ha per housing unit, bracketed between urban-suburban development and rural housing, which is primarily associated with agriculture.

Figure 6.2 A forested landscape perforated by housing in Massachusetts. Source: From Thompson et al. (2014). Photograph by David Foster 2009. Reproduced with permission of Harvard Forest Archives, Harvard Forest, Petersham, MA, USA.

by moving into the suburbs and then into the countryside, exurban development has increased dramatically. Theobald calculated that there were 1,25,729 km^2 in urban and suburban housing in the coterminous United States in 2000 versus 9,17,090 km^2 in exurban development—more than a seven-fold difference. Model projections (which carry a host of sociological, economic, and political assumptions) suggest that exurban development will continue to increase.

The low density of exurban development tends to perforate rather than fragment the existing land cover. When this occurs in undeveloped landscapes such as forests, the interface between people and wildlands is increased (Figure 6.2). Peoples' houses are at greater risk of wildfire and encounters with wildlife may be more frequent, which may either enhance or diminish support for conservation depending on whether songbirds and squirrels or bears and mountain lions are encountered. In either case, management and conservation may be shifted from core habitats to the interface, because that's where people and their properties are.[11]

[11]Recognizing the growing importance of this interface, the United States Forest Service has initiated a "Forests on the Edge" project, which addresses the interactions between people and development with both private and public forested lands; see http://www.fs.fed.us/openspace/fote/index.html.

Much of the concern of ecologists and conservationists about the effects of exurban development or other land uses has traditionally been with the loss of native ecosystems and the preservation of their remnants. The reality, however, is that most of the native cover in North America, Europe, and other parts of the world was converted to human uses long ago, an inevitable consequence of human needs for food and fiber. Attention has therefore expanded from a focus on remnants of native habitats to include lands already being used—the places where people live and work. The emphasis is shifting from a black-and-white, protected or not protected perspective to one emphasizing landscape mosaics of varying shades of gray that depict multiple land uses with multiple conservation values (see Essay 24, *The dangers of black-and-white conservation (2007)* (page 203)). The mosaic, however, is dynamic, with land uses changing in response to urbanization or changing crops and agricultural practices. Documenting and understanding these dynamics and assessing their effects on landscapes is the focus of landscape ecology, which I consider in Essay 12, *Landscape ecology: The science and the action (1999)* (page 88).[12]

Changes in land use do not occur without a reason. Farmers make decisions about which crops to plant or whether to let land lie fallow based on many factors—soil condition, water availability, equipment on hand, tradition—but mostly on economics. Land uses are dynamic in part because the economic drivers change. In response to growing global demands and high prices for almonds, for example, some farmers in California's Central Valley replaced annual crops with almond orchards, which require more water throughout the growing season. Now, with California in the midst of a historically unprecedented drought, water uses are sharply restricted. Some farmers have uprooted trees they planted only a few years ago—a clear example of the interaction between land use and climate.

Land-use decisions are made at a local scale by individual landowners, who tenaciously defend their property rights.[13] Land-use policies are also set (in the United States, at least) locally. Yet the economic factors that drive these decisions and policies are increasingly global in scope. Soybean exports from South America influence farming practices in North America and support factory farming of livestock in Europe (and spur more deforestation to plant more soybeans in South America). The establishment of biofuel targets by the U.S. Congress in 2005–2007 affected corn prices in the Midwest, prompting some farmers to

---

[12]Like landscape ecology, conservation fuses science with action. In this essay I warned against the splintering of landscape ecology that might follow from its growth as a discipline. Conservation biology has grown to have perhaps an order of magnitude more participants than landscape ecology, so it is even more vulnerable to this spontaneous fragmentation. Although that hasn't happened yet, there are clear signs of internal dissonance (see Chapter 24).

[13]Eric Freyfogle explores the evolving relationship between individual property rights and the common good in *The Land We Share* (2003).

convert lands reserved for conservation to corn production.[14] And the demands of China for almost everything are altering economies and land-use practices nearly everywhere.

The challenges that land use and land-use change create for management and conservation, then, stem from the multiple scales of factors that influence decisions and policies and from the pivotal role of individual landowners in making decisions about the land. In most places that support agriculture, forestry, or grazing, the landscape is a mosaic of ownerships, fragmenting decision-making about land uses. This makes it difficult to develop a local or regional consensus about broad goals or priorities for land use, fostering piecemeal rather than comprehensive conservation efforts.

Land use, however, is only part of the picture. It determines what is there to support species. But whether the species are there depends on their distributional dynamics as well. Distributional changes, then, represent the third force of change confronting conservationists and ecologists, which I consider next.

[14] Tiffany (2009) and Dale et al. (2010); escalating prices also affected the availability of corn to make tortillas in Mexico.

ESSAY 11

Will land-use change erode our conservation gains? (2007)*

I made several points in this essay that reappear time and again throughout this book. One is the need to consider the lingering imprints of past land uses on what we see now. These anthropogenic legacies from decades, centuries, or millennia past mingle with current practices and natural disturbances to create a mélange of patterns that color and distort what we regard as "nature" and strive to conserve. Another is the need to consider what is likely to happen in the future and plan accordingly. Including climate change and its effects is essential, of course, but conservation is as much a societal activity as an ecological or scientific one, so sociopolitical factors and globalization must be part of the equation. Finally, almost as an afterthought, I drew attention to the landscape matrix in which protected areas—a traditional focus of conservation efforts—are embedded. Usually, most of this matrix is unprotected, but that doesn't mean that it lacks conservation value. Conservation in one form or another can be applied (nearly) everywhere.

Picture a preserve—a "last great place" (the marketing label of The Nature Conservancy), or even a pretty good place. Conservationists identify these places and prioritize them through rigorous planning protocols (see, e.g., Groves 2003), but how can we ensure that these places will retain their conservation value over the long haul, or even that they will continue to be the right places? The world, after all, is dynamic.

Climate change and its ecological and societal impacts are (finally!) receiving widespread attention. I'd like to focus instead on a related and arguably more urgent issue: land use and land-use change. Human land uses are the major agents of habitat loss throughout the world, and rates of land-use change are accelerating. Examples are legion. Some 4.5 million ha of wetlands disappeared in the United States between the 1950s and the 1970s, chiefly through conversion to agriculture. Urban land cover increased 348% in the Greater Yellowstone Ecosystem between 1975 and 1995. In Western Australia, some 93% of the native vegetation had been converted to agriculture by the early 1960s, leaving only scattered woodland

*Wiens, J.A. 2007. Will land-use change erode our conservation gains? *Bulletin of the British Ecological Society* 38(3): 39–40. Reproduced with permission of the British Ecological Society.

Ecological Challenges and Conservation Conundrums: Essays and Reflections for a Changing World, First Edition. John A. Wiens.

remnants. In semiarid regions of New Mexico and Idaho, grazing and fire have passed thresholds and produced essentially irreversible changes in land cover from grassland to shrubland (or vice versa). The list goes on.

It is naïve, of course, to think that only recent land uses affect the integrity of the habitats we strive to conserve. Seemingly "natural" tropical forests in Puerto Rico bear the imprints of agricultural land uses centuries ago, and the species richness of plant communities in some French forests still reflects the effects of a brief period of agricultural land use during the Roman occupation in AD 50–250. The widespread dominance of oak and hickory in forests of the Midwestern United States likely had its origins in the activities of Native Americans, and Aborigines have used fire to shape Australian landscapes for many millennia. What we see today contains the legacies of past land uses. "Natural" is a relative term.

Understanding the effects of past land uses on natural habitats and biodiversity is useful only insofar as it helps us select the places we wish to protect and manage the places we don't protect. What is really needed is knowledge that can help us anticipate and adjust for future land-use changes. This requires an understanding of the underlying socioeconomic and political forces that contribute to land-use change. In Eastern Europe, for example, the breakup of the former Soviet Union led almost immediately to the replacement of the extensive monocultures of collective farms by smaller, more diverse single-family farms, enhancing the heterogeneity and wildlife value of local landscapes. The recent expansion of soybean agriculture in parts of Brazil was fostered in large part by global economics and advances in agricultural technology. The current rush to embrace biofuels as energy sources is causing massive cropland conversions in the central United States and a cascading array of ripple effects as corn prices escalate. And we are now seeing shifts in the forest industry from traditional timber harvesting to the development of forest farming for fiber production using exotic or engineered tree species, accompanied by a movement into areas in the southern hemisphere previously used for agriculture or grazing. In the future, demographic changes such as the retirement of the baby-boom generation, technological advances such as telecommuting, and economic expansion in developing countries (witness China!) will all contribute to major shifts in land uses. Globalization means that land uses in one part of the world are increasingly influenced by factors elsewhere in the world, and those pesky thresholds in ecological responses to land-use change portend that we are likely in for some rude surprises.

My emphasis here on land-use changes is not intended to leave climate change out in the cold (so to speak), Climate change undoubtedly will influence land-cover patterns at a variety of scales, but the reverse may also be true. A "top story" on the NASA website, for example, was headlined "landcover changes may rival greenhouse gases as cause of climate change" (www.gsfc.nasa.gov/topstory/2002/20020926landcover.html). Clearly, consideration of either climate change or land-use change in isolation from the other provides only a partial picture of the changing context of our conservation actions. How we deal with the two, however, is likely to be quite different. Considerations of climate change tend to start at the global level and work down, whereas land use is essentially local, and the effects of land use and land-use change amplify upwards to broader scales. Consequently, while there may be some hope of dealing with climate change through broad multinational policy accords, land use is likely to remain immersed in a morass of local and idiosyncratic policies and politics, particularly in the United States.

So, will land-use change compromise conservation planning efforts if we ignore it or just pay it lip service? You bet! Think of what the surroundings of protected places will look like in 10 or 20 or 50 years. Think of what the effectiveness of conservation will be *within* those places if we ignore the surroundings. Think of the multiple ways in which conservation progress may be thwarted by the accelerating pace of economic globalization and changing land values and land uses in many of the biodiversity "hotspots," or even the "coldspots," of the world.

What can we do? Actually, quite a lot. Many conservation strategies are aimed at influencing land uses or lessening their impacts on biodiversity, so there is a strong foundation of experience on which to build. We could bolster some of this experience by using imagery to assess the form and magnitude of land-use changes surrounding protected areas since they were placed under conservation protection. Have they become islands? Have the connections across the landscape disappeared? But we need also to couple such assessments with analyses of the potential consequences of different scenarios of future land-use changes. We need to develop ways of assessing "conservation futures."

So what's the bottom line of this ranting and raving?

- If we don't look to the future, we probably won't be very effective in protecting our present conservation investments or investing wisely.
- Climate change and land-use change will affect these investments, at multiple scales.
- We are moving to consider climate-change impacts much more aggressively; unless we do the same for land-use change, success at dealing with climate change will be illusory.
- Most of our attention is being given to "protected lands"; we must also recognize that the matrix of unprotected lands—the places where people live and work—can contribute to the goal of preserving biodiversity.

ESSAY 12

Landscape ecology: the science and the action (1999)*

I've always been interested in spatial patterns and relationships in ecology, so when landscape ecology emerged as a discipline with an identity and name in North America in the early 1980s, I began calling myself a landscape ecologist rather than (or in addition to) an ornithologist or ecologist. One thing led to another, and, in 1996, I was elected President of the International Association for Landscape Ecology (IALE), the umbrella organization for the discipline. This provided a platform for pontificating beyond the academic classroom, so I wrote this essay.

Since its beginnings in Europe, well before it reached the shores of North America, landscape ecology has blended science with action. As the discipline has grown, centripetal forces have now and then threatened to fragment landscape ecology into basic or applied subdisciplines. I cautioned against such splintering, and so far it hasn't happened—landscape ecology continues to attract scientists and practitioners with widely disparate interests.

I also emphasized what might be called a biocentric approach to landscapes. We create and manage landscapes according to human interests and perceptions, but the ecological effects of these actions are mediated through biological organisms, populations, and ecosystems. Applications of landscape ecology in conservation may require adjusting the scales on which landscape patterns and processes are considered.[1] This was (and still is) the perspective of one who came into landscape ecology from "mainstream" ecology; those entering the discipline by other avenues—agriculture, landscape architecture, geography, forestry, economics—had (and still have) a more anthropocentric perspective. Which is fine. This is what gives the discipline its vitality.

More than most disciplines, landscape ecology bridges the gap between science and practice, between question-driven research and problem-driven applications. It does so in a wide variety of guises. Browse any volume of *Landscape Ecology* or attend any conference of landscape ecologists and one cannot help but be amazed by the diversity of topics and approaches represented. People such as Zev Naveh,

*Wiens, J. A. 1999. Landscape ecology: the science and the action. *Landscape Ecology* 14: 103. Reproduced with permission of Springer Science+Business Media.

[1] Indeed, "landscape" is generally defined in terms of human perceptions of a countryside or land area. I encountered considerable resistance (bordering even on scorn) when I suggested that landscape ecology could apply at the scale of beetles or ants; see Wiens and Milne (1989).

Ecological Challenges and Conservation Conundrums: Essays and Reflections for a Changing World, First Edition. John A. Wiens.

Monica Turner, Rob Jongman, or Lenore Fahrig are interested in different problems, use different approaches, and perceive landscapes in different ways. They are drawn together by their shared interest in the spatial dimensions of their problems.

The diversity of landscape ecology is a great strength. The sharing of problems, perspectives, and procedures among individuals leads to insights that otherwise would remain hidden. But this diversity also poses a threat. Landscape ecology is growing rapidly, and the diversity of approaches and perspectives that gives the discipline its vitality contains the seeds of polarization. As potential subdisciplines reach critical mass they may break away; landscape ecology may become as fragmented as the landscapes we study.

To avoid such splintering will require a conscious and concerted effort. The challenge, I believe, is to link the *science* of landscape ecology with the *action* of landscape ecology. This linkage is necessary to maintain the unity of the discipline, but there are more important reasons to forge this union. If the action of landscape ecology is to make valuable and lasting contributions to such areas as land-use planning, environmental management, or natural resource conservation, it must be firmly anchored in the science of landscape ecology.

Landscape ecology deals with the structure, dynamics, functioning, and scaling of landscapes. Because landscapes are such an integral part of human history and culture, we are predisposed to view these elements of landscapes through a filter that is tinged by human perceptions and values. The actions of landscape ecology are usually conducted in the context of these perceptions and values. We manage landscapes or develop land-use policies in relation to goals that ultimately are defined in sociological, political, and economic terms. But the subjects of these actions—landscapes and the biota or resources they contain—are not members of the same cultural system as humans, and our goals or values may often be irrelevant or antithetical to their well-being or persistence. They may not play the game according to our rules.

Developing a strong foundation for our actions thus requires that we shed our anthropocentric perceptions of landscapes and their values and study landscape structure, dynamics, functions, and scaling on their own terms. This is where the science of landscape ecology comes in. As with any science, landscape ecology is characterized not so much by the questions that are asked as by the way they are answered, not so much by the tools that are used as by the way they are used. As scientists, landscape ecologists now use a variety of sophisticated tools—spatial statistics, GIS, spatially explicit computer models, and experimental model systems to name but a few. These tools are applied within a framework that emphasizes research conducted with quantitative rigor and objectivity, regardless of the problem posed or question asked. Only through such a framework can we develop the predictions and principles needed to guide actions along reliable and productive pathways.

In her book *Placing Nature: Culture and Landscape Ecology* (Island Press, 1997), Joan Nassauer observed that "landscape ecology should be as much about doing as it is about thinking." If landscape ecology is going to deliver on its promise, however, the doing and the thinking must involve a coupling of action and science. This is the theme of the 5th World Congress of the International Association for Landscape Ecology (IALE), to be held in the Rocky Mountains of Colorado in late summer 1999. The Congress will provide an opportunity to see how this coupling is to be accomplished, and what insights and initiatives emerge.

CHAPTER 7

Distributional changes: invasive species

One consequence of a changing world is that species don't stay put. Range maps depict where a species occurs as a static blob, fixed in space and time. However, most species' distributions are dynamic. Cattle egrets, native to tropical and subtropical areas in the Old World, first appeared in North America in the 1940s, apparently having flown across the Atlantic Ocean to South America and then moved northward. They were breeding in Canada by the early 1960s, and have now expanded to California and much of Latin America. Barred owls, denizens of forests in eastern North America, followed riparian corridors westward during the last century and became established in western Canada and the Pacific Northwest by the 1970s.

These and many other species may shift distributions of their own accord in response to land-use changes, weather, climate, population dynamics, species interactions, and a multitude of factors. In other cases, the range shifts are consequences of human actions. People have been moving plants and animals around for as long as people have been moving around, which has been a long time indeed.[1] Polynesians carried a variety of plants and animals on their voyages among the Pacific islands long before European contact, introducing taro, yams, coconuts, bananas, dogs, pigs, and rats. Most of the plants and animals that now occur at lower elevations on the Hawai'ian Islands have been introduced (intentionally or not) by people. The lush vegetation, exotic-looking birds, and mongooses that tourists encounter bear little resemblance to the flora and fauna that evolved there. Lowland Hawai'i is an unmanaged arboretum, a menagerie of aliens that have displaced or eradicated native species.

Human-aided introductions of species are inherently neither bad nor good; this depends on the intent of the introduction, its outcome, and who is making the judgment. The introduction of *Opuntia* (prickly pear) cactus to Australia to serve as fencerows around pastures turned out to be bad when the plants escaped cultivation and spread widely; the introduction of *Cactoblastis* moths to control the cactus was good, because it worked (Figure 7.1). Following the successful biological

[1]This is one reason for thinking that the Anthropocene began well before the Industrial Revolution.

Ecological Challenges and Conservation Conundrums: Essays and Reflections for a Changing World, First Edition. John A. Wiens.

Figure 7.1 An area in Queensland, Australia, formerly covered with prickly pear cactus (*Opuntia stricta*). The cactus was introduced to the region in 1926, and 3 years later the moth borer (*Cactoblastis cactorum*) was introduced as a biological control agent to reduce populations of the cactus. Source: Encyclopedia Britannica (http://www.britannica.com/EBchecked/media/138479/Area-in-Queensland-Australia-covered-with-prickly-pear-cactus-an?topicId=336811) © The State of Queensland, Department of Agriculture & Fisheries 2016.

control of *Opuntia* in Australia, however, moths were introduced to control exotic or indigenous *Opuntia* species in South Africa, Mauritius, Hawai'i, and several Caribbean islands, with some success. Not content to stay where it was introduced, however, *Cactoblastis* dispersed to other islands in the Caribbean and, in the early 1990s, to Florida, where it threatens several native *Opuntia* species as well as commercial stocks. *Cactoblastis* is now treated as a pest in the United States because of its potential to devastate areas in the southwest and Mexico that are rich in endemic *Opuntia* species. Several *Opuntia* species are also important agricultural and subsistence resources in Mexico. Should the moths disperse to these areas, they would cause major ecological and economic harm. Suggestions that the moths might be controlled by releasing parasitoids have raised concerns that the parasitoids might also reduce populations of other pyralid moths that currently keep native *Opuntia* species in check, resulting in population increases and invasions by the native *Opuntia*.[2] In ecology, nothing is as simple as it first seems.

In many instances, however, the potential for an introduced species to have devastating impacts on ecological systems is not in doubt—they are invaders, *invasive* species. In *The Ecology of Invasions by Animals and Plants*, the British ecologist Charles Elton drew attention to the havoc that species coming from elsewhere could wreck on native, well-balanced communities and ecosystems.[3] There are many well-known examples. Brown tree snakes colonized the Pacific island of Guam a few years after World War II, probably as stowaways in ship or air cargo. In the absence of predators and with abundant food, the snake population exploded, leading directly to the extinction of a dozen native bird species and indirectly reducing vegetation diversity through their effects on native pollinators. The ecology of Guam has been utterly transformed by this one species.[4]

"Invasive" species, "invasions," and like terms clearly portray such species as the enemy. Elton made liberal use of such military metaphors, writing:

> *I have described some of the successful invaders establishing themselves in a new land or sea, as a war correspondent might write a series of dispatches recounting the quiet infiltration of commando forces, the surprise attacks, the successive waves of later reinforcements after the first spearhead fails to get a foothold, attack and counter attack, and the eventual expansion and occupation of territory from which they are unlikely to be ousted again.*[5]

Invasive species can alter ecosystems through multiple pathways. Water hyacinths were introduced from tropical South America to the United States in 1884 as gifts to visitors at the New Orleans World Fair. The plants spread quickly, and within a few years many waterways in the South were choked by the plant, depleting oxygen in the water, killing native aquatic plants and fish, and impeding shipping.[6] In Australia, cane toads were introduced from South America

---

[2]Zimmerman et al. (2004) provide a comprehensive review of *Cactoblastis* and its effects.
[3]Elton (1958). Elton wrote in the mid-1950s, when notions of balanced, equilibrial communities were still in vogue. The ecological research and thinking spawned by Elton's work were recognized in *Fifty Years of Invasion Ecology. The Legacy of Charles Elton* (Richardson, 2011). Elton's book is full of insights that apply today, and is well worth a read.
[4]See Fritts and Rodda (1998).
[5]Elton (1958: 109).
[6]In what is surely one of the more imaginative proposals for control of an invasive species, a measure was introduced in 1910 in the U.S. House of Representatives to release hippopotamus from Africa into choked waterways in Louisiana; the hippos would eat the water

in 1935 to control native beetles that damaged sugar cane crops. The toads were ineffective control agents but spread from Queensland along the eastern and northern coasts, and are continuing their march around the continent. The toads produce poisonous toxins in their parotid glands, which have severe impacts on mammals and reptiles that prey on them. Water buffalo, introduced into northern Australia in the mid-1800s as a meat source for isolated settlements, spread into swamps and wetlands when the settlements were abandoned a century later. Their wallows, trails, trampling, and consumption of native vegetation have devastated floodplain ecosystems.[7]

These are but a few of many examples of what can happen when a species colonizes or is introduced to a new place. They are immigrants; like Robert Heinlein's "stranger in a strange land,"[8] they are aliens faced with new challenges and new opportunities. Most fail to become established or settle down and integrate into the new setting—they become naturalized, as it were. However, a few run rampant through an ecosystem.

To become invasive, a species first must be a successful colonizer: it needs to get to a place and become established. Getting there is increasingly facilitated by people as transportation links even remote places with global networks. The expanding networks of global shipping or airfreight routes have the potential to spread invasive colonists in ballast or cargo far and wide. Once introduced, the likelihood of an immigrant becoming established is enhanced by the attributes that ecologists have long associated with good colonizers of disturbed lands: high reproductive output, rapid maturation, broad ecological tolerances, and high dispersal capacity. Invasive species are also often aggressive competitors, and the likelihood that they will cause major ecological disruption increases if they are effective predators (such as brown tree snakes) or large herbivores (such as water buffalo).

Not every species that will eventually become invasive plays its hand immediately, however. It may take some time after a species reaches an area to gain a foothold (or put down roots). In some instances, an introduced species that initially seems benign and well integrated may begin causing harm only when changing land use alters the ecological setting, perhaps by removing predators or other factors that control its population or by providing new resource opportunities. The story of *Cactoblastis* and *Opuntia* illustrates how a species believed to be beneficial can become an invasive factor in a different setting.

In some situations where an invasive species has effectively replaced a native species, other species may come to depend on the replacement. This can create conflicts if the other species are themselves subjects of conservation or

hyacinth while also providing meat for people. The "American Hippo bill" fell one vote short of passage in the House.

[7]Tim Low describes the particular vulnerability of Australia to invasive species in *Feral Future* (2001).

[8]Heinlein (1961).

management actions. For example, saltcedar (tamarisk), which is native to Asia and the Mediterranean, was introduced into the southwestern United States toward the end of the 19th century to control wind and soil erosion. It rapidly spread through riparian habitats, choking out native willows and cottonwoods, altering streamflows and groundwater, and increasing soil salinity—a classical invasive species. As the native riparian vegetation disappeared from many areas, however, southwestern willow flycatchers turned to nesting (successfully) in saltcedar thickets. The flycatcher is a federally endangered species, however, creating conflicts with the (also successful) efforts to remove saltcedar using a biological control agent, saltcedar leaf beetles, introduced from Russia in 2001. Bowing to the legal requirements of the Endangered Species Act, releases of the beetles in saltcedar thickets have been suspended, to the great dismay of landowners and land managers.[9]

Such conservation conflicts are created by the complexity of ecological systems, which opens the door to multiple priorities and agendas and differing views about what is important—values. I'll consider such conservation conundrums later on in the context of setting priorities (Chapter 17) and in the broader context of values (Chapter 24). Because values derive from human cultural and social norms that differ from place to place and change over time, however, we should consider sociopolitical factors as another force of change confronting ecology and conservation—the subject of the next chapter.

[9]Nonetheless, the beetle has now dispersed to Arizona, which is currently in the grips of a drought. Farmers and ranchers who claim that saltcedar sucks up precious water welcome the beetle, but others caution that the vegetation that might replace saltcedar along watercourses could consume even more water. In any event, the declining water flows in rivers are more likely consequences of climate change and drought; saltcedar is a convenient scapegoat for what is destined to become a recurrent problem.

CHAPTER 8

Societal, cultural, and political change

Ecology and conservation are not conducted in an idealistic vacuum, free of external influences. They are guided and constrained by the social, cultural, and political attitudes and agendas of a particular place and time. Because conservation is concerned with the effects of human actions on ecological systems and with marshaling efforts to reduce those effects, it lies more directly at the crossroads between science and society than does ecology, although both are ultimately manifestations of culture and society.[1]

Whether a species is useful, benign, or harmful depends on what it benefits, what is harmed, and how it is valued by people, either as a component of nature or as a resource or commodity. The conversion of *Cactoblastis* from useful to harmful that I described in Chapter 7 had little to do with the moth doing what moths do, but with whether people viewed *Opuntia* as something bad to be controlled or eradicated (as in Australia) or as something good to be protected (as in Florida or Mexico). Feral pigs, first introduced to the Hawai'ian Islands by Polynesians, have wreaked ecological havoc by uprooting native vegetation, creating wallows, and spreading invasive weeds and epizootic diseases. Pigs have also become important to recreational and subsistence hunters, however, and because pigs have been in Hawai'i for more than a millennium they have become part of the culture. Consequently, control efforts are often resisted. Depending on whose values are considered, feral pigs are either bad or good. Striped bass introduced into the California Delta are either good or bad, depending on whether you are a sport fisherman angling for the bass or someone concerned about imperiled native fish species that are eaten by the bass.

It's all a matter of values. The difficulty, of course, is that society is diverse—there are multiple values, each of which is claimed to be preeminent by those who hold

[1]At least as a basic science, ecology is more concerned with finding out how and why things are rather than how they should be or how to keep them that way. But ecology is not immune to social, cultural, and political influences, perhaps most directly through government funding priorities, which affect which questions are (or are not) asked by researchers. The culture of the science itself also has a strong effect, through the power of the paradigm I discussed in Chapter 2 and Essay 4.

Ecological Challenges and Conservation Conundrums: Essays and Reflections for a Changing World, First Edition. John A. Wiens.

it. The Endangered Species Act is held in great reverence and is fiercely defended by conservationists and environmental advocates, but it is opposed with equal passion by many landowners and developers, who think it restricts their property rights or places insignificant species that "don't matter much" above the economic interests of people. Such conflicts can become intense or even violent when they involve water, as I discuss in Essay 13, *Wildlife, people, and water: Who wins? (2012)* (page 100).

Laws, regulations, or cultural prohibitions are developed by societies to resolve conflicts among values, or at least to establish which values dominate and will govern individual or collective actions. Historically, many cultures had restrictions on harvesting or hunting native plants or animals. Raul Valdez recounts how hunting was tightly restricted in the Mongol Empire of Genghis Kahn and his descendants during the 13th century (and by Middle Eastern cultures several millennia before that), and Matthew Brown and Robert Haworth have described how cultural conventions among the Walpiri people in the Tanami Desert of central Australia governed the use of bush tucker and management of the land.[2] Although not everyone agreed with the strictures, they had the force of law, often because they were couched in religious or spiritual terms or, in the case of Genghis Kahn, power.[3] As modern societies have become larger and more complex, the regulations and restrictions governing how people relate to the natural world have mushroomed, aided and entrenched by their codification into laws.

Laws are intended to ensure certainty and stability in human actions, telling people what they can and can't do and specifying penalties for the latter. Philosophically and pragmatically, however, this emphasis does not mesh well with the dynamic and uncertain nature of ecological systems. The stipulations of environmental laws often set boundaries that restrict flexibility in management or conservation. The Endangered Species Act, for example, imposes restrictions not only on what farmers, ranchers, and developers can do that might jeopardize a threatened or endangered species, but also on what managers or scientists can do to help the species. The use of adaptive management, which could enable managers to respond to the rapid and unanticipated changes that happen in dynamic and incompletely understood systems, may be precluded by the inflexibility of legal restrictions and regulations.[4]

---

[2]Valdez (2013) and Brown and Haworth (1997). Such cultural restrictions are perhaps the earliest form of conservation.

[3]Any taboos that ancient Polynesians may have had on overharvesting of native plants and animals, however, apparently did not apply to the indigenous flora and fauna they encountered as they spread across the Pacific two millennia ago; the record of extinctions accompanying the appearance of Polynesians on islands is extensive. See Steadman (2006).

[4]Angelo (2008) and Allen et al. (2011) discuss the incompatibility of adaptive management with legal systems.

Laws, although intended to be firm and fixed, are not immutable. In a democracy, laws are made by legislative bodies populated by politicians who are elected and who are therefore (in principle) responsive to public opinion. And public opinion as well as the social and cultural forces that drive it (e.g., advertising, information and social media, religion), is anything but firm and fixed. Attitudes and values change, and the changes may affect how laws are interpreted by the courts or whether they are vulnerable to legislative repeal as the balance of political power shifts.

Recent events in Australia provide a sobering example. Australia has the dubious distinction of being among the world's leaders in CO_2 emissions per capita. After years of political haggling, the Labor government joined with the Green Party in 2012 to pass a tax on carbon emissions. This was widely hailed internationally as path-breaking progress, and the approach seemed to be working. However, costs of gasoline and electricity went up and public support for the carbon tax, which was always weak, faded. The (conservative) Liberal Party won a decisive victory in the 2013 election, and in 2014 the Prime Minister, Tony Abbott, followed through on his pledge to repeal the carbon tax, leaving the country without any clear climate policy.[5] What the political process gives, it can take away[6].

Clearly, changes in societal and political attitudes can affect how conservation is implemented and what is conserved. Setting aside nature reserves or restricting activities that might imperil at-risk species may generate broad public and political support in good times, but under economic or environmental pressure this support may rapidly crumble. Faced with mounting budget shortfalls, for example, California first cut back on the maintenance of state parks and then, in 2011, announced that 70 parks (nearly a quarter of the entire system) would be closed. Public outcry was immediate, and several nonprofit organizations and local governments stepped into the breach to ensure that some of the parks stayed open. Although in the short term, this (mostly) solved the problem, it cannot be a long-term solution. The episode illustrates the vulnerability of protected areas, and of conservation more generally, to the ebb and flow of economic and political tides.

Faced with a crisis, economic strain, or climate change, politicians and the public may question not only the value of environmental laws and the need for conservation, but also the science on which management and conservation are based. And, although conservation relies on science, it banks on public support. Without the engagement of the public as protectors, users, and inhabitants of the environment, conservation would flounder. But "the public" is not some homogeneous mass of people. Rather, it is a complex amalgamation of races, ethnicities, interests, beliefs,

[5]Not one given to subtlety, Tony Abbott once described the science behind climate change as "absolute crap" and any link between climate change and an increased frequency and severity of wildfires as "complete hogwash."
[6]An update: In September 2015 Tony Abbott was ousted as leader of the Liberal Party and replaced as Prime Minister by Malcolm Turnbull. How this will affect environmental policies is yet unclear.

employments, wealth, and much else, all of which affect attitudes and perceptions about "nature" and its values, and thus about conservation.

The amalgamation that makes up a society changes over time. As the ethnic or racial composition, age structure, wealth distribution, or other demographic attributes of a society change, the collective perspective on the environment also changes. These changes, in turn, slowly but surely work their way through the political process to affect how government funds are allocated or which laws and regulations are created, amended, or overturned. Funding for the National Park Service, for example, is based in part on visitation rates. Racial and ethnic groups differ in their use of and support for parks, so changes in the make-up of a population may alter the support base for parks and, consequently, their role as protected areas for conservation. If trends for children to be increasingly disengaged with nature hold into adulthood, support for conservation may change as the age structure of a population changes. However, other societal and cultural factors may also be in play. For example, Oliver Pergams and Patricia Zaradic attributed a decline in recreational use of parks and public lands in part to "videophilia," the increased attachment of people to video games and the internet.[7] Conservation in the digital age may need to change its message and appeal.

The sociopolitical context of conservation and ecology, then, is an ever-changing tableau of attitudes, perspectives, interests, laws, and agendas that is grounded on a changing base of population demographics. Which brings us to the fifth force of change: population growth.

[7]Pergams and Zaradic (2006, 2008).

ESSAY 13

Wildlife, people, and water: who wins? (2012)*

The history of the western United States is written in many ways—religion, gold (itself a religion of sorts), conquest, exploration, blood—but perhaps especially water. Westward expansion during the 19th century was limited by water, and government programs to harness and direct water ("reclamation," it was called) galvanized the agricultural and urban development that followed.[1] Demands for water soon exceeded supplies, leading to conflicts and, in turn, a byzantine array of water rights, rules, and regulations over who gets the water.

In this essay I touched on the labyrinthine interplay of forces that can be unleashed when there is not enough water to go around. I wrote this essay in the summer of 2012, when California and southern Oregon were in the beginnings of a drought. Since then the drought has deepened, even further impeding efforts to resolve water issues in the Klamath Basin and California Delta. In the Klamath Basin, a hard-won agreement that would remove dams to restore river flows, enhance salmon migration, provide water for farmers and ranchers, and restore treaty rights to tribes languished in the U.S. Congress, a victim of conservatives who argued that it favored fish over people. Then, in 2013, the priority rights of Klamath tribes "since time immemorial" were finally recognized by the government and courts, allowing them to allocate water for the protection of endangered and spiritually valued fish. As the drought intensified during that summer, the tribes exercised their rights, leaving several hundred farmers and ranchers without irrigation water. Faced with the prospect of a renewal of hostilities, the tribes, government agencies, and irrigators forged a new agreement in the spring of 2014. This agreement also requires Congressional approval (and funding) and is not embraced by all the tribes (nor all the farmers and ranchers), so its fate is uncertain. Meanwhile, the drought goes on.

In the California Delta, the issues are if anything even more complex. The water that passes into and through the Delta is used by some 25 million people and supports agriculture that provides a substantial share of the nation's produce. As the current drought reaches a severity not seen for at least the past 1,200 years, the ecological, social, political, and economic effects are being felt more widely and more deeply. Because the situation in the Delta illustrates so clearly how conservation is at the mercy of multiple forces of change, I discuss it in greater detail in Chapter 10.

*Wiens, J.A. 2012. Wildlife, people, and water: who wins? *Bulletin of the British Ecological Society* 43(4): 54–55. Reproduced with permission of the British Ecological Society.
[1]Recounted with flair in Marc Reisner's book, *Cadillac Desert* (1993).

Ecological Challenges and Conservation Conundrums: Essays and Reflections for a Changing World, First Edition. John A. Wiens.

Through all the debate, quarreling, shouting, and negotiating, however, one reality remains inescapable: the amount of rain or snow that falls, glaciers that melt, or water remaining in underground aquifers is finite—we must learn to manage, store, use, and replenish what nature provides. People who believe that the solution to water scarcity is creating more water (as some politicians have argued) are drilling a dry well.

When water in the Klamath Basin of southern Oregon and northern California is scarce, farmers who need it for irrigation get first rights. What they don't need next goes to three endangered fish species, and only after that to a complex of national wildlife refuges that provide critical stopover habitat for migrating waterfowl. As a result of low winter precipitation or drought, water flowing to the refuges can be cut off—as it was in early 2012.

The result was the driest conditions in the Lower Klamath refuge in 70 years. Portland's daily newspaper, *The Oregonian*,[2] picked up the story:

> *A cut-off of water supplies to a key Klamath Basin national wildlife refuge contributed to the deaths of 10,000 or more birds this year, the most in a decade.*

Without water, wetland habitat was reduced by a half and avian cholera spread, killing the birds (mostly snow geese, *Chen caerulescens*; Figure 1).

Water issues in the Klamath Basin have always been contentious, pitting water users against one another and against the needs of species and ecosystems. In 2001, another drought year, fish got the water instead of farmers, prompting public outrage, protests, and political actions. The next year farmers got most of the water, and there was massive fish mortality. Conflicts bubble to the surface whenever water becomes scarce, as may happen more often in the future.

Such conflicts are not restricted to the Klamath Basin. In California, the epicenter of water issues is in the Delta where the Sacramento and San Joaquin rivers meet before emptying into San Francisco Bay. The Delta provides water to two-thirds of California households. Delta water also irrigates millions of hectares of farmland, which account for nearly half of the nation's domestically grown fresh produce.

The Delta is one of the most heavily altered ecosystems in western North America (Figure 2).[3] Dams on upstream tributaries have regulated water flowing into the Delta, changing natural streamflows. Two massive water projects have created a network of diversions and pumping stations to move water to agricultural and urban areas elsewhere in California, and flows are further channeled by a labyrinth of levees, many built over a century ago. The extensive tidal wetlands that once covered the Delta have been replaced by farmland, much of which has subsided and is now well below sea level. Because the levees are old and lie in an

---

[2]Learn (2012)

[3]Additional information about the Delta can be found in Lund et al. (2010), Hanak et al. (2011), and National Research Council (2012a).

Figure 1 Snow geese on the Lower Klamath Wildlife Refuge in the Klamath Basin in a good water year. Source: U.S. Fish and Wildlife Service and Tupper Ansel Blake.

Figure 2 A waterway in the California Delta, channelized by levees. Note the subsidence of the farmland on the left. The water is at sea level and subject to tidal effects. Source: California Department of Water Resources.

earthquake-prone region, they are susceptible to failure, threatening the farmland and the reliability of water supplies.

The Delta is also home to several critically endangered fish species. Populations of these and other species have declined to critical levels as a result of the altered flow regimes, loss of wetland habitat, mortality at the pumping stations, the effects of invasive predators, disruption of trophic webs, and changes in salinity due to tidal intrusion. Legal mandates to protect the endangered fish and the salmon migrating through the region now limit the amount of water that can be exported from the Delta. As in the Klamath Basin, the conflicts over who gets the water—people or wildlife—are escalating as demands for water grow.

In an attempt to resolve the continuing disputes between water users and environmentalists and address concerns about future changes in the Delta and its water, the California legislature passed the Delta Reform Act in 2009. This act establishes "co-equal goals" for the future of the Delta: "providing a more reliable water supply for California and protecting, restoring, and enhancing the Delta ecosystem." It's a remarkable piece of legislation. The ecological integrity of the Delta is given equal footing with the needs of 25 million people in California. The devil, of course, is in the details of how these goals will be met. One possible solution, talked about for decades, is to construct massive upstream water intake facilities and tunnels or a peripheral canal to convey the water for people under or around rather than through the Delta. The hope is that this would allow wetland habitats and more natural flow regimes within the Delta to be restored. The costs of this or any of the other alternatives being debated would be enormous; that they are even being considered speaks to the imperative of finding a long-term solution.

Doing nothing is no longer an option. For decades, the emphasis has been on maintaining the status quo: ensuring the structural integrity of the levees and the flow of fresh water to the primary users and controlling salinity intrusion into the Delta. Now, however, the Delta is facing both physical and ecological collapse. Subsidence of farmlands continues, increasing the risk of catastrophic levee failures and permanent flooding. Sea-level rise and an increasing frequency and duration of high-water events are further weakening levees. Tides and storm surges are pushing saline water farther into the Delta, threatening water quality for both people and fish. Climate change is altering the mix of rain and snow in the mountains that feed the Delta, which will increase the intensity of winter floods and change the seasonality of runoff. Collectively, these changes may push the Delta past a tipping point, to become expanses of brackish water fringed by tidal marsh. Such habitat changes are likely to favor invasive plants and animals at the expense of the natives, impacting first those species teetering on the brink of extinction.

If policies and practices remain unchanged, the water supply from the Delta will become less reliable and advocacy positions will harden even more, rendering compromise more remote. By establishing water availability and ecological integrity as co-equal goals, the California legislature has forced parties to recognize the legitimacy of different interests and reconcile competing demands. Plans to chart a course for actually doing this are nearing completion. Those concerned about water issues in the Klamath Basin and elsewhere would do well to consider a similar policy.

As I write this, over half of the continental United States is in the grip of severe drought. Crops are withering in the fields and food prices are expected to increase, sending ripples through the economy when we can least afford them. The effects on wildlife and ecosystems may be equally severe. Increasing conflicts over water are inevitable. Resolving such conflicts will require breaking away from the "me first" approach of the past. If the needs of people and the environment are not somehow balanced, the environment will suffer. And so, in the long run, will people.

CHAPTER 9

Population growth

All of the forces of change are tied, in one way or another, to the growth of the human population and the economy. How the size, growth, and resource demands of the human population change in the future will determine whether conservation will be successful and what "success" will look like.

In 1900, the world population was around 1.6 billion people. Writing in 1930 (without the benefits of comprehensive demographic data or computer models), H.L. Wilkinson[1] projected that the global population might reach 4 billion by the turn of the 20th century and perhaps 8 billion by 2070. In 1953, with the memory of the Second World War still fresh and the demographic impacts of the subsequent baby boom not apparent, Fairfield Osborn[2] forecast a world population of 3.6 billion by 2000. In fact, the global population was over 6 billion in 2000, and it stood somewhere in excess of 7 billion people in 2013. Although growth rates have slowed, they are applied to an increasingly large number (Figure 9.1). The United Nations projects that, given current demographic trends, global population may reach 8.5 billion by 2030, 9.7 billion by 2050, and 11.2 billion by 2100.[3] Beyond this, projections range from a stabilization of population at around 9 billion to numbers in excess of 25 billion. Whatever the number, it is very, very large.

Because the global population is changing in tandem with increasing expectations for economic growth, accumulation of wealth, and individual well-being, demands on the world's natural resources and energy reserves continue to expand. At some point, the relation between numbers of people and per capita resource demands will become unsustainable. Speculating on where that point might lie and what might happen when it is reached has occupied scholars and scientists for more than two centuries. Most famously, Robert Malthus, in the ponderously titled *An Essay on the Principle of Population; or, a view of its past and present effects on human happiness; with an enquiry into our prospects respecting the future removal or mitigation of the evils which it occasions,*[4] drew attention to the inevitable collision between population growth in times of plenty and the onset of population

[1] Wilkinson (1930).
[2] Osborn (1953).
[3] *World Population Prospects: The 2015 Revision.* http://esa.un.org/unpd/wpp/publications/files/key_findings_wpp_2015.pdf.
[4] Malthus (1803).

Ecological Challenges and Conservation Conundrums: Essays and Reflections for a Changing World, First Edition. John A. Wiens.

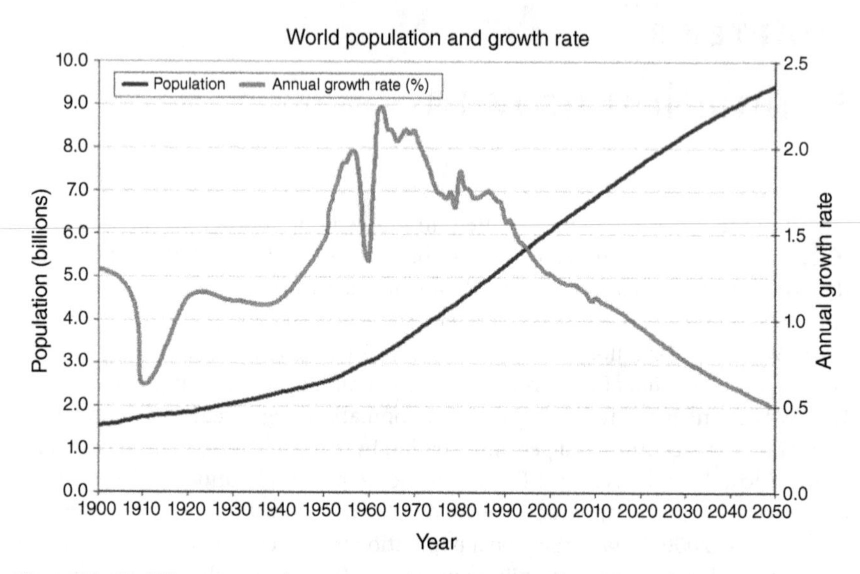

Figure 9.1 World human population (ascending curve) and annual growth rate (variable curve), 1900–2050 (projected), based on United Nations data. Source: DSS Research.

limitation by famine and disease when demands outstripped supplies.[5] Malthus explicitly argued that improvements in agriculture could not remove limits to population growth. The Industrial Revolution and its effects on economics, politics, agriculture, and virtually all elements of society in most of the world seemed to invalidate Malthus' thesis, although in reality they only reset the limits rather than removing them.

Later writers have drawn on Malthus and on subsequent scientific advances (especially in theoretical population biology) to recast the issue, although the underlying theme of finite limits to the earth's capacity remains. In *The Population Bomb* (1968),[6] Paul Ehrlich contended that the limits had already been passed and massive starvation and social upheaval lay just ahead.[7] Donella Meadows and her

---

[5] Although my suggestions in Essay 6, *Fat times, lean times and competition among predators (1993)* (page 33) were cast in the context of niche overlap driven by interspecific competition rather than population limitation driven by famine and disease, I realize now that they have a distinctly Malthusian ring.

[6] Ehrlich (1968).

[7] Ehrlich's assertion led to a wager with the economist Julian Simon over whether the prices of several global commodity metals would rise (as Ehrlich predicted) or fall (Simon's expectation) between 1980 and 1990. Ehrlich lost the bet (although he would have won had the wager been over three decades rather than one). The debate helped to draw attention to the broader issue of the collision between human population growth and resource scarcity, whenever that might occur.

colleagues offered a more hopeful prospect in the first edition of *Limits to Growth* in 1972, but by the time of the second edition in 1992 they concluded that the global population had already overshot the earth's limits; the 2004 update reinforced this conclusion and bemoaned the ineffective international response to the crisis.,[8,9] Joel Cohen synthesized many of the projections and debates in the appropriately titled *How Many People can the Earth Support?*, and the topic has received more recent attention in a report from the National Research Council of the National Academy of Sciences.[10]

The human population is not only just growing, but it is also redistributing. Arguments have been offered that migration from densely populated regions to more sparsely settled areas could provide at least a temporary solution to regional population problems. For example, in *The World's Population Problems and a White Australia*, H.L. Wilkinson proposed that Australia could provide a destination to alleviate "overpopulation" in Europe, particularly Britain, by providing "a home for the European race and civilization."[11] Fairfield Osborn offered a similar argument, with less obvious racial overtones, in 1953.[12]

More apparent has been the shift in population distribution from rural to urban settings. In the early part of the 20th century, only 2 of 10 people lived in urban areas; by 2010, more than half the world's population was urban. This shift was largely driven by economic forces. Although projections suggest that this trend will continue, particularly in the cities of developing countries, future redistributions may increasingly be driven by environmental factors. The dust bowl of the 1930s sent a wave of environmental refugees ("Okies") from the Great Plains to California.[13] The effects of future climate change and sea-level rise may lead to far greater movements of people fleeing low-lying coastal areas and places where agriculture has become marginal in search of greener pastures (or any pastures) elsewhere.

The consequences of this unrelenting growth and redistribution of the global human population are the topics of unending speculation. Will there be a "soft

---

[8]Meadows et al. (1972, 1992, 2004).

[9]The international response seems to have largely been to gather people together in conferences to talk about the "population problem" and issue reports—what might best be characterized as the blah-blah-blah approach to dealing with a problem.

[10]Cohen (1995) and National Research Council (2014). The global population was 5.7 billion when Cohen wrote his book. Understandably, he did not provide a definitive answer to the question; rather, he elaborated how complex the question really is. The answer, not surprisingly, is that "it all depends."

[11]Wilkinson (1930: 315). Wilkinson's book was published in 1930, when the "White Australia" policy was in force. This policy gave preference to immigrants from Britain and excluded those from Asia and Japan, ostensibly to prevent competition for jobs. Immigration policies were gradually relaxed following the Second World War, and the White Australia policy was dismantled in 1966. It was not until 1978, however, that references to country of origin were officially eliminated from immigrant screening criteria.

[12]Osborn (1953).

[13]The subject of John Steinbeck's *The Grapes of Wrath* (1939).

landing" in which population numbers stabilize and the use of resources becomes sustainable? This is the hopeful vision expounded in *Limits to Growth*. Alternatively, will there be an unleashing of the Four Horsemen of the Apocalypse (conquest, war, famine, and death) upon the world, as Malthus and the New Testament foretold (and Paul Ehrlich predicted)? Even in the most optimistic scenarios, the consequences for the environment and the earth's biodiversity will be profound. This, and the changes already wrought by past human population growth, is what creates the moral imperative and the urgency for conservation. All of this matters because where people are and what they do circumscribes the options for conservation.

Conservation is value-driven. Values, and therefore the goals of conservation, change as society changes. One consequence of population change and redistribution is that the world one generation experiences is unlike that which previous generations experienced or subsequent generations will experience. My father, an immigrant from Russia, experienced war and famine in his youth, hardships I have never had to endure. I remember as a child hiking with my parents and uncle through forested hillsides in West Virginia that no longer exist, destroyed by strip mining and then removal of entire mountaintops to get at buried coal deposits. The Elvis Presley and Sputnik that were part of my growing up are quaint historical references to my children, who instead are experiencing changes in society and civil rights that were talked about in hushed tones, if at all, when I was young. Everyone's life brackets a segment along a continuum of change, some of which is undeniably progress, some not so much. But it all represents changes in what theoreticians and modelers term "initial conditions," the starting point from which individuals perceive whatever follows.

This segmenting of the forward progression of time by individual lifetimes leads to a "generational ratcheting." Each generation recalls, often with nostalgia,[14] how things were when they were young—their formative impressions of the world about them. This affects perceptions of what is "normal" in the environment. What I regarded as normal for the hills of West Virginia is quite different from what a child today would see as "normal." As the environment changes, each generation encounters it in a different way, and the baseline of "normal" shifts. Societal values and expectations, and the context for conservation, shift as well.

[14] But not always with nostalgia; I don't recall my father ever being nostalgic about his childhood in Russia.

CHAPTER 10

Linkages among changes

Just as population growth and redistribution, escalating use of resources, and societal perceptions and valuations of the environment are linked, so also are all the other forces of change. It may be useful to consider them separately, as I have done, but in reality they interact in complex ways. As climate change unfolds, for example, it is affecting which agricultural crops will grow where, and land uses are changing as a result. Land uses also change with population and economic growth, and societal and economic forces pull people into cities or spread them out in exurban developments. Climate change may also facilitate the spread of invasive species. And whether a species is regarded as an invasive to be controlled or is desired for recreation or subsistence depends on societal values and on who is doing the valuing.

Examples of the web of linkages among forces of change could be given for virtually any ecosystem. Because it illustrates these linkages particularly clearly (and because I know something about it), I'll highlight water issues in the Sacramento–San Joaquin Delta in California as a case study,[1] expanding on my comments in Essay 13, *Wildlife, people, and water: Who wins? (2012)* (page 100).

The Delta is all about water—how much of it there is, where it goes, and who gets it. The water comes chiefly from rainfall and winter snowpack in the mountains to the north and east (Figure 10.1). The amount of water entering the system can vary tremendously among years, from occasional floods to severe droughts.[2] To ameliorate this variation, reduce flood risks, provide water for agricultural and domestic uses, and regulate releases of stored water, most of the upstream tributaries were dammed during the past century. More recent climate changes, however, have led to increased winter runoff and earlier snowmelt

[1]The Delta lies where the Sacramento and San Joaquin rivers meet before emptying into San Francisco Bay. It encompasses roughly 3,00,000 ha.· The watershed extends over a much larger area, from the eastern slopes of the Coast Range to the western slopes of the Sierra Nevada, sending nearly half of the state's average annual streamflow toward the Delta. Lund et al. (2010) and the National Research Council (2012a) provide additional detail about water uses and issues in the Delta, and Whipple et al. (2012) recount the historical ecology of the Delta.

[2]Moyle et al. (2010); Moyle and his colleagues offer several recommendations for achieving the variability and complexity that create an ecologically productive estuary.

Ecological Challenges and Conservation Conundrums: Essays and Reflections for a Changing World, First Edition. John A. Wiens.

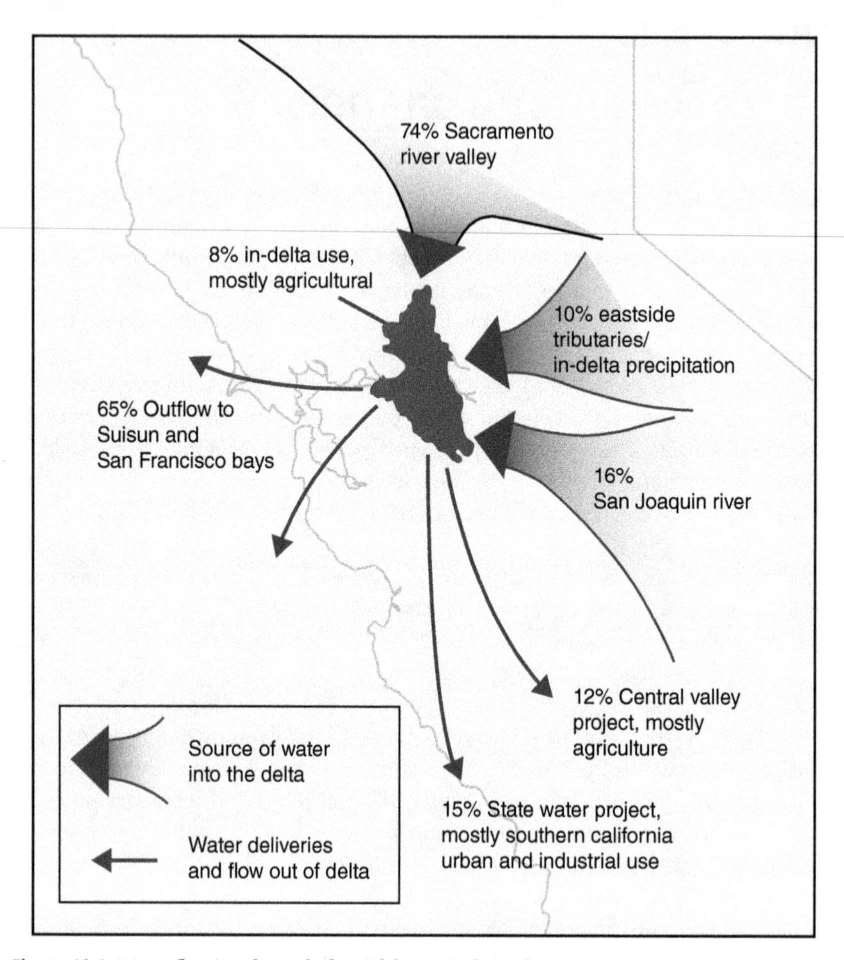

Figure 10.1 Water flowing through the California Delta is the main source of supply for two major California water delivery projects, the State Water Project and the federal Central Valley Project. Millions of Californians rely on water from the Delta for their drinking water, and one-third of the state's cropland uses water flowing through the Delta. Source: California Legislative Analyst's Office (http://www.lao.ca.gov/2008/rsrc/water_primer/water_primer_102208.aspx).

in the spring, altering hydrological flows into the Delta and compounding management challenges. Climate models project that this shift toward earlier runoff will continue, exacerbated by even greater variability among years.

The Delta has been dramatically altered by land uses. What was a vast tidal wetland two centuries ago has been converted into a labyrinth of channels hemmed in by some 1,800 km of levees (see Essay 13, Figure 2, page 102). Most of these levees were originally built over a century ago for protection from floods and tides, to allow the lands behind the levees to be drained for farming, and then to provide water for irrigation of the crops. When the wetlands were drained and farmed, the organic peat that had been deposited over millennia (and which makes farming so productive) was exposed and oxidized. The combination of the loss of peat with groundwater extraction to supplement irrigation has led to land subsidence. A substantial area of the Delta now lies more than 4 m below sea level, making it vulnerable to flooding should levees fail (as has happened).

The levees also channel water through the Delta to two massive water-conveyance systems that divert water from the Delta to the San Joaquin Valley and areas west of the Delta (mostly for agriculture) and to cities in southern California and the Bay Area (for the 25 million people who live there). Collectively, human uses of water within the Delta and through the two conveyance systems consume slightly more than one-third of the water entering the Delta (Figure 10.1). These uses have a major impact on how the flows of the remaining two-thirds are managed.

The Delta was originally an estuarine system, and its waterways are still subject to tidal flux and salinity intrusion. Agricultural and urban users, however, require fresh water. To keep tidal intrusions at bay and ensure water quality, outflows from the Delta are regulated to create a hydraulic barrier at the mouth of the Delta. As sea level continues to rise,[3] threatening water quality (as well as the aging levees), more water flow will be needed to maintain this barrier, affecting how much water can be allocated to other uses.

The Delta is not only the water hub for much of California's agriculture and most of its people, but it is also ecologically important. Most of California's fish species occur in the Delta, and it is a major overwintering area for millions of migratory birds. As California's human population has grown and the agricultural economy has mushroomed, conflicts have welled to the surface, pitting people from northern California (where the precipitation falls) against those from southern California (where the people are) and the interests of the environment against those of farmers and city dwellers.

Enter the Endangered Species Act. The Act stipulates that all federal agencies are prohibited from authorizing, funding, or carrying out actions that "destroy or adversely modify" critical habitats.[4] Consequently, agencies have the responsibility to maintain sufficient flows in federally managed waterways to sustain populations of endangered fish, such as Delta smelt or winter- and spring-run Chinook salmon. Since 2007, legal mandates to maintain flows for fish have led to significant reductions in water exports for human uses. Smelt and salmon numbers have continued to decline (Figure 10.2), increasing the prospects of further export

[3]Sea level has risen by 20 cm in San Francisco Bay over the last century; future projections suggest an additional rise of 92 cm over the remainder of this century; National Research Council (2012b).
[4]Endangered Species Act Section 7(a)(2).

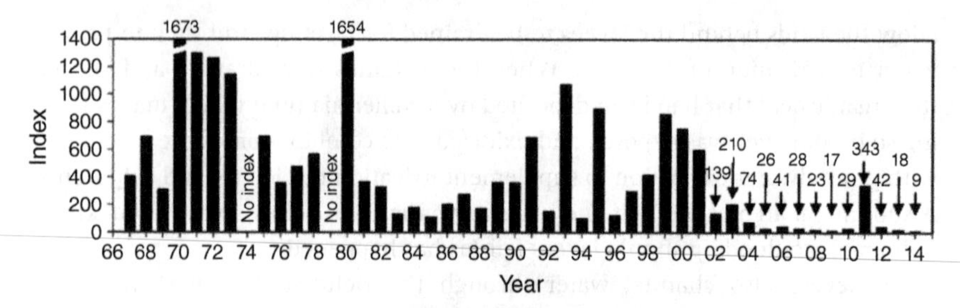

Figure 10.2 Fall mid-winter trawl Delta smelt annual abundance indices, 1967–2014. Note that the 2014 survey recorded only nine smelt, the smallest number ever recorded. This occurred despite court-ordered measures to protect the smelt by ensuring adequate water flows at critical times of year. Source: California Department of Fish and Wildlife.

reductions and additional releases from upstream dams to provide cool water for fish downstream and escalating conflicts over who gets the water.[5]

Changing climate, sea level, land use, species' abundances and distributions, societal demands, legal requirements, and numbers of people have all contributed to bringing water issues in California and the Delta to a breaking point—there is simply not enough water to go around.[6] Now add in the current drought affecting California (2012–2015), which by some accounts is the most extreme in the past 1,200 years. In summer 2014, several major upstream reservoirs were at less than a quarter of their capacity (Figure 10.3). Long-time water rights could not be honored. Denied surface water to meet their needs, farmers and municipalities drilled deeper wells to tap groundwater, threatening depletion of those reserves, further land subsidence, and weakening of levees. Crops withered and farmland lay fallow, costing jobs and affecting the economy well beyond California. And yet, the court-mandated flows to sustain endangered fish continued—that's the law. Predictably, some politicians proposed changing the law to suspend or cancel provisions of the Endangered Species Act that allocate scarce water to fish, arguing that the Act allows water to "flow out to sea and be wasted." Climate models project that such extreme droughts are likely to become more frequent and last longer in the future. It's not a pretty picture.

Conservationists and ecologists face a dilemma. We know that these forces of change interact and that to consider any one divorced from the others would be folly. But we also know full well the difficulty of trying to put them all together

[5]Smelt numbers continue to drop. A survey in March 2015, when the smelt usually form spawning aggregations, recorded only six fish. This led Peter Moyle, who has studied the smelt for over 40 years, to warn that the species will likely be extinct in the wild in another year or two.

[6]In fact, California has allocated water rights to five times more surface water than is available in a year of good precipitation; Grantham and Viers (2014).

Figure 10.3 The South Fork of the Feather River feeding Lake Oroville, California, on September 5, 2014, during the third year of unprecedented drought. Lake Oroville is the second largest reservoir in California. When full, the lake elevation is 736 ft (224 m); when this photograph was taken the lake elevation was 678 ft (207 m). Photograph by Kelly M. Grow, California Department of Water Resources.

when we have barely begun to understand the parts. Our inclination is to follow the recipe that has worked so well in science: decompose a complex problem into manageable parts and address the primary factors that may be directly involved.

Fine. But the different forces of change operate at different scales, which renders some more amenable to direct conservation or management action than are others. Land-use change and the arrival and spread of invasive species are most immediately evident at local scales of a few hectares or tens of square kilometers. These are the scales where ecological research is customarily conducted, management is applied, and conservation actions are undertaken and where the effects of these activities may be most direct. Other forces of change—climate change and population growth and demography—are driven largely at regional to global scales. Although conservation and ecology may be effective in influencing national and multinational policies at these scales by pointing out problems and their consequences, they are unlikely to have direct effects on the forces themselves. Sociopolitical forces span a range from the local scale of landowners and local governments to the global scale of the World Bank and the United Nations Environmental Program. Conservation and management are affected by these forces at multiple scales, with differing consequences.

Because of these differences in scale, there may be advantages to focusing on the forces of change separately. The underlying risk, of course, is that by doing so we will forget about the linkages with other factors and the ways they can compromise the simplified explanations and conclusions we draw. Actions taken on the basis of such forgetful science may be ineffective or worse. There is no easy solution. Inevitably, most scientific research and most conservation action must concentrate on one or a few factors that seem overwhelmingly important to the questions being asked. It would be well worth the effort, however, to spend time thinking through how those factors may interact with others, how they coincide or differ in the scales on which they affect things, and especially how the dynamics of change are likely to play out.

Whether they are considered individually or collectively, the forces of change I've discussed in this part create a variety of conundrums for conservation—confusing problems that defy simple solutions.[7] These are the topics of the next part.

[7] I was curious, so I looked up the meaning of "conundrum." The one I've given is the second definition; the first is "a riddle whose answer involves a pun," such as "What is black and white and read all over?" (a newspaper, although that example is really outdated because most newspapers are now printed in color and few people read them all over, or at all).

PART III
Conservation conundrums

Change is ubiquitous. But it's not just change that challenges ecologists and conservationists. It's that the rate, magnitude, and nature of change are themselves changing. At the same time, the window through which ecologists and conservationists view environments has also changed. I've noted before how screening out extreme variations and thresholds (the anomalies) might bolster faith in the order and stability of nature (see Figure 11.1A in the next chapter). With the recognition by ecologists of nonequilibrium dynamics and by the public of more weird weather, the window is shifting and expanding (Figure 11.1B). More of the variation, extremes, trends, and thresholds can now be seen. The challenge for ecology and conservation is to understand how these changes in dynamics affect ecological systems.

It is beyond the scope of this book, and certainly beyond my experience and thinking, to consider all of the ecological and environmental consequences of the boisterous changes in climate, land use, species' distributions, sociopolitics, and human populations that are now underway. Instead, I'll consider several topics that pose particularly thorny problems—conundrums—all of which are consequences of variation, thresholds, and uncertainty. Variation can lead to unanticipated thresholds and nonlinearities, which in turn create mounting uncertainties. Taken together, these effects make it difficult to understand the dynamics of ecological systems, and thus to conserve and manage them successfully.

As before, I'll not offer a review or synthesis, but a personal perspective on some topics I've thought about, bouncing back and forth between reflections, commentaries, and essays.

Ecological Challenges and Conservation Conundrums: Essays and Reflections for a Changing World, First Edition. John A. Wiens.

CHAPTER 11

Variation and history

We see variation in nature when we open the window of time. If the variation seems to be slight or well-contained (Figure 11.1A), we may be content to treat the system as essentially stable. Variation exists, but conditions remain unchanged over the long run, with no obvious trends. This is what mathematicians, statisticians, or water-resource managers mean when they talk about stationarity. But it's turning out that variation in the environment and its effects on ecological systems are greater than we had envisaged. Variation is often so great and unpredictable that it may befuddle our efforts to understand, manage, or conserve natural systems—"stationarity is dead."[1]

Whether stationarity and similar equilibrium-like concepts are actually dead[2] or simply moribund, a focus on variation may change how ecologists and conservationists go about their work. Ecologists may need to reconsider how they use statistics to lend credibility to their findings. Statistics are generally used to partition variation so that the core patterns (e.g., differences between means, trends, associations among variables) can be detected. Too much variation may obscure those patterns. This is one reason why outliers are often removed from statistical analyses (see Essay 18, *Black swans and outliers (2012)* (page 160)). To be sure, those core patterns in data are important; they are what we construct explanations and theories about. But it may be just as important to consider the patterns in the variance component of the data and to develop explanations or hypotheses about what caused the variation. The variations in data are real, but the mean is simply an arithmetic derivative of the variations over the arbitrary extent of the data. The greater the variation, the more meaningless the mean.

Variation may also affect how an ecological study is designed. If variation in a study system is small, a short-term study in a single place can probably capture enough of it to characterize the system and its dynamics adequately. As variation increases and becomes more erratic (as in the middle part of Figure 11.1), however, such studies will encounter only a small slice of the variation. Any conclusions or inferences about the actual dynamics of the system may be incorrect if extrapolated more broadly. Of course, the greater and more unpredictable the variation,

[1]Milly et al. (2008).
[2]Bringing to mind Mark Twain's response on reading his obituary in the *New York Journal*, that "reports of my death have been greatly exaggerated."

Ecological Challenges and Conservation Conundrums: Essays and Reflections for a Changing World, First Edition. John A. Wiens.

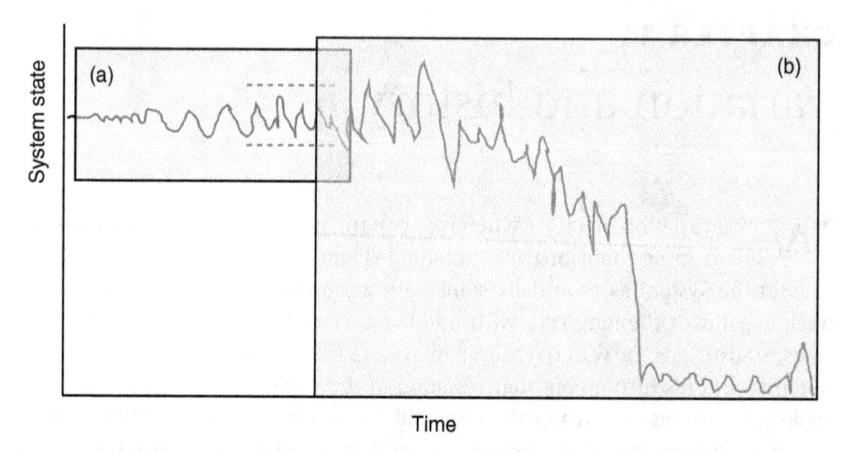

Figure 11.1 The conceptualization of kinds of environmental change shown in Figure 3.1, indicating (roughly) the windows through which ecologists and conservationists traditionally viewed system dynamics under a presumption of equilibrium (a) and of nonequilibrium, variation, trends, and thresholds (b).

the longer it will take for a study to understand the system. This raises the vexing question of when to conclude a study—how long is long enough?[3]

I first became aware of the confounding effects of variation and the difficulty of deciding when enough is enough during our studies of breeding bird communities in the grasslands and sagebrush shrubsteppe of western North America.[4] What began as a 3-year study (the normal duration of a research grant) turned into a 17-year odyssey as my students, postdocs, and I tried to understand what was going on in a system in which every year seemed different from the last. Eventually, we stopped trying to fit the square peg of our data into the round hole of prevailing (equilibrium-based) theory and recognized how environmental variation kept the communities from reaching the stability we had been expecting.[5] Along the way, we conducted a back-of-the-envelope modeling exercise to ask what sorts of patterns we might have detected had we sampled each of our study areas just once instead of multiple times. Sample sizes would of course have been reduced, making it more difficult to detect statistically significant patterns. Nonetheless, it was apparent that short-term sampling over a range of study sites could lead to differing conclusions, depending on when and where the surveys were conducted.[6]

---

[3]The long-term study of reproductive success of Cassin's auklets on the Farallon Islands described in Essay 17, *Tipping points in the balance of nature (2010)* (page 144) provides a good example.
[4]Summarized here and there in *The Ecology of Bird Communities* (Wiens 1989a).
[5]Wiens (1974, 1977).
[6]Wiens (1981).

So, how did we decide when to stop? Certainly not when we had thoroughly understood the system or had answered all our questions! Rather, I moved to a different university, far from the study sites; we had difficulty obtaining grant funding to conduct "the same old stuff"; the students and postdocs moved on; and I got interested in (or distracted by) other questions. I suspect that many studies end for similar reasons.

Variation affects conservation by making the outcomes of actions less certain. A centerpiece of efforts to protect or enhance the recovery of imperiled species, for example, involves identifying the factors threatening a species and then acting to reduce or eliminate those threats. Things get complicated, however, if the threats change. This is what has happened with attempts to recover populations of northern spotted owls in the Pacific Northwest of the United States. Based on a massive amount of research that identified loss of old-growth forest habitat as the primary threat to spotted owls, the Northwest Forest Plan of 1994[7] attempted to balance protection of the owl and old-growth forests while ensuring continued timber harvesting at a reduced level. The compromise pleased no one. The logging industry continued to lose jobs while the environmental community wanted even greater reduction in timber harvest.

Then, while managers and conservationists were trying to sort this out by focusing on the management of old-growth forests, a new threat emerged. Barred owls, a closely related but more aggressive species that had moved in from eastern forests, began to displace spotted owls competitively from a suitable habitat.[8] Conservation and management approaches have shifted to include control of barred owls as well as habitat protection, Yet, spotted owl numbers continue to decline.

This is not an isolated example. As climate change shuffles the distributions of species, the target species of conservation will be exposed to new webs of interactions, so such changes in threats are likely to become more frequent. Conservation and management will need to recognize these changes and adapt.

That's in the future, something we can only guess at. The window through which we actually see variation opens only on the past, giving us a view of what has happened but not what lies ahead. In other words, history. But which history is important, told by whom, over what span of the past? Ecologists of different stripes have different perspectives on the past. To those who conduct short-term studies, the here-and-now, not history, is what's important. Others take a long view, encompassing hundreds to tens of thousands of years. Somewhere in between lies the "historical range of variation" (HRV) that is incorporated into

[7]Rapp (2007).
[8]Wiens et al. (2014). Plants to control barred owls are described at http://www.fws.gov/oregonfwo/species/data/northernspottedowl/.

a good deal of natural-resource management in the United States.[9] The different perspectives on history include differing amounts of environmental variation, which will influence how one considers an ecological system at a particular point in time (e.g., "now"). "Historical variation" depends on how one views history.

I consider some facets of how history plays into ecology and conservation in the two essays included here (Essay 14, *The eclipse of history? (2008)* (page 122) and Essay 15, *From our southern correspondent(s): Which history? (2013)* (page 125)), so I'll only offer some brief additional comments here, along with another digression (Chapter 12).

History is intertwined with time. But while time is an uninterrupted stream, always moving ahead, history proceeds in fits and starts, moving jerkily from the past to the present and into the future. It is a chronicle of singular events that can change the trajectory of what follows, rather like the butterfly effects of chaos theory.[10] Human history is a series of events that "changed the world," or at least altered some part of it. The assassination of Archduke Franz Ferdinand and his wife Sophie in 1914 triggered World War I; the terrorist attacks of 9/11 changed how governments dealt with security threats and altered air travel for everyone; and the *Exxon Valdez* oil spill changed the economy and culture of Alaskan coastal communities. Each event reset the trajectory of history from what it had been before.

Ecological history also follows a wandering trajectory full of legacies of a past no longer with us. Sometimes we can reconstruct what might have happened. We know, for example, that the disappearance of American chestnut from eastern deciduous forests in the United States permanently changed the species composition of the forests over a large area.[11] A forest fire or beetle infestation can leave an imprint on a landscape that can last for centuries; if environmental conditions have changed during the interim (as is likely), any replacement forest that eventually develops will be different. The sequoias or bristlecone pines that we admire for their size or longevity are "living ghosts," remembrances of the different environments in which they germinated many centuries ago.

Human actions may also leave enduring legacies in ecological systems. Gold was discovered in the Sierra Nevada of California in 1848, precipitating the California Gold Rush. By 1853, miners were using hydraulic mining, in which a jet

---

[9] HRV is the starting point for the discussions about how history is incorporated into conservation and resource management contained in the contributions to Wiens et al. (2012).

[10] The butterfly effect refers to how a small change in the initial state of a system can result in a large change in the subsequent state of the system. It was formalized theoretically by Edward Lorenz in 1963, although the idea that something as small as a butterfly could set off ripple effects that would affect later history was developed in a science-fiction short story (*"A Sound of Thunder"*) by Ray Bradbury in 1952.

[11] Although arguably not the ecological functioning of the forests; whether the disappearance of the chestnut created a major change or a minor ripple in the forests depends on whether species composition or ecological functioning is deemed to be important—a matter of values.

of water under high pressure was directed at streamside cliffs to extract the gold that was buried in sediments. Entire hillsides were washed away, the sediments and debris going downstream and eventually into San Francisco Bay. By the time hydraulic mining ended some 30 years later, vast amounts of sediments released by the mining operations had raised streambeds, exacerbated flooding, decreased tidal range, and destroyed spawning grounds for fish. Along with the sediments, mercury used in the mining operations continued to move downstream into the Sacramento–San Joaquin Delta and beyond for over 130 years after the hydraulic mining ceased—a continuing impact of a human activity long past.[12]

Here's another, more personal, example of historical legacies. In 1991, I visited the Luquillo Experimental Forest in eastern Puerto Rico, a little more than a year after hurricane Hugo had hit the area with full force. Although forest regeneration was already well underway, evidence of the destructive effects of the hurricane was not hard to find. But the effects on the forest were patchy, some places devastated, other nearby places seemingly untouched—not what one would expect from a hurricane. I asked Ariel Lugo, who was showing me around, why. Ariel told me that, although topography, exposure, and soil substrate were important, they discovered after some sleuthing in old Spanish land records that some places in the forest had been coffee plantations early in the 17th century. The plantations were later abandoned, reverting to forest. The regenerated forests, however, differed in species composition and structure from the native forests that had not been cleared for cultivation. When Hugo passed through, the forests in the former plantations suffered greater damage than those in the uncultivated areas. Previous land uses were important.[13]

These examples illustrate that things that happened in the past can have a profound effect on what we see today. Ignoring the past may therefore be a mistake, a consequence of failing to open the window wide enough to see the variation. Because of variation and change, however, it may also be a mistake to think that the past provides a baseline for comparison with the present. Time, then, for another digression, on baselines and targets.

[12] Whipple et al. (2012) recount some of this history.

[13] Lugo (2008) provides a comprehensive review of hurricane effects on forest ecosystems at Luquillo and elsewhere.

ESSAY 14

The eclipse of history? (2008)*

Ecologists have always been a bit ambivalent about the history of the systems they study, and resource managers and conservationists even more so. On the one hand, everyone knows that what happened in the past affects the present, and therefore history shouldn't be ignored. On the other hand, most scientists are interested in understanding how things are now and most managers face the task of dealing with what they have to work with now, so dwelling on the past can be an unwanted distraction. Those who cling to a balance-of-nature or equilibrium view of the world may argue that the present is pretty much like the past anyway, so it's unnecessary to consider history. Others suggest that nature was in a much better condition in the past than it is now, so we should use history as a guide to returning to the "good old days." To people who see variation everywhere they look, identifying useful reference points in the stream of history can seem futile—what one sees depends entirely on the scope of history considered. After looking at past tree-ring records, for example, the paleoecologist Steve Jackson concluded that "no two years within a decade are alike, no two decades within a century are alike, and no two centuries within the last millennium have been alike."[1]

Through much of my professional career I fell into the last group—aware that history must be important but unsure what to do about it. Then one day I got caught up in conversations about the relevance of history with colleagues from the U.S. Forest Service. They were required to manage forests within the historical range of variation (HRV[2])—the envelope of variation in forest conditions over some period of recent history. Some scientists and managers in their agency, however, were beginning to question whether HRV, or history in any form, could be at all relevant to a world undergoing climate change.

So we decided to organize a conference to discuss the issue, which eventually led to a book on historical environmental variation and its use in conservation and resource management.[3] Along the way I wrote this essay and (with Richard Hobbs) the following one. None of these efforts resolved how history should be used in studying, managing, or conserving nature, but all of them speak loudly of the need to understand history and learn from it.

*Wiens, J.A. 2008. The eclipse of history? *Bulletin of the British Ecological Society* 39(4): 63–64. Reproduced with permission of the British Ecological Society.
[1]Jackson (2012: 95).
[2]Being a federal agency, the use of acronyms in the Forest Service spreads like a virus.
[3]Wiens et al. (2012).

Ecological Challenges and Conservation Conundrums: Essays and Reflections for a Changing World, First Edition. John A. Wiens.

Can history tell us anything useful in a world undergoing rapid environmental change, where the future will be unlike the past? This question is at the heart of an ongoing debate among ecologists and resource managers. For example, hydrologists and water managers have based their work for decades on the assumption of stationarity—the notion that natural systems fluctuate within an unchanging window of variability. This assumption established a framework for managing water supplies, regulating floodplain development, and evaluating flood risks. The magnitude of recent and projected anthropogenic and climate-induced changes in river basins and streamflows, however, has led some scientists to declare that the concept is no longer useful, that "stationarity is dead" (Milly et al. 2008).

Similarly, foresters have relied on assessing the "historical range of variation" to establish targets for managing forests with respect to fire or harvesting practices, for example, and the concept is included in planning directives for the United States Forest Service. Yet some scientists and managers are now asking whether the concept and guidelines are irrelevant in the face of climate change. They argue that, as species disappear or shift distributions and biotas mix, we are likely to experience "no-analog" futures—assemblages of species unlike any we have seen before (Williams & Jackson 2007). In such a world, knowing how things have varied in the past may provide little guidance about how they will be in the future.

If history can't help us manage proactively, does this mean that history is irrelevant? Certainly not! The problem is not with history, but with the use of history to establish targets or "desired conditions" for management. This is how stationarity and historical range of variation have been used. In other contexts, the target of ecological restoration is usually a return to some previous, more natural, condition, and the laws governing the assessment of damages from environmental accidents in the United States define recovery as a return to what the status of a species would have been had the accident not occurred. Past history is used to provide the reference criteria for management or restoration.

Such approaches recognize that natural systems are not static, that variation in time is a property of nature. Yet they all express, in one way or another, adherence to a balance-of-nature perspective. While acknowledging that nature varies from year to year, the window of variation over many years is assumed to be fixed. The shorter the time period considered (the smaller the window), the smaller the range of variation. Because most ecological studies are constrained to last only a few years by the realities of grant cycles (or the imperative for graduate students to complete their degrees), the range of natural variation that one sees is often small. It is easy to regard this variation as background noise, something to be removed through statistical analysis.

If the window is expanded, however, more of the real variability of nature emerges, and we begin to see more extreme events. If the window is large enough (say, many decades or a century or two), it becomes apparent that the "historical range of variation" is anything but stable. Using an even larger window on history, paleoecology has shown us time and again that the current distributions of species and composition of communities were very different in the past. Paleoclimatology and paleontology indicate that climates in the distant past may have approximated those predicted to develop over the next century, although the species and biological communities were entirely different (they were "no-analog pasts"). Clearly, our perceptions of stability ("stationarity") in natural variation is a figment of using an unrealistically narrow window on the past.

So what does this say about the relevance of history to ecology? Even if we acknowledge that ecological systems are not in steady-state equilibrium and that the range of variation has itself varied in the past, how will this help us to understand or predict how these systems may vary in a no-analog future?

We can begin by recognizing that the future is contingent on the present. The pathways into the future are constrained by where we are now—which species are present in an area, which land uses are dominant, how landscapes are connected or fragmented. But the present is contingent on the past—on history. What we see now when we look at nature bears the legacies of past species' distributions, the ghosts of past species interactions, the impacts of extreme events that may have occurred centuries ago, and the fingerprints of human alterations of landscapes that happened in past millennia. The sorts of changes in climate or disturbance that have recently pushed some systems across thresholds into alternative states undoubtedly occurred in the past, setting trajectories of change that have led to what we see today. History determines a distinctly nonrandom range of possibilities, from which the present has emerged.

Historical ecology can also help us to understand how ecological systems came to be as they are. It can reveal which species have persisted and which have not, and which attributes characterize the survivors or the losers. It can show us how systems have responded to extreme events, which the climate modelers tell us are likely to become more extreme and more frequent in the future. It can tell us something about past patterns of colonization, adaptation, and extinction associated with past climate changes. It can provide the information needed to test models of the future—if a model can't "backcast" the past (about which we know something), how confident can we be about its forecasts of the future? And all of this can contribute to a better understanding of the determinants of ecological resiliency, which will be critical to the sustainability of ecological systems in a rapidly changing future.

It has been said that those who ignore history are condemned to repeat it. Given the changes now underway, it's unlikely that we'll be repeating anything that has occurred in the past. It's also true, however, that those who ignore history cannot benefit from its lessons. Historical ecology can help us deal with the future through a better understanding of the past.

As frameworks for setting fixed targets for management or restoration, stationarity and historical range of variation, if not dead, should at least be eclipsed, relegated to obscurity. But history? History lives on. It's what got us here, and it's what will set our pathways into a different and uncertain future.

ESSAY 15

From our southern correspondent(s): which history? (2013)*

I mentioned before how Richard Hobbs and I periodically meet at Little Creatures in Fremantle, Australia, to ponder what seemed at the time to be pressing issues in ecology, restoration, and conservation. Essay 7 (page 44) was the first product of these musings. The experience (and the beers) proved to be so rewarding that we decided to do it again; this is the second of these essays from Little Creatures.

Richard and I got to talking about history, and how different histories, such as those of Australia and North America, can affect how one thinks about ecological systems and their dynamics. Australian ecologists, for example, have a different perception of the variability of environments than North American ecologists, which has affected such things as how they think about equilibrium or how long they study a system before feeling that they understand it. But it's not just that. History is in the telling, and who tells the history can make all the difference. This is true whether the tellers are Australian or American, British or French, European explorers or Aborigines, or (for palynologists or paleontologists) about the record by a pollen profile or in rock strata. All tell selective stories of how things were once upon a time. It is these stories—versions of history—that help to identify historical baselines or restoration targets. Something worth keeping in mind as people contemplate the "desired state" of nature and the conservation or management actions that will get them there.

Another year, another visit by John to Western Australia to collaborate with Richard and, inevitably, to visit Little Creatures Brewery, a place where the two colleagues have mused over things ecological while drinking beer.

Richard and John have a long history of meeting in various places, and the meetings usually start with a disagreement about whose turn it is to buy the beers, margaritas, or whatever. The trouble with history is that it happened in the past and is usually based at best on partial evidence and selective recollections. Hence, it is open to radically different interpretations. Indeed, Richard frequently uses

*Wiens, J.A. and Hobbs, R.J. 2013. From our Southern correspondent(s): which history? *Bulletin of the British Ecological Society* 44(1): 54–57. Reproduced with permission of the British Ecological Society.

Ecological Challenges and Conservation Conundrums: Essays and Reflections for a Changing World, First Edition. John A. Wiens.

selective memory loss to conclude that it must be John's round, and John counters with equally certain but different recollections of the past. Which history is correct?

The problem is that "history," as in the recounting of past events rather than what actually happened, must be chronicled by someone, and there's the rub. The history of conflicts is usually written by the winners. The histories of countries are often written by the European colonists or their descendants. The history of whose round it is, is often written by the fastest one to say "It's your turn." "History" is a product of who's telling the story, which is why accounts of the same events (e.g., who brought last time) can differ so markedly.

The context of the time also affects perceptions of history. In sciences such as ecology, the history of ideas often emphasizes the concepts in vogue at a particular time. For some time, equilibrium concepts bolstered the view that ecological systems varied within a relatively narrow range and ecological succession was an orderly, predictable process. Such concepts influenced the questions asked and what was regarded as the usual state of nature. The continuing grasp of past thinking—an unwritten "sense of history"—permeates any science, ecology included, and may strongly influence or even impede its development.

Thinking about history over beers at Little Creatures, we got to wondering how history plays into ecology in North America and Australia (see *BES Bulletin* 42(1): 49–51 for our earlier musings on this topic). Australia is an ancient land, parts of which have been largely undisturbed, geologically, since Gondwanan times. The soils are similarly ancient and generally poor in nutrients. Australia is also, over much of the continent, one of earth's driest places, and has been so for many millennia. North America, by contrast, has been geologically active and subjected to multiple glaciations. It has been a more fertile and wetter place, leading to a greater variety of major vegetation types spread over the continent as a whole. The ebb and flow of vegetation over the North American landscape over the past 10,000 years have been documented through a rich record of pollen in lakebed deposits, a recording of history that is much more scattered and fragmentary in the dryness of Australia.

These are the histories written by geologists, paleontologists, and palynologists to piece together the deep environmental histories of the continents. Other more recent histories recorded the conditions experienced by the first European explorers and settlers. The first European immigrants to North America encountered environments that were largely familiar, and (after waging several wars and displacing indigenous populations) they set about converting the woodlands and prairies to farmlands. The colonization of Australia occurred two centuries later, and those who came found a strange land containing unusual animals, strikingly unfamiliar vegetation, and Aboriginal cultures they did not understand. The "Europeanization" of Australia through introductions of plants and animals, extinction of a range of native animals, and establishment of farmlands and sheep and cattle production fundamentally altered Australian landscapes, at least in the less arid regions away from the desert interior. In both continents, much of the history taught in schools dates from the arrival of Europeans. The effects of indigenous peoples on the flora and fauna—which occurred over thousands of years in North America and tens of thousands of years in Australia—are just beginning to be appreciated by ecologists and anthropologists, if not yet by the general public.

The point of all this, we realized, is that "history" can be quite different depending on where you are, how far back you look, and whose version of history you use (not to mention how many beers you've had). The differences become important as ecologists, managers, and conservationists set targets or baselines. Consider the restoration of degraded ecosystems. Some in North America argue that the aim should be to return the environment to its "original" state, before Columbus landed on Hispaniola. This goal is not only unattainable, due to the massive changes humans have inflicted upon the landscape since then, but it also ignores the earlier impacts of indigenous populations, stretching back for additional centuries. The history of human alterations of the environment did not begin with Columbus in North America, or with the Dutch explorers who first set foot on Australia a century later.

Or, consider the management of disturbance-prone ecosystems. For decades, management of fire in forests and bushlands emphasized suppression: unchecked fires damaged valuable resources, threatened human dwellings, and disrupted the inherent stability of the ecosystems. Historical variation in fire frequency and severity has now been incorporated into fire-management policies in North America, although there are debates about what span of history should be used. In northern Australia, the ways fire was used by Aborigines over millennia have been incorporated into fire-management practices in savanna ecosystems.

Or, consider the recovery of imperiled species. The recovery goals for endangered species in both continents aim to attain self-sustaining populations. The demographic attributes required to meet this goal, however, differ depending on the scope of history used. Is it the current environment, one that existed when a species was officially listed (generally when it was perilously close to extinction), the time prior to European colonization, or something else?

Or, here's a more immediate example. Several decades ago a road was built on the Oregon coast by filling in an area of the Salmon River estuary as part of an amusement park development that later failed. The U.S. Forest Service is now restoring the estuary, removing the road and the trees that have grown alongside over the decades. The history that determines the ecologists' desired state goes back before the road, when the trees were not there but the estuary was. For many of the public, however, the relevant history is more recent. The trees are part of the desired state, so people object to their removal.

So what point in the flow of history determines the desired state? Is it when the only disturbances affecting the system were natural rather than anthropogenic? Is it when it was first realized that there was a problem requiring restoration, management, or conservation? Is it what we fondly remember from our childhood? Is it the blurry vision of the past that emerges after too many pale ales at Little

Creatures? If a major disturbance occurs at an identifiable point in time (the eruption of Mt. Saint Helens, the *Exxon Valdez* oil spill, or the onset of mining in the jarrah forest in Western Australia come to mind), it may seem obvious that the historical reference point is whatever occurred just before the disturbance. But even then the undisturbed environment would probably have changed over the interim, shifting the target. Ecosystems are dynamic; stability or steady-state is an illusion quickly destroyed by time.

This becomes all the more obvious when we consider the cascading effects of future climate change on ecosystems, landscapes, and people. In many cases, systems may move beyond their historic range of variation (however defined) into novel configurations and dynamics not previously encountered. If the future is so uncertain and "history" is so open to differing interpretations depending on whose version of history is used, how far back it goes, and where it occurred, one might question whether history has anything important to contribute to restoration, resource management, or conservation.

Yet, history cannot be ignored. History is what tells us how we got to where we are now, and that in turn sets some boundaries on where we are likely to go in the future. Knowing how different species have responded to environmental changes in the past, for example, can reveal much about their resiliency or vulnerability to future changes. Understanding how the ecological and life-history attributes of species have influenced community assembly following past disturbances may help us predict responses in the future. Determining why particular restoration or management practices have or have not worked in the past may provide hints about which to employ under the changed conditions of the future. Knowing which invasive species have disrupted ecosystems and which have not can help to direct control efforts more effectively as invasions become more frequent. In all of these situations, history offers important insights. The trick is to be judicious in how history is used, to learn from it rather than to be its captive.

Having cleverly had the foresight to jot the preceding thoughts down on a napkin and put it in his pocket for later, John used history judiciously and pointed out that it was Richard's round. In characteristic fashion, Richard was equivocal about the role of history in reaching conclusions like this, but, in the absence of solid information either way, obligingly trotted off to the bar to order more beers.

CHAPTER 12

A digression on baselines and targets

Baselines and targets are essential to any applications of ecology, whether they be resource management, restoration, or conservation. Baselines tell you where a system has been, in case you want to get back there again. Targets tell you where you would like to go, the goals of your actions.

Clearly stated baselines or targets enable people to gauge whether they are making progress and provide transparency so that others can evaluate whether the actions and goals really make sense. Scientists like baselines and targets to be objectively determined, but the reality is that they are value-driven, either by the scientists' own sense of values or by societal values. For example, decisions about the goals of restoring a damaged ecosystem, such as a drained wetland in Nebraska or a strip-mined hillside in West Virginia, involve a complex balancing of what is feasible with what is desirable. What is feasible often depends on how much money is available to undertake the restoration action, and what is desirable depends on whether the perspective is that of a mining corporation, a governmental regulator, or an environmentalist. Establishing targets or baselines isn't easy. Yet both are essential.

Baselines are usually needed if one is undertaking a historical comparison of how things are now in relation to how they were at some time in the past. They are often used in assessing the impacts on species, communities, or ecosystems arising from natural or human disturbances, and then in determining how recovery is progressing toward that baseline condition (i.e., the baseline becomes the target for recovery efforts).

Oil spills, mentioned previously (Essay 9, *Oil, oil, everywhere ... (2010)* (page 60)), provide a good example. In the United States, the assessment of damages and recovery from an oil spill is governed by the regulations of a Natural Resource Damage Assessment (NRDA) under the Comprehensive Environmental Response, Compensation, and Liability Act (CERCLA). If oil or some other hazardous substance is released into the environment, injury to a natural resource is documented by an adverse change from the previous state of the system; recovery is then defined as "the return of injured natural resources and services to baseline," where baseline is "the condition of the natural resources and services that would have existed had the incident not occurred."[1] These are

[1]U.S. Code of Federal Regulations 15 CFR §990.30.

Ecological Challenges and Conservation Conundrums: Essays and Reflections for a Changing World, First Edition. John A. Wiens.

the regulations that were used to assess the impacts of the *Exxon Valdez* oil spill on species and natural communities.

The problem is in interpreting what the regulations actually mean and determining what sorts of measurements are appropriate. Because the *Exxon Valdez* spill became embroiled in litigation that lasted for decades,[2] different interpretations and approaches turned into hardened positions that fueled continuing controversy. The spill was not a subtle event, so establishing injuries to natural resources was generally not too difficult. Most disagreements centered on how the baselines or targets for recovery should be assessed.

There are two ways to interpret what might be expected "had the incident not occurred." One is to assume that conditions prior to the spill would have remained unchanged, so recovery would be signaled by a return to pre-spill conditions. Quantitative information on the status of a species or community immediately before a disturbance is rarely available, so historical records must be used (if such exist). For the *Valdez* spill (which occurred in 1989), pre-spill surveys of sea otters had been conducted in 1973, 1974, and 1984 using a variety of survey methods; seabirds had been surveyed in 1972–1973 and 1984–1985. The marine environment where the spill occurred is highly variable, so it is not surprising that these surveys produced different results. Depending on which numbers were used as a baseline, conclusions about impacts and recovery differed.

The second way is to acknowledge that conditions varied, both before and following the spill, and to incorporate such variation into the assessments. This proved to be tricky because the environment varied both in time and in space, so simply surveying a few places at a few times would not yield an adequate sampling of the dynamics of the system. After pondering this problem for some time, Keith Parker and I concluded that there really is no fixed baseline in such a variable environment; rather, recovery must be assessed by conducting statistical comparisons between impacted and reference sites, incorporating variations in natural resources in both time and space and in other environmental factors that could confound the analyses.[3] This approach, like nearly everything about the *Exxon Valdez* spill, fueled disagreements.

The problem with baselines, then, is that they move around. No time is the same as another and no place is the same as another. Stepping back to be general enough that the variations don't matter leaves one with targets such as "avoiding extinction," "ensuring persistence," or "avoiding further degradation." Although these are laudable and important goals, they are too open-ended. While it may ultimately be true that nothing is permanently conserved or restored, the notion of having targets for actions is to enable an assessment of how well the actions are working and when "success" has been achieved, at least for a while.

[2] See Wiens (2013a) for a review and synthesis of the scientific studies and issues.
[3] Wiens and Parker (1995) and Parker and Wiens (2005).

For a long time, the goal of ecological restoration has been to return a damaged or degraded place or ecosystem to some former state.[4] Often, this former state referenced conditions prior to explorations or settlement by Europeans.[5] Some conservationists have argued for similar, "back to nature" targets. Such views of a pristine nature or "naturalness"[6] as something we should strive to re-create are hopelessly idealistic and naïve. Using such targets is a recipe for failure.

One way around this problem is to establish generic rather than specific targets. Instead of attempting to replicate the species composition (much less the relative abundances of species) as the target for ecological restoration of a wetland, grassland, or some other habitat, the goal might be to re-establish ecological functioning in the restored area. Obviously, the target or baseline for an action, and the actions that are undertaken, will differ depending on whether the objective is to restore a species, community, habitat, ecosystem function, ecosystem service, or something else. Each of these poses definitional problems. Although we may be reasonably confident about the identity of a species, determining what a habitat or ecosystem service is may be more subjective, and thus potentially more contentious. This is part of the broader issue of determining what it is we are trying to conserve, which I consider in Chapter 20.

None of this means that past conditions are irrelevant to establishing restoration or conservation targets. History may not enable one to define *specific* targets, but by indicating how systems have functioned in the past it can help to guide current actions. Historical analyses, for example, have shown that hydrologic flows were quite different in different parts of the California Delta before the system was engineered with a vast network of levees and channels.[7] The engineering has only submerged these differences, however, not erased them. Consequently, applying a particular habitat-restoration approach may work in some places in the Delta but not in others. Some assessments of the effects of the *Exxon Valdez* oil spill also suffered from this assumption that all locations in the spill zone could be treated as if they were the same.

---

[4]The everyday meaning of "restore," after all, is to repair a painting, antique furniture, or a historic building, returning it to its former glory.

[5]What is often called a "pre-Columbian" baseline, as if the purported discovery of America by Columbus marked a demarcation between pristine nature and nature altered by humans. While environmental change took on a different trajectory once Europeans entered the picture, pre-European indigenous people had been having a significant effect on the environment for millennia. The Columbian benchmark is also decidedly North American; it is largely irrelevant to events in Europe, Africa, Asia, or Australia, all of which had different histories of European contact. Mann (2006, 2011) explores the settings before and after Columbus' landing on Hispaniola in 1492 in some detail.

[6]A notion considered from multiple perspectives by the contributors to Cole and Yung (2010).

[7]Whipple et al. (2012).

The problem, then, is that variation in time and space confounds attempts to establish firm targets or baselines that can be used to direct management, restoration, or conservation actions. Time marches on, continuously changing things. While the marching image doesn't fit spatial variation very well, the message is the same: different places have had, continue to have, and will have different dynamics. We are left with what Steve Jackson termed "George Webber's dilemma"[8]: you can't go home again, and even if you could you have no idea how long ago or far away "home" is. In a variable environment, you will encounter something different depending on how far back you go and how far away you go, making any specification of "baseline" or "target" scale-dependent.

If we can't specify fixed baselines or targets, how can we know where we are going or when we're getting close? Perhaps if we can't establish fixed baselines and targets we should create variable ones. John Hiers and his colleagues[9] have suggested using a "dynamic reference concept," in which areas to be restored are simultaneously compared with a set of reference sites over time. Differences among the reference sites capture spatial variation, and changes over time record temporal variation; together, they define the "moving target" for restoration. The approach is similar to the design we used to assess recovery of seabirds from the effects of the *Exxon Valdez* oil spill.[10] In either case, the challenge is in the selection of appropriate reference sites. Because the sites vary in space and time, comparisons must be statistical, requiring quantification of key variables and continued monitoring. And what is "appropriate" depends on the specification of the desired state, be it restored sites or recovered populations, and on who makes the determination.

This approach is based on contemporaneous comparisons. But the pace of environmental change is accelerating, which means that by the time the data for a comparison are collected and analyzed and appropriate restoration or conservation actions designed, the environment will have changed again. The actions may be appropriate for a time that has already passed. I comment on the need to anticipate future conditions in Essay 16, *Shooting at a moving target (2011)* (page 133).

Given all these problems and complexities, it's easy to see the allure of a fixed baseline or target, be it sometime in the past, now, or even off in the future. Having seen the variation, however, we can no longer ignore it.

---

[8]After the protagonist in Thomas Wolfe's novel, *You Can't Go Home Again* (1940); Jackson (2012).
[9]Hiers et al. (2012).
[10]Wiens and Parker (1995).

ESSAY 16

Shooting at a moving target (2011)*

In 2011, I found myself embedded a large and varied assortment of debates with restoration ecologists in the Yucatán. There was lots of talk about the tools and techniques of restoration and its social, political, and ecological contexts. There was also a widespread recognition that environmental variation could complicate efforts to achieve the goals of restoration projects. No one was arguing (openly, at least) that restoration should aim to re-create conditions of long ago, but many of the projects that were described had fixed, stationary targets. There were heated discussions about climate change (perhaps because it was so hot and humid), but not many ideas about how to incorporate it into restoration, which almost by definition looks to the past.

But what use is it to restore a habitat, only to see the gains wiped out by climate change? I suggested that restoration needs to be more facile and forward-looking, anticipating what the environment will be like when a restoration project is completed, and thereafter. What I called anticipatory adaptive management.

Thinking about the future is hard, because it's couched in so much uncertainty. Everyone—perhaps especially scientists—likes as much certainty in their lives as possible. The lesson of the Mayan cultures of the Yucatán is that nothing about the future is certain. But we do have ways of assessing future probabilities that the Mayans lacked. Perhaps we can do better.

And, of course, the Mayans didn't disappear, nor did their culture. It's still there, just not in its former splendor.

Late August in the Yucatán is hot and muggy. What better place, Richard Hobbs and I thought, to renew our discussions about the future of all things ecological—over beer and margaritas, of course. To justify this junket, we organized a symposium for the 4th World Conference on Ecological Restoration, hosted by the Society for Ecological Restoration in Mérida, México. The conference drew some 1000 participants from over 60 countries, who spoke on topics ranging from cultural landscapes to coral reefs to ecosystem valuation to water and peace in Mesopotamia, all in the context of restoration. Our symposium addressed restoration ecology in a changing world—hence the title, shooting at a moving target.

*Wiens, J.A. 2011. Shooting at a moving target. *Bulletin of the British Ecological Society* 42(4): 55–56. Reproduced with permission of the British Ecological Society.

Ecological Challenges and Conservation Conundrums: Essays and Reflections for a Changing World, First Edition. John A. Wiens.

I'm not a hunter, but the metaphor will be obvious to anyone who is. When aiming at a flying duck, partridge, or wood pigeon, one must lead the target. Aim where you anticipate it will be when your shot reaches the target, not where the target is now. In other words, one must look into the future to have any chance of success. With a flying bird, it is a matter of projecting the flight trajectory a few seconds ahead. Hitting the target of restoring an ecosystem may require anticipating what the environment will be like in several decades, and the trajectory of change is rife with uncertainties.

This challenge of identifying (much less hitting) moving targets in an uncertain future confronts all areas of conservation and resource management. Until recently, targets were often established in relation to some set of past conditions. Hydrologists used stationarity—the assumption that ecosystems vary about a stable long-term mean—as a basis for defining 100-year floods. Forest managers aimed to maintain forest conditions within the historical range of variation, and many restoration ecologists and conservationists (at least in North America and Australia) set the targets of their actions in terms of some idealized notion of how things were before Europeans arrived on the scene. But all this is changing. Hydrologists have recognized that "stationarity is dead," foresters are questioning the value of historical range of variation, and most restoration ecologists and conservationists have come to accept that pre-Columbian conditions may not be suitable targets after all (and never mind that the scene Europeans first happened upon in North America, Australia, and elsewhere was already profoundly altered by millennia of occupation by indigenous people). Thomas Wolfe's admonition that "you can't go home again" applies with full force to management or conservation that aims to re-create the past.

This realization was evident everywhere in the Mérida conference. There was lots of hallway talk about "novel ecosystems," "no-analog futures," and "the new ecological world order." No one was denying that climate change, land-use change, invasive species, economic globalization, and other social and political forces were gathering into a perfect storm of change and uncertainty. The problem, of course, was that no one knew how to deal with all of it, or even with the pieces that restoration ecologists worry about. Aiming for targets of the past is not likely to produce the lasting results we want, but how should one aim for targets in a future that we know will differ from the present and past in so many ways?

Perhaps we should turn to the hunting metaphor again. When aiming at a flying duck, one can set the shot pattern to be broad (e.g., cylinder bore) or narrow (full choke). With a broad pattern the shots are scattered, increasing the probability that a few will hit the target, although with uncertain results. If one's aim is true and the target's trajectory is predictable, a narrow, more focused pattern will be better, maximizing the number of shots hitting the target and increasing the probability of the desired result.

Which approach is best in conservation and resource management? If the future status of the target is uncertain, using a variety of methods (a shotgun approach) in the hope that the one that will work not only has a low probability of success but will also scatter the efforts and dilute the resources available to address the problem. A more focused approach may be better, particularly if we can adjust the aim if it seems we're not hitting the target, which may have veered in unexpected directions.

This is where adaptive management comes in. The premise of adaptive management is to try something out; if it doesn't work (because it's an ineffective approach, the target has shifted, or the environmental setting has changed); adjust the approach and try again, or try something different. Unlike duck hunting, there is often the opportunity to take more than one shot, adjusting on the basis of past results. But the essence of shooting at a moving target—leading the target—suggests something more. Waiting for the results of a management

Figure 1 An unrestored Mayan ruin at Sayil, Yucatán, México. Sayil was settled around AD 800. The city reached its greatest extent around AD 900, when it covered an area of approximately 5 km^2 and had a population of perhaps 10,000 in the city itself with an additional 5,000–7,000 living in the surrounding area. This population probably exceeded the limits of the agricultural capacity of the land; Sayil began to decline around AD 950 and the city was abandoned by AD 1000. Photograph by John Wiens.

action to decide how or whether to adjust the approach may not work when the target is moving rapidly. To be effective in a changing world full of novelty and surprises, one may need to anticipate where the target will be, not where it has been or is now. One needs to use what I call *anticipatory adaptive management*. This will not be easy. It requires foresight and a capacity and willingness to change management practices quickly (not the forte of most government agencies). Judged by the presentations at the Mérida conference, the challenge of ecological restoration in a changing world is on everyone's mind, even if it's not clear what to do about it.

Amid all this talk of changing targets, there was one conspicuous irony of the Mérida conference. Centuries ago, the Yucatán was a center of Mayan culture, and the area is sprinkled with hundreds of archaeological sites (Figures 1 and 2). Some of these, most famously Chichén-Itzá and Uxmal, are being actively restored

Figure 2 The restored Pyramid of the Magician at Uxmal, Yucatán, México. Maya chronicles say that Uxmal was founded about AD 500. Uxmal was the capital of a Late Classic Mayan state around AD 850–925. Toltec invaders took over around AD 1000 and most building ceased by AD 1100, although Spanish colonial documents suggest that Uxmal was occupied into the 1550s. Construction of the first pyramid temple began in the 6th century AD and the structure was expanded over the next 400 years. The pyramid fell into disrepair after AD 1000 and was looted during the Spanish conquest of the Yucatán. Restoration of the pyramid was initiated in the mid-19th century, began in earnest in the 1970s, and continues to this day. Photograph by John Wiens.

to their former grandeur. Re-creating the structures the Mayans inhabited many centuries ago can provide a glimpse of the richness of the Mayan culture. The target of restoration is rooted in the past. The environment surrounding these structures, however, is not the same as it was when the Mayans thrived, and it will not be the same in the future. We cannot hope to restore the past environment, any more than we can hope to re-create the Mayan culture by restoring Chichén-Itzá and Uxmal.

CHAPTER 13

Ecological thresholds

To a conservationist or manager, and to the organisms that live in an ecological system, the most challenging changes may be those that occur suddenly and without warning and alter the system in dramatic ways. Examples are legion.[1] Florida Bay, a shallow estuary in southern Florida, was for many decades a clear-water system low in nutrients (oligotrophic), that was dominated by seagrasses. In the space of a few years in the late 1980s to early 1990s, it changed to a turbid, nutrient-rich system dominated by phytoplankton blooms (eutrophic), largely as a result of increased nutrient inputs. Reversing these changes will be extremely difficult.

Other examples: In the semi-arid savannas of northern Australia, intense grazing by cattle has resulted in a loss and increased patchiness of grass cover, reducing the frequency of the periodic fires that maintained the grasslands. The shift to a shrub-dominated system has been rapid, bare areas have expanded, and erosion has increased, degrading soil quality. An opposite shift has occurred in the western United States, where removal of sagebrush cover by grazing and wildfires in shrub-dominated landscapes has led to invasion by exotic grasses (primarily cheatgrass), which provide fuel for recurrent fires that forestall the recovery of the slow-growing shrubs. In both cases, the shift in dominance by grasses or shrubs may be permanent.[2]

These are examples of ecological thresholds or tipping points—abrupt changes in the state of an ecological system, often in response to small changes in environmental conditions. Look again at the figure from Chapter 3 (reproduced here as Figure 13.1 so you don't have to page back to find it). Thresholds imply a big, sudden change that takes the system into a different state, such as oligotrophic to eutrophic or grassland to shrubland—an alternative stable state.[3] In many cases, it

[1]The Resilience Alliance maintains a web site listing examples of thresholds in both ecological and social systems. As of February 2015, it has listed 103 examples. See http://www.resalliance.org/index.php/database.
[2]Bagchi et al. (2013). A special issue of *Frontiers in Ecology and the Environment* (Volume 13, Number 1, 2015) considers multiple examples of such state transitions in arid landscapes.
[3]Referring to a condition or state as "stable," while appropriate for theoretical discussions, obscures the reality that the alternative state may exhibit substantial variation, as shown in the left portion of Figure 13.1.

Ecological Challenges and Conservation Conundrums: Essays and Reflections for a Changing World, First Edition. John A. Wiens.

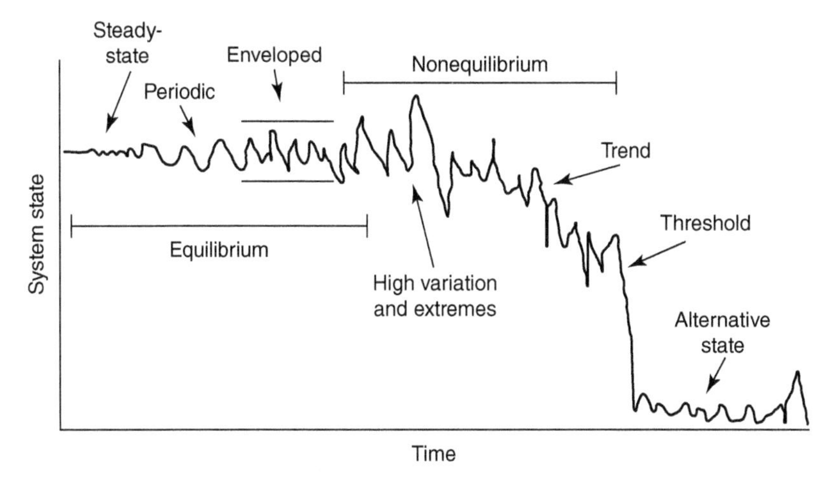

Figure 13.1 The conceptualization of the kinds of changes that may occur in ecological systems shown in Figure 3.1. Relevant here is the transition from a variable state to a threshold and alternative system state.

may be difficult to return to the former state without massive management efforts, and even then the interventions may move the system to some state other than the one desired. Because thresholds occur suddenly, there is often little warning that the system is approaching the precipice, resulting in surprise when the system escapes management and shifts to something else. This is why thresholds are so important in conservation and management.

Despite the intuitive appeal of the concept, the broad theoretical support, and the popularity of Malcolm Gladwell's book, *The Tipping Point*,[4] the idea of thresholds is difficult to apply in practice.[5] To begin with, there is the issue of whether thresholds are only abrupt changes or whether the changes must also be irreversible (or require a huge effort to reverse). For example, a flood in a river may be sudden and cause a change in state when it breaches the river banks, inundating the adjacent floodplain. However, the river system and floodplain return to something approximating their former state when the floodwaters recede. Energy, in the form of floodwater flows, is needed to cross the banks into the floodplain, but a reduction in energy is sufficient to bring the river back within its banks. Given the transitory nature of the state change, I think of this as an example of an extreme event (Figure 13.1) rather than a threshold.

---

[4]Gladwell (2000).

[5]Groffman et al. (2006) and Guntensbergen (2014) have summarized some of the difficulties, and Levin and Möllmann (2014) have developed an approach for incorporating thresholds (what marine scientists call regime shifts) into ecosystem management.

Conservationists are most concerned about thresholds that produce long-lasting, potentially irreversible changes, as in the above examples. This issue of irreversibility is a thorny and contentious one. Carolina Murcia and her colleagues have claimed that there is little evidence of irreversibility of system changes, and that "no proof of ecological thresholds that would prevent restoration has ever been demonstrated."[6] There does seem to be clear evidence of irreversible changes in some ecological systems (see Footnote 1), but I suppose time will tell whether the system changes are ultimately reversible or not. The sudden change in reproduction by Cassin's auklets that I describe in Essay 17, *Tipping points in the balance of nature (2010)* (page 144), for example, seemed to be a clear example of a threshold, only to turn into something else as time went on.

There is also the question of what is meant by a "sudden" change. Where is the threshold if a system changes gradually to a different, potentially irreversible state? There is probably no better example of this than the saga of the Aral Sea in Central Asia. At one time it was the fourth largest lake in the world. Beginning in the 1960s, the former Soviet Union initiated a major project to divert the rivers flowing into the Sea in order to irrigate fields for (mostly) cotton agriculture in Kazakhstan, Uzbekistan, and Turkmenistan. The desert bloomed and the Sea shrank, splitting into two lobes by 1995 and then a single remnant by 2014 (Figure 13.2). Efforts to save what remains of the Sea have focused on constructing a dam to hem in the waters of the western lobe; the eastern lobe (now completely dry) has been given up for lost. The change was irreversible, but it occurred over a half-century—gradual from our perspective, but sudden in terms of the long-term dynamics of the Aral Sea ecosystem.

The issues of irreversibility and suddenness largely relate to how one defines a threshold. Other issues are more contextual, and more confounding. For example, a particular species of interest is likely to exhibit different thresholds for different environmental factors based on the physiological tolerances or habitat selection of individuals. If the focus is instead on population dynamics, the same species may experience other thresholds. Because environmental factors are themselves interrelated to varying degrees, how a species responds to changes in one factor may depend on how other factors vary. For instance, a sudden, massive die-off of piñon pine in the southwestern United States was initiated by drought, which stressed the trees. This made them vulnerable to an outbreak of bark beetles, which finished them off.[7] In other situations, the presence or absence of keystone predators can create threshold dynamics that ripple through the ecosystem. In Alaskan nearshore environments, sea otters feed on sea urchins, which in turn feed on kelp. Where otter numbers have plummeted (as in the Aleutian Islands), the urchin populations, no longer constrained by predation, have

[6] Murcia et al. (2014: 5).
[7] Breshears et al. (2009).

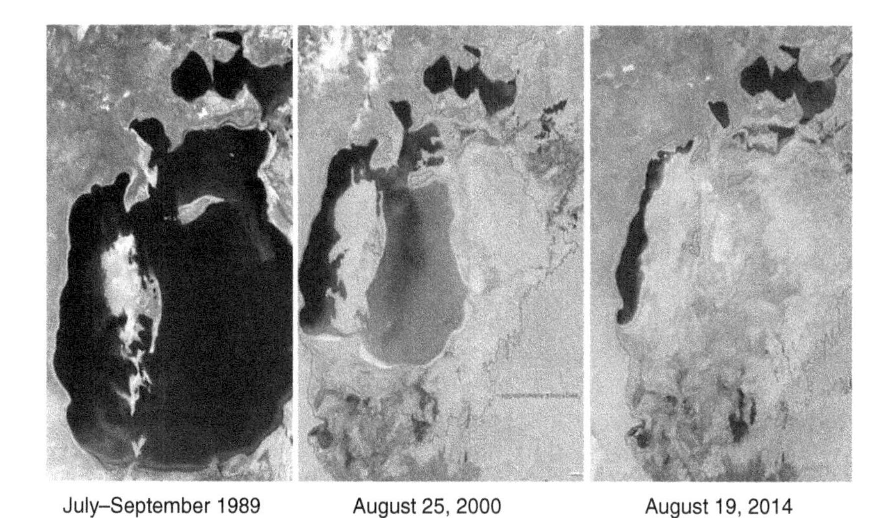

Figure 13.2 The disappearance of the Aral Sea, Central Asia, 1989–2014. Source: http://riverdiary.umwblogs.org/files/2014/04/AralSeaShrinking.jpg.

exploded, grazing the kelp nearly to a vanishing point. This has removed food and cover for fish, which have also declined.[8]

Threshold changes in habitat (chiefly vegetation) have received particular attention, partly because there is a strong theoretical basis for expecting thresholds to occur and partly because the amount of habitat in an area is something that concerns conservationists and can be measured and managed. Theory suggests that a random loss of habitat from a continuous coverage (e.g., a forest or grassland) will initially perforate the habitat (as shown in Figure 6.2). When habitat loss exceeds a certain value, however, connections are broken and habitat is fragmented—a threshold has been passed. Various landscape models predict much the same thing. In both approaches, the amount of habitat loss that triggers a fragmentation threshold varies depending on the model assumptions and the scale of movement or "neighborhood size" of the species or process of interest.[9] Empirical, experimental, and modeling studies suggest that habitat connectivity may suddenly decline when coverage in a landscape falls below 10–30%, although some studies place the threshold as high as 60%. In an intensive study of the Atlantic Forest in Brazil, Christina Banks-Leite and her colleagues found that the compositional integrity of communities of mammals, birds, and amphibians was maintained until forest cover declined to 24–33%, below which integrity declined sharply as

[8]Estes and Palmisano (1974). Now orcas (killer whales) have entered the picture, developing a taste for sea otters that can forestall otter recovery (Estes et al. 2004).
[9]With (2002).

forest-specialist species were replaced by disturbance-adapted generalists.[10] Whatever the actual value, there is general agreement that breakage of connectivity can produce abrupt, threshold changes in how disturbances, nutrients, or organisms move across a landscape, altering ecosystem processes and population dynamics.[11]

History creates additional complications. When a system changes suddenly, one usually looks for immediate causes—what just happened that could have made the system change? Because of time lags and legacy effects, however, a threshold change that we see now may have been set in motion by an abrupt environmental change in the past. Or we may mistakenly conclude that an environmental change that we see now has not pushed a system over a threshold, only to see the response emerge at some later time—a delayed response. Because changes in environmental factors may have synergistic and cumulative effects (as in the piñon pine example), what we see as the final "tipping point" trigger may be only the most recent of a series of changes leading to the threshold. It is rare indeed to have the historical data and the ecological insight to put the chain of events together, much less to be able to incorporate them into conservation or management plans.

Changes in the environment or in the responses of systems to the environment are not the only thresholds that affect conservation and management. The sociopolitical context in which decisions are made can lead to thresholds in environmental actions. The decision to list a species under the Endangered Species Act, for example, is not directly related to the species or its habitat having passed some biological threshold that launches it into a sudden and precipitous decline. Rather, it is based on a variety of factors, some scientific (e.g., population viability analyses, remaining population size), some pragmatic (e.g., feasibility of implementing recovery actions), and some political (e.g., avoiding of listings that would have large economic impacts). The listing decision, however, has ramifying effects on agency priorities, the agendas of conservation organizations, and the attitudes and practices of landowners—all threshold responses.

All of these issues and realities present formidable challenges to incorporating thresholds into conservation and management in anything more than a conceptual way. We know that thresholds exist in ecological systems, and that more

---

[10]Banks-Leite et al. (2014). They used the survey information, in combination with economic cost scenarios for forest restoration, to conclude that an annual investment equivalent to 6.5% of what Brazil spends on agricultural subsidies would revert the ecological systems across farmlands to levels found in protected areas, thereby enhancing ecosystem services.

[11]Turner et al. (1989), Ludwig et al. (2000), and Reiners and Driese (2004). Recall, however, that metapopulation theory requires that there be some degree of breakage of connectivity in order for subpopulations to vary largely independently of one another, creating the setting for metapopulation dynamics and stability. Connectivity (or dispersal among patches) must be just right: too little and too many local populations may suffer extinction, too much and the patch subpopulations no longer function independently of one another. See Hanski and Gaggiotti (2004) for details.

thresholds will be encountered more often as the pace and magnitude of environmental changes increase. We generally can't detect thresholds until they are passed.[12] We are increasingly aware of the idiosyncratic, situation-specific nature of threshold dynamics but are yet to develop any useful generalizations. If we know or suspect where thresholds might lie in the coupling between ecological systems and the environment, however, we might want to focus attention on situations that are close to the threshold, where efforts might be successful in forestalling abrupt changes. Alternatively, we might want to concentrate on state changes that could be reversed with a cost-effective investment. Alternatively, if a system has entered a state space in which it is so degraded that it cannot be put back together again (as the Aral Sea), it may not merit conservation or management attention. In other words, triage. I'll comment further about triage in conservation in Essay 22, *Talking about triage in conservation (2015)* (page 188).

[12]Progress is being made on developing methods to anticipate approaching thresholds, based on changes in the variance of key indicators; Scheffer et al. (2009).

ESSAY 17

Tipping points in the balance of nature (2010)*

I seem to keep coming back to the theme of a balance of nature, rather like a recurrent dream or nightmare. Perhaps this is because the belief is so deeply ingrained in Western thought and therefore, inescapably, in ecology and conservation. Comforting as the concept is, I think it stands in the way of making ecology and conservation relevant in a changing world.

Perhaps the greatest affront to a balance of nature belief is the notion that natural systems may undergo abrupt, unpredictable changes into something quite different—the antithesis of a balanced nature. Such thresholds are the subject of a rich body of theory in nonlinear systems dynamics, chaos, cognitive psychology, and a host of other disciplines, and references to "tipping points" have become part of everyday discourse.[1]

I initially planned to use the example of annual variations in reproduction by Cassin's auklets on the Farallon Islands off San Francisco to illustrate threshold dynamics. As is often the case, however, the example illuminated several other points. Yes, there was an apparent threshold (certainly a nonlinearity) when reproduction totally failed in two successive years. This, of course, led to speculations about possible causes. Because information was available on other attributes of the marine ecosystem (including the krill that the auklets eat), the speculations had strong empirical support. This isn't usually the case when something unanticipated happens. Scientists at PRBO (now Point Blue) Conservation Science had been monitoring auklet reproduction for nearly four decades, so the sudden plunge in reproduction could be weighed against long-term dynamics. This record showed variation in annual reproduction increasing in the years before the plunge, as some theoretical arguments have suggested should happen.

Because the Point Blue scientists are nothing if not persistent, they have continued to monitor the auklets in the years following the plunge. Jamie Jahncke and Pete Warzybok provided additional information, which showed reproduction returning to exceed the long-term average within three years, and then increasing to an unprecedented level. I've updated the graph from the original essay below. The peak in 2010 was associated with a high proportion of successful second broods, allowing many auklets to produce more than one chick per pair. The high success coincided with strong oceanic upwelling and high krill abundance throughout the summer; juvenile survival and recruitment were also high during this period.

*Wiens, J. A. 2010. Tipping points in the balance of nature. *Bulletin of the British Ecological Society* 41(1): 68–69. Reproduced with permission of the British Ecological Society.

[1]Thanks in large part to Gladwell's book, *The Tipping Point: How Little Things Can Make a Big Difference* (2000).

Ecological Challenges and Conservation Conundrums: Essays and Reflections for a Changing World, First Edition. John A. Wiens.

However, there's an additional twist to the story. In 2014, auklet productivity was high at many colonies along the West Coast. At the same time, a warm-water incursion along the coast led to the disappearance of krill from large areas. So, as large numbers of auklets fledged and went to sea, they were immediately faced with poor foraging conditions. Many emaciated carcasses of young auklets washed up on shorelines, attracting considerable public attention to a species most people never see (or even know about). The shift to warmer, less productive conditions in the California Current is unprecedented in the historical record and may be long lasting.

Whether or not the auklet example illustrates a tipping point, it certainly demonstrates the complex and unpredictable dynamics of the system, and it makes one wonder why auklet reproductive output was so stable during the 1970s and 1980s and so variable thereafter. But we wouldn't even be wondering about such things if the long-term monitoring had not been done (see Essay 29, Is "monitoring" a dirty word? (2009) (page 241) *and Essay 30, The place of long-term studies in ornithology (1984)* (page 244)*). Look at different 3- to 4-year segments of the long-term record and you can make up a different story for each. This may be the reason why we have such a proliferation of stories in ecology, and why we have not seen tipping points more often (Figure 1).*

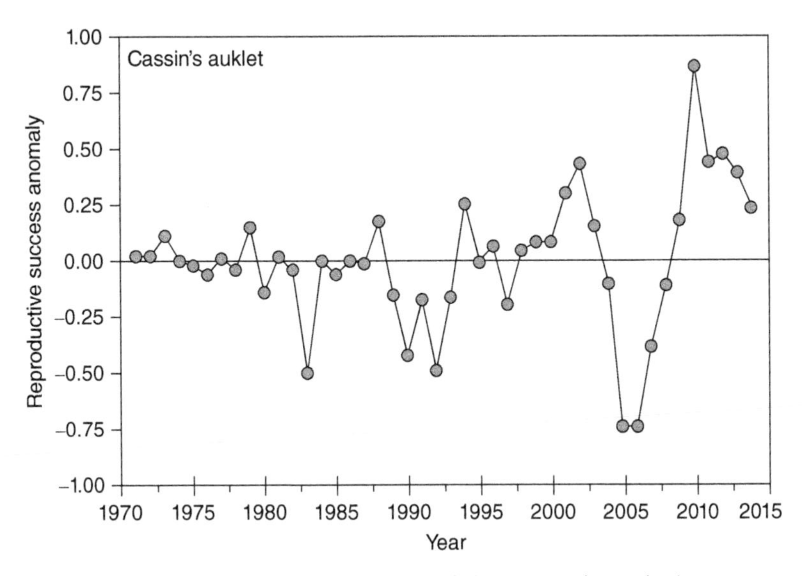

Figure 1 Standardized reproductive success anomaly (mean annual reproductive success–long-term mean) for Cassin's auklets on Southeast Farallon Island, 1971–2014. The zero line represents the long-term mean. The points represent the deviation from the long-term mean for individual years in terms of annual mean number of chicks fledged. Source: Reproduced with permission of Point Blue Conservation Science.

The notion of a balance of nature is an article of faith in many cultures. It is founded on a desire for things to stay the same, or if they change, to do so gradually and predictably, and by not very much.

Of course, the concept of nature in balance is no longer accepted as unquestioningly by ecologists as it was 30 or 40 (or even 20) years ago. Nature is variable, and models based on stability have given way to arguments based on "windows of variation," "stationarity," or "state space." However, the expectation remains much the same: nature may vary, but she does so only within relatively narrow bounds.

All of this may be about to change. Since the onset of the Industrial Revolution, humans have been altering nature at an accelerating rate. Scientists increasingly talk of thresholds or tipping points in global systems, points at which there are sudden (and perhaps irreversible) shifts in the ways atmosphere–earth–ocean systems function together. Within decades, ocean acidification may reach the point at which many marine organisms will be unable to form shells or protective exoskeletons, fundamentally altering marine food webs. Earth system scientists are warning of sudden changes in the stability of the Antarctic ice sheet or of changes in ocean circulation patterns that may weaken the flow of the Gulf Stream, affecting climate, growing seasons, and societies throughout Western Europe.

Concerns about tipping points and thresholds are not new (Robert May theorized about such abrupt state changes in population dynamics nearly 40 years ago), but we've made little progress in figuring out how to anticipate them. Tipping points surprise us; we usually recognized them only in retrospect. If we pay attention, however, we may see a clue in the variance of a system. Chaos theory predicts that variations will increase in size and frequency as a bifurcation boundary—an irreversible tipping point—is approached.

Let me offer an example that may provide a clue about at least one species, the Cassin's auklet (Figure 2) (*Ptychoramphus alueticus*). For the past 38 years, scientists from my organization, PRBO Conservation Science, have been recording reproductive performance of this small seabird on the Farallon Islands in the Pacific Ocean, 45 km west of San Francisco, California. The mean annual fledging success (chicks per pair) is shown below.

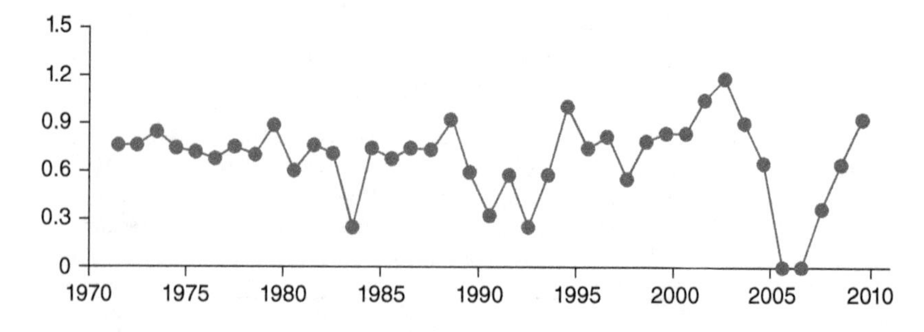

Several things about this long-term record bear on the topic of this essay. Over the first half of the study, reproductive output was remarkably stable (except in 1983, an El Niño year). If the work had stopped then, it would stand as a good example of nature in balance. Variation then increased during the 1980s, and became even greater during the late 1990s and 2000s. In 2005 and 2006, there was

Figure 2 Cassin's auklet. Photograph by Ron LeValley. Reproduced with permission of Point Blue Conservation Science.

a sudden and complete reproductive failure. If the work had stopped then, it would stand as a good example of a threshold. Over the following 3 years, reproductive output returned to previous levels, something some might consider evidence of population resilience. These details have emerged only after decades of study, however. The value of long-term data in an era of rapid environmental change can scarcely be overstated.

As our studies of seabirds on the Farallon Islands have expanded (Figure 3), it's become apparent that these dynamics are mostly driven by shifts in the distribution and abundance of krill (largely *Euphausia pacifica*). Krill, in turn, are responsive to broad-scale changes in ocean conditions. The increasing variance in reproductive output of the auklet may be warning of a tipping point to come. And if we're right about the sensitivity of the birds to marine conditions, we may be approaching an irreversible shift not just in seabird populations on the Farallones, but also in the dynamics of marine food webs.

Tipping points are likely to become more frequent and common in ecological systems as climate-driven changes accelerate. In some cases (such as the auklet reproductive failures seen up to now) the system may bounce back. Of greater concern are the "tipping points of no return," where a system has been so fundamentally altered that bouncing back across the threshold is not possible. The risks of such irreversible tipping points are too great for us to be optimistic or complacent.

Figure 3 Southeast Farallon Island. Photograph by John Warzybok. Reproduced with permission of Point Blue Conservation Science.

The solution is not more theories, more models, or more business-as-usual ecology. Most ecological studies are too short to separate normal variation from sudden, threshold shifts, fostering doubt about the real-world relevance (or existence) of tipping points. Even the Intergovernmental Panel on Climate Change, in their 2007 report, avoided using the terms "threshold" or "tipping point," instead referring to an increased risk of "large-scale discontinuities." Some scientists have argued that statements about tipping points in Earth's climate are misleading and unsubstantiated, and could lead to an erosion of public and political support for emissions controls: if we've already passed tipping points of climate change, why bother going out of our way to take action?

If we are to maintain some semblance of a balance in nature, however, managing ecological systems to avoid such thresholds will be imperative. Unfortunately, many of the coming changes (e.g., ocean acidification, loss of high-elevation ecosystems, and salt-water intrusion into coastal wetlands) may be beyond management. In these situations, we must aim to delay or mitigate the change as best we can by anticipating tipping points. We must also learn how to navigate across thresholds and prepare for what lies on the other side. I must confess that I have no idea how to do this for auklets on the Farallones, but if we think about tipping points only as something that might happen sometime in the distant future, we are sure to be unpleasantly surprised.

A new world full of thresholds and surprises awaits us. We need to get ready.

CHAPTER 14

Ecological resilience

Ecological resilience is the flip side of thresholds. Simply stated,[1] resilience is "the capacity of a system to absorb disturbances and still retain its basic function and structure."[2] "Resilience" is rapidly becoming a buzzword in ecology, conservation, management, and policy, perhaps because it offers the prospect of avoiding thresholds and unwanted changes in the state of a system. The hope is that by capitalizing on the inherent resilience of nature or enhancing resilience through management, ecological systems will be able to deal with the affronts of the Anthropocene and persist. Resilience means that a system bends but doesn't break. The status quo, or something close to it, is maintained.

Leaving aside any ambiguities of what qualifies as a "disturbance" in the definition (see Chapter 4), the concept has a common-sense appeal. A rubber band has resilience: pull on it and then let go and it will snap back to its former state. A grassland that is moderately grazed may change in some details of species composition and nutrient cycling, but if grazing is stopped it can return, in time, to something approximating its "basic function and structure" (whatever that means). Grazed too heavily, however, the system may be so degraded that it cannot recover—a threshold has been passed and the system moves to an alternative state, such as a shrubland or barren desert. Its resilience has been exhausted.

Because the concepts of thresholds and resilience are derived from systems theory, it's easy to lapse into systems-talk about feedback controls and such. But what are these "systems?" In ecology and conservation, most discussion of resilience is focused on ecosystems, especially the nexus where ecology meets society.[3] The examples given by Brian Walker and David Salt in their book, *Resilience Thinking*,[4] are all large-scale ecosystems—the Florida Everglades and the like. However, any ecological system displays some form of resilience (or, to use a different term, adaptability) to change, and ecologists have been talking about such things for many decades. Frederic Clements' notions of ecological succession centered on

[1] Something that seems extraordinarily hard for ecologists to do.
[2] Walker and Salt (2006: xiii).
[3] This is the domain where the staunchest advocates of resilience thinking—Buzz Holling, Lance Gunderson, Brian Walker, Steve Carpenter, and others in the Resilience Alliance—have advanced the concept most vigorously.
[4] Walker and Salt (2006).

Ecological Challenges and Conservation Conundrums: Essays and Reflections for a Changing World, First Edition. John A. Wiens.

the capacity of ecological communities to return to a climax community (the "basic function and structure") following disturbance.[5] Much of the discussion among ecologists about population regulation during the 1950s and later was cast in terms of density-dependent feedbacks that would act to stabilize population numbers about some level ("carrying capacity") determined by resources and the environment.[6] Models of community assembly such as those developed during the 1970s by Jared Diamond or Philip Grime[7] posited a process that would lead to a community composition determined largely by competitive interactions (the feedback mechanism), with some variations depending on initial conditions (i.e., priority effects—which species happened to get to a disturbed site first). Although applying the resilience label to ecological dynamics is relatively recent, the general idea has been about for some time.

These examples might lead one to think that resilience is what maintains equilibrium in a system, and that the concept is therefore closely tied to equilibrium thinking and a balance of nature philosophy. According to my arguments earlier in this book, the equilibrium premise is either passé or simply wrong. If in fact environments vary on multiple scales in space and time (as they do), then the ecological systems that persist under such conditions must be able to cope with the variations; natural selection (at whatever level) will have seen to this. This is why, ultimately, all ecological systems have some degree of resilience. It is also why resilience thinking explicitly recognizes that environments vary, sometimes substantially and erratically, and that systems may change in response as they attempt to track a moving target (see Essay 16, *Shooting at a moving target (2011)* (page 133)). In fact, the process of continuously adjusting to changing environmental circumstances bolsters the resiliency of a system.

I said above that all ecological systems have at least some degree of resilience. This may not be so for systems under intense management. To optimize the production or harvesting of resources, people attempt to control sources of variation ("disturbance") and remove threats or stressors. This reaches its zenith in agriculture, where extensive plantings of monocultures such as corn or soybeans require applications of chemical fertilizers or pesticides to enhance productivity and protect the crops from disease.[8] In the process, natural resilience is sacrificed, replaced by engineered (and heavily subsidized) resistance. Brian Walker and David Salt

[5]Unless the disturbance was continuing or recurrent, in which case the community tended toward a disturbance-maintained climax, what Clements called a "disclimax" community—something we might now call an alternative state; Clements (1916).
[6]Lack (1954) and Andrewartha and Birch (1954).
[7]See Weiher and Keddy (1999) for a review.
[8]Thus contributing to pollution of streams and rivers that receive runoff from the fields. Nassauer et al. (2007) have described how intensive agriculture in the corn belt of the midwestern United States contributes to nutrient loads in the Mississippi River and a "dead zone" in the Gulf of Mexico.

have argued that this approach may work well so long as nothing goes wrong, and it may well seem to be cost-effective to foster greater productivity over the short term at the expense of running down resilience. However, there is no such thing as an optimal state of an irregularly dynamic system. As new threats emerge (as they inevitably will), additional nutrient inputs or stronger pesticides may be needed to maintain productivity levels, plunging the system into an arms race (e.g., pesticides vs pests) that further erodes resilience and increases the vulnerability of the system to unusual stresses or sudden environmental changes, such as a severe drought.

Concerns about resilience are also relevant to other human uses of the environment. For example, development often comes at the expense of clearing and simplifying landscapes. Commercial development on tropical and subtropical coastlines provides a particularly compelling example. In their "natural" state, many of these coastlines are fringed by dense thickets of mangroves in tropical and subtropical areas or salt marshes in temperate and high-latitude locations. The vegetation provides habitat and spawning areas for wildlife and buffers shorelines against wave action. When these shorelines are developed, the fringing vegetation is often removed to create better access to beaches or improve views from coastal properties, at a cost of reduced system resilience. Coastal properties and habitats on the Gulf Coast of the United States, for example, were decimated by the storm surge accompanying hurricane Katrina in areas lacking fringing marshes, while damage was much less where the vegetation had not been cleared. Evidence of the resiliency provided by mangroves comes from areas affected by the Indian Ocean tsunami in 2004. I recall seeing photos from the Sri Lankan coast showing one beachside hotel standing undamaged while the neighboring hotels had been utterly destroyed. The owner of the surviving hotel had left the mangroves untouched, while the others had cleared them. In places like Aceh in Indonesia that had long since lost the protection of mangroves, the tsunami wreaked devastation. There are limits, however: any resilience that mangroves might have provided would likely have been overwhelmed by the 5- to 8-m tsunami waves.

Resilience is tied to the variability, diversity, heterogeneity, and connectedness of an ecological system. Rather like a boxer bobbing and weaving, an ecological system that is variable in time and space is less likely to be decimated by a single disturbance—it is harder to hit one moving than one standing still. A large monoculture is more susceptible to the spread of a disease than interspersed plantings of different crops, and a continuous expanse of forest of the same age and type is more likely to burn over a large area or suffer a widespread beetle outbreak than a multi-aged, species-diverse, or patchy (even fragmented) forest. A population that is divided into subpopulations with limited connectivity (i.e., a metapopulation) may be more likely to persist than a single population of the same size. The value of variability, diversity, heterogeneity, and the right amount of connectivity in reducing vulnerability to disturbances or their spread has long been recognized by ecologists[9]; what is (arguably) new is making explicit the link to resilience and a reduced likelihood of crossing thresholds.

Although the resilience concept has considerable intuitive appeal, it may be difficult to apply in practice.[10] Rachel Standish and her colleagues concluded that the concept "remains vague, varied, and difficult to quantify."[11] Part of the difficulty is contingency: resilience is defined in terms of response to disturbance, but the specifics of a disturbance may lead to differing conclusions about resilience. Resilience is also species- or system-specific[12] and may depend on the current state of a system (e.g., whether the population is large and widely distributed or small and isolated) as well as its past history (e.g., whether it has been subjected to repeated disturbances over a short time, weakening its resilience). Managing a system to enhance its resilience requires an understanding of which attributes of the system may confer resilience, how much diversity or heterogeneity is necessary to provide sufficient resilience, and how resilience relates to thresholds. None of that comes easily.

Sometimes, however, what looks like a shift to an alternative state and a loss of system resilience may be misleading, and the system may still have the capacity to rebound if conditions change. The threshold that seemed apparent when reproductive success of Cassin's auklets suddenly declined to zero (described in Essay 17, *Tipping points in the balance of nature (2010)* (page 144)) lasted only 2 years; reproduction resumed once oceanographic conditions changed. Despite appearances, the system retained its resilience. However, if reproductive output were to remain very low for a longer time, say 5–6 years (6 years is the maximum recorded lifespan for a Cassin's auklet), the population would no longer have the capacity to bounce back—no replacements for the dying adults would have been produced.

So where does this leave us? Yes, there are problems in documenting the factors that determine the capacity of ecological systems to absorb changes without reaching a tipping point. There is an implicit assumption that current conditions are somehow the best, perhaps because that's what we are familiar with. If it

---

[9]den Boer (1968) and Turner et al. (1989).

[10]Notwithstanding the attempt by Walker and Salt (2012) to develop guidelines for practical applications.

[11]Standish et al. (2014).

[12]Benson (2012) has argued that the concept of resilience should apply only to ecological systems, which have feedback controls and the capacity to self-organize that species lack. Consequently, if resilience thinking is to be incorporated into conservation measures such as the Endangered Species Act, management will need to shift from a species-centered approach to one that is based on systems and system dynamics. It is not yet clear (to me, at least) how resilience thinking should be infused into conservation, but I'm pretty sure that "intelligent tinkering" (Leopold's phrase, which Benson has applied here) with the Endangered Species Act is not the best way to do this. As many have warned, tinkering with the Act may open the door to wholesale amendments that weaken or destroy its effectiveness. And there are plenty of people poised to do just that.

is less than what we desire (i.e., "degraded"), we believe we can capitalize on the resilience capacity of the system to fix that through management or restoration. These are incorrect assumptions and false hopes. Nonetheless, the concept of resilience is useful. It draws attention to the reality that there are limits to how much alteration or change a system can take before it becomes something else, even though we may not know what those limits are. And realizing that systems have some inherent capacity to respond to change may allow us to adopt more flexible management practices, mobilizing intense actions when we suspect that the system is approaching a threshold or its resilience capacity is threatened, and using a lighter touch when we think the system is in its comfort zone.

We should no longer think of conservation or management as aiming at some finite, ideal state of the system, but rather at a range of conditions that encompasses environmental variations while allowing the systems to ebb and flow in response. And if the system should cross a threshold into an irreversible alternative state? Well, so be it. We may be better off shepherding the system in its new setting and configuration, rather than mounting a Herculean (and perhaps futile) effort to return it to its former (historical) state.

Life beyond a threshold isn't always bad; it's just different, a new setting for a new domain of resilience.

CHAPTER 15

Dealing with novelty

Novelty presents a conundrum to conservationists and ecologists because it says that things are different, new, and unusual, and perhaps even interesting. If they are very different, the concepts, approaches, and tools that have served well in the past may no longer be effective. It's not surprising, then, that ecologists and conservationists have ignored novelty for a long time.

The expectation that ecosystems or ecological communities should have a consistent and predictable composition and structure ("nothing novel here") runs deep in ecology. In his classic paper on the strategy of ecosystem development in 1969, Gene Odum described ecological succession as an orderly process that "culminates in a stabilized ecosystem in which maximum biomass ... and symbiotic function between organisms are maintained ... "[1] In 1975, Martin Cody and Jared Diamond were confident in asserting that "similar [natural] selection by similar environments should produce similar optimal solutions to community structure," with competition among species being the driving force.[2] And although island biogeography theory admitted a role for turnover in species composition, the overall species richness and niche composition of communities were expected to stabilize when colonization balanced local extinction—an equilibrium.[3] In these conceptualizations, there was no room for novelty, except perhaps as outliers to be set aside—Kuhn's anomalies. I consider the importance of outliers in Essay 18, *Black swans and outliers (2012)* (page 160).

Not everyone ignored novelty, however. Even as the Clementsian paradigm of succession was gaining force, Henry Gleason was proposing that communities developed by the chance colonization or disappearance of species. To Gleason, a "fluctuating and fortuitous immigration of plants and an equally fluctuating environment" would result in an assemblage that was "merely the fortuitous juxtaposition" of species, rather than the integrated, organism-like entity envisioned by Clements.[4] Lots of novelty. The flurry of null models of community assembly that emerged in the 1980s reached similar conclusions. In a changing environment, the ebb and flow of species could generate so much turnover

[1] Odum (1969: 262).
[2] Cody and Diamond (1975: 7).
[3] MacArthur and Wilson (1967), revisited by Losos and Ricklefs (2010).
[4] Gleason (1926: 8–10, 23). See Chapter 23 for a bit more about this debate.

Ecological Challenges and Conservation Conundrums: Essays and Reflections for a Changing World, First Edition. John A. Wiens.

that any community was compositionally unlike any other—they would be "no-analog" assemblages.[5]

Paleoecologists, whose perspective on time encompasses massive environmental changes, were among the first to draw attention to such no-analog assemblages. Steve Jackson noted that species shifted distributions independently of one another in response to the retreat of continental glaciers in North America over the past 20,000 years. This resulted in a parade of species combinations that changed continuously—sometimes gradually, sometimes rapidly—producing novel assemblages unlike anything seen before or since.[6] The emergence of such no-analog assemblages, however, is not something seen only over geological time spans. Keith Kirby described substantial turnover in species composition associated with millennia of human activities in Wytham Woods in England.[7] In California, alteration of flows in rivers and streams over the past century has led to the disappearance of some native fish species and invasions by exotics, creating new, novel assemblages.[8]

These changes are in the past. Local, regional, and global climates are shifting into increasingly extreme conditions, and it seems likely that future conditions may fall outside the historical range, creating new, no-analog climates.[9] As the climate changes, habitats disappear from some places and emerge anew in other places, and species are already shifting distributions in response. Because such changes have until recently been gradual, they have scarcely been noticed, but historical comparisons provide a clear picture of change. In 2003–2008, Morgan Tingley and his colleagues surveyed breeding bird species on elevational gradients at several locations in the Sierra Nevada of California, repeating the surveys of Joseph Grinnell in 1911–1929.[10] In the time between the surveys, the climate at these locations had gradually become warmer and wetter. Of the 53 species, 91% had moved (primarily upslope) to maintain their association with historic climate conditions, their "climate niche." Because the species responded differently to the climate changes, however, the composition of local communities was re-shuffled.

As the effects of climate change on species and landscapes play out in the future, shifts in species' distributions are likely to become more frequent. Re-shufflings in communities will be more rapid and pronounced, creating novel combinations of species that neither we nor the species have seen before. Based on projected changes in the distributions of breeding birds in California under different scenarios of climate change, Diana Stralberg and her colleagues suggested that

[5] Williams and Jackson (2007).
[6] Jackson (2012).
[7] Kirby (2012).
[8] Moyle (2013).
[9] Wiens et al. (2011).
[10] Tingley et al. (2009); Moritz et al. (2008) conducted a parallel analysis of changes in small-mammal communities in Yosemite National Park.

perhaps half of the state might contain no-analog bird assemblages by 2070.[11] This represents a substantial increase in community novelty, especially at the local scales most relevant to conservation and management.

One particular expression of novelty—novel ecosystems—has recently generated considerable attention, and with it debate. Richard Hobbs and his colleagues define novel ecosystems as "non-historical species configurations that arise due to anthropogenic environmental change, land conversion, species invasions, or a combination of the three."[12] The emphasis is on system functioning, not just species composition, and the changes are irreversible (i.e., a threshold has been passed). There is more to novel ecosystems, then, than the change in species composition from a historical or contemporary baseline that characterizes no-analog assemblages.

Recognizing the distinctiveness of novel ecosystems—that a threshold has been passed and a return to previous conditions may not be possible—may compel managers and conservationists to realize that things are indeed different and different approaches may be needed (a topic I ponder in Essay 19, *Moving outside the box (2009)* (page 164)). Because novel ecosystems are often regarded as degraded, attention may also be drawn to the potential value of managing or restoring such places. For example, the Conservation Reserve Program in the United States reimburses farmers who allow fields to go fallow, producing habitats that look like grasslands but bear little floristic resemblance to the native prairies that were historically there—novel ecosystems. In fact, because human activities have had such a profound effect on the environment in so many places, novel ecosystems may be common and widespread. Based on a global analysis, Michael Perring and Erle Ellis concluded that roughly one-third of the land surface of the Earth[13] currently contains novel ecosystems. If conservation is restricted to places that have close fidelity to historical conditions, there will be few places to do conservation.

I said at the beginning of this chapter that novelty poses a conundrum because it says that things are different. Why? What difference does it make? Potentially, a lot. Consider protected areas, for example. Most protected areas have been established to provide habitat for particular species or assemblages, usually those that were there in the past. Protected areas are fixed in space but not in time, however, so they are subject to the ebb and flow of both native and exotic species as they wash across the landscape, creating no-analog communities or novel ecosystems as they go. Consequently, protected areas may no longer protect what they were

[11] Stralberg et al. (2009) and Jongsomjit et al. (2013) extended the analysis of climate-change effects to include projections of future land-use change (housing development). The combined effects on bird distributions could be large; in some instances, species projected to expand their distributions with climate change actually lost ground when future land use was included in the analysis.
[12] Hobbs et al. (2013: 17).
[13] Perring and Ellis (2013).

intended to protect, the conservation value of the places may be altered, managers may need to change their practices to deal with species and assemblages beyond their previous experience, and public support for conservation of those places may diminish as they are perceived to be "unnatural." Protected areas may not be the conservation panacea we thought they would be (see Chapter 18).

As no-analog communities or novel ecosystems develop, existing webs of species interactions may be disrupted, casting both resident species and newcomers into relationships they have not previously encountered. A recent report, for example, described how a fish native to Eurasia, the round goby, has invaded the headwaters of the Danube, the Rhine, the Baltic Sea, and the Great Lakes of North America, probably transported in the ballast water of ships.[14] The goby outcompetes native fish for food and alters invertebrate prey communities, transforming the communities into previously unknown combinations of species and reducing ecosystem biodiversity. This is a particularly graphic example of the effects that one new species can have on an ecosystem, but the literature on invasive species is replete with other examples. The underlying message is that the development of no-analog communities involves more than just species turnover—*which* species enter a system may make all the difference.

Perhaps because the notions of no-analog and novel ecosystems threaten the foundations of traditional ecology and conservation, they do not sit well with some people, generating debate.[15] Three issues deserve mention because they bear on topics I consider elsewhere in this book. First, a system is novel or no-analog only in comparison with a baseline—the "normal" or desired state. Usually, the baseline is the historical condition of a place or ecological system. For example, in dismissing the concept of novel ecosystems, Carolina Murcia and her colleagues argue that the goal for conservation and restoration should be "to reestablish—or emulate, insofar as possible—the historical trajectory of ecosystems, before they were deflected by human activity" rather than to embrace "invasion-driven" novel ecosystems as a new normal.[16] I address the latter point below, but as I discussed in Chapter 12, such historical targets, especially if they are in centuries or millennia past, are pipe dreams. The farther back in history one goes to establish a baseline, the more likely that the contemporary systems will be no-analog or novel, so how the baseline is set determines whether one regards the new systems as novel or not, good or bad—a matter of values.

Second, there is the issue of thresholds. Hobbs et al. define novel ecosystems in terms of crossing an irreversible threshold to become something different, making it impossible to return to the baseline (historical or otherwise).[17] This demarcation separates novel ecosystems from what Hobbs et al. call "hybrid ecosystems"—systems that fall outside of the historical range of variation (e.g.,

[14]Brandner et al. (2013).
[15]See Murcia et al. (2014), Hobbs et al. (2014a, 2014b), and Simberloff (2015).
[16]Murcia et al. (2014: 549).
[17]Hobbs et al. (2013).

no-analog assemblages) but that could possibly be returned to the historical state. The distinction is important, for it can indicate when efforts to restore a system may be worth undertaking and whether traditional conservation and management approaches are likely to be effective. In practice, however, it may be difficult to determine whether or not a system has crossed a threshold and whether the crossing is actually irreversible (see Chapter 13), not to mention establishing what constitutes the historical range of variation for the baseline state of the system (or how far into the past "history" goes).

The "historical," "hybrid," and "novel" states of systems are categories, which may be more useful for management than a continuous spectrum.[18] Given the definitional difficulties mentioned above, however, I prefer to think of novelty as a continuum of conditions ranging from (for the sake of completeness) absolute historical fidelity at one extreme to completely different at the other. No-analog assemblages or hybrid ecosystems span situations that depart from historical or reference conditions to varying degrees, and novel ecosystems are those that differ from reference conditions in both their functional and compositional properties, whether or not a threshold can be identified. The emphasis, then, is on the degree to which an assemblage or ecosystem does or does not correspond to what has been seen in some reference time period. The challenge is to determine how far a system falls outside our previous experience (i.e., "novelty") and under what conditions traditional approaches to conservation and management should be rethought.

The third issue is what to make of invasive species. The label "invasive species" may include native species from elsewhere that have expanded their range, such as barred owls in the Pacific Northwest (see Chapters 7 and 11), or exotics from far away that have been either intentionally or inadvertently introduced. No-analog communities and novel ecosystems often include some of both; neither would have occurred in the historical or reference ecosystems. To those who value naturalness and strive to re-create historical conditions, invasive species (especially the exotic ones) are an abomination. Like the round goby, or the striped bass I mentioned in Chapter 8, they can push out (or eat) native species and transform entire communities and ecosystems, creating novelty almost single-handedly. Critics argue that the novel ecosystem concept may be a "Trojan horse" to conservation, leading to an acceptance of invasive species so long as they provide vital ecosystem services.[19] Proponents contend that not all invasive species are equal—some can blend into a system with few effects, others can be eradicated or controlled through management, while yet others may move the system over a threshold, foreclosing traditional management and conservation options. Because novel ecosystems by definition include species from elsewhere,

[18] Managers, like most people, feel more comfortable dealing with categories, to which they can apply standardized approaches, than with gradients of continuous variation, which may require more nuanced, situation-specific approaches.
[19] Simberloff (2015).

debates are interwoven with those about invasive species, which I consider in Chapter 23.

Beneath the surface of debates about no-analog communities, novel ecosystems, and invasives—how to deal with novelty, really—lies a deeper philosophical concern. The issue is whether those who care about nature—conservationists, restoration ecologists, environmentalists, and others—should continue to strive to preserve or enhance the ecological communities and ecosystems that have come to be regarded as normal. Alternatively, should we accept that the changing world will produce more and more novelty and recognize that places that aren't "normal" or "natural" may nonetheless have conservation value. Such places are degraded only by reference to some baseline, and values and perceptions determine what is an appropriate baseline. To incorporate novelty fully into the conservation agenda will require a change in public perception about such things as "naturalness" and "degradation," which will only happen when conservationists, restoration ecologists, and others re-examine their own perceptions about what their goals should be.

Novelty is upon us. We can wish it weren't so, but better that we recognize the opportunities it presents, figure out how to manage and conserve it, and even embrace it.

ESSAY 18

Black swans and outliers (2012)*

As I said in Essay 2, I'm a bird person—have been since childhood. So naturally, I was attracted to Nassim Nicholas Taleb's book about black swans. My disappointment at finding that it actually had little to do with black swans (the birds) was overcome by the use of swans as a metaphor for outliers. This appealed to my inordinate fondness for variance and extremes (and my love of metaphors). It reinforced my concern that too much science is swayed by the siren song of the average, the search for what is "normal."

Dan Simberloff once said (I don't recall where) that what is noise to a physicist (unwanted and confounding variation) is music to the ears of an ecologist. Daniel Botkin's book, Discordant Harmonies,[1] *may have best captured the kind of music Simberloff had in mind. My belief has always been that average conditions are boring; the real action and interesting things are in the variance. Especially, as Taleb argued, in the extremes. And as climate change increases the occurrence of extreme heat, rainfall, droughts, floods, and the like, today's outliers may become tomorrow's normal. We'll have a better chance of being prepared if we start paying more attention to outliers now.*

Since the time of the Roman Empire, black swans had been used as an example of something that could not exist. So when Dutch explorers first encountered black swans along the coast of western Australia in the 17th century, they were astounded (Figure 1). Swans, after all, were supposed to be white. To the Aborigines who had occupied Australia for many millennia, however, black swans were the norm, so much so that they figured in their narratives of the Dreamtime.[2]

I'm drawn to think about black swans not because of my ornithological leanings, but because I've been reading Nassim Nicholas Taleb's fine book, *The Black Swan: The Impact of the Highly Improbable.*[3] Taleb uses the black swan as a metaphor for unexpected, improbable occurrences that have hugely disproportionate impacts on our lives—on our economics, politics, social systems, and perceptions of reality. Black swans shake the foundations of certainty and predictability that make people comfortable. They can alter the trajectories of human affairs. The world

*Wiens, J. A. 2012. Black swans and outliers. *Bulletin of the British Ecological Society* 43(1): 42–44. Reproduced with permission of the British Ecological Society.
[1]Botkin (1990); updated in Botkin (2012).
[2]Parker (1897).
[3]Taleb (2010).

Ecological Challenges and Conservation Conundrums: Essays and Reflections for a Changing World, First Edition. John A. Wiens.

Figure 1 A real black swan (*Cygnus atratus*), Yanchep National Park, Western Australia, 2010. Photograph by John Wiens.

now is not as it was before the terrorist attacks of 9/11, and the political and social changes unfolding in the Arab world were initiated when Mohammed Bouazizi set himself on fire in front of a local municipal office in Sidi Bouzid, Tunisia. History is dominated by the impacts of such unexpected events (which is what makes history interesting).

Taleb's thesis is that it is in the outliers—the highly improbable occurrences—where the real action is. If this is so, we should view outliers as opportunities to test ideas or learn something really interesting, rather than as impediments to statistical analysis and predictability, something to be washed (or wished) away. I'm reminded of two examples. In 1964, when I was in graduate school, a colleague was studying intertidal communities in Prince William Sound, Alaska, when the Alaska earthquake elevated some of his study plots by several meters. Seizing the opportunity provided by this natural experiment, he applied for a revision in his funding to examine how the communities would respond to this disruption of their intertidal zonation. His application was denied, largely because the disturbance was not replicated (although his study plots were) and was not part of the original study plan.

Some years later, one of my graduate students was investigating competitive interactions between hummingbirds and hymenopterans (bees and wasps) at floral nectar sources in montane meadows. The hymenopterans generally dominated the hummingbirds and excluded them from flowers. One day, a sudden, unseasonable cold front suppressed hymenopteran activity, and the hummingbirds swarmed over the floral resources. When the weather warmed the following day, the hymenopterans returned and the hummingbirds were again relegated to foraging at flowers on the sly. It was a perfectly good natural experiment. When the results were submitted for publication, however, rejection was immediate: the observations were anecdotal, there were no real controls, and proper statistics could not be used.

The difficulty faced in both of these examples was that they were dealing with extreme events—the outliers—rather than documenting what was normal. People like to focus on what is "normal" because it is predictable. Unpredictable events, the black swans, can be unsettling, disruptive, or worse. Often, explanations are created after the fact to reconcile the occurrence with what was expected, preserving what Taleb calls "retrospective predictability." (Labeling an event a "100-year flood" seems to provide some reassurance that another won't occur for another century.)

Scientists (including ecologists) are no different. We study the behavior of individuals, the composition of communities, or the dynamics of ecosystems to determine what is "normal" and predictable under a given set of conditions. Thomas Kuhn's thesis in *The Structure of Scientific Revolutions*[4] was that scientific paradigms explain a range of observations and foster predictability; they guide investigators to delve ever deeper into the normal (Kuhn called it "normal science"). Anomalous observations (the black swans) are ignored or explained by post facto adjustments of the paradigm—Taleb's retrospective predictability. Applied ecology is based on using "normal" relationships to predict the outcome and effectiveness of management actions. The statistics we use are focused largely on measures of central tendency. If there are outliers in a data set, rules and protocols are at hand for removing them to facilitate "clean" statistical analyses.[5]

But times are changing. Ecologists are more inclined to recognize the opportunities provided by infrequent extreme events (my colleague and student might find a more receptive audience now). More to the point, however, those infrequent extreme events are becoming more frequent. Record heat waves, record droughts, record floods—what used to be outliers are becoming the new normal.[6] The opportunities and challenges provided by outliers are growing. Black swans are showing up everywhere. How should ecologists respond?

---

[4]Kuhn (1970).

[5]For example, Liu et al. (2004).

[6]See http://www.columbia.edu/~jeh1/mailings/2011/20111110_NewClimateDice.pdf for an informative analysis by Hansen, Sato, and Ruedy.

First, learn to love outliers, because there will be more of them. Extreme events can push systems to or beyond ecological thresholds, informing us about the resilience of natural systems. Every outlier has something to tell us, but we have to ask.

Second, shed the yoke of statistics. Statistics can be a hugely useful tool, but one that can also be restrictive. Ignoring outliers to cleanse data for smoother statistical analysis may be justified in some situations, but not without first probing more deeply into why (or whether) the outliers are in fact outliers.

Finally, remember that extreme events or outliers are recognizable only against the backdrop of what is "normal." Black swans were as startlingly improbable to the European explorers, who knew only white swans, as white swans would have been to Aboriginal Australians. Recognizing exceptions requires a baseline, which historical ecology and long-term studies provide.

These steps, though simple, have not been part of mainstream ecology. Taking them will entail broadening our focus beyond the "normal" to embrace the outliers and extremes, paying greater attention to the variance component of statistical analyses, and dredging through long-term data sets to uncover those telling but infrequent events. If we admit to the importance of black swans (metaphorically), we should seek them out.

ESSAY 19

Moving outside the box (2009)*

It seems that I can't escape my penchant for referring to Thomas Kuhn and his ideas about paradigms and normal science. They seemed particularly germane to the topic of this essay—the need for a shift in thinking and approaches in conservation to deal with a rapidly changing world. I cast the challenges to conservation in terms that will be familiar to readers by now: variation, thresholds, resilience, and other aspects not normally part of the conservationist's toolbox. My concluding metaphor, to ancient maps labeling vast unknown territory as "Terra Incognito," is still apt. The concepts of no-analog assemblages and novel ecosystems are now more fully developed. We realize that, for better or worse, such systems are already with us and will become more common. But conservationists, restorationists, managers, and ecologists have yet to determine what this means—will it entail tweaking of the existing paradigm (the box) about the edges, or require new and novel approaches that lie outside the box? It should be an interesting time.

Among the conservation practitioners and natural-resource managers I talk with, the reality of climate change is no longer at issue. They ask instead "What should I do?" And they expect ecologists and other scientists to provide answers, as we have about other issues in the past.

Increasingly, however, projections of climate change and its consequences are revealing a future that may be profoundly different from what we have experienced during the past century, when much of our thinking and theory in ecology took form. Models that track shifts in species' distributions in relation to changing climate predict that "no-analog" assemblages—combinations of co-occurring species we have not seen before (and that have not seen each other before)—may become commonplace as existing communities are torn apart and reassembled in new ways. Our own analysis of shifts in the distributions of birds in California, for example, suggests that as much as half of the state may contain assemblages that lack a modern counterpart by 2070. As the composition of assemblages changes, webs of interactions among species will be transformed. The new invasives may be not exotics introduced from afar, but native species expanding into new territory as

*Wiens, J. A. 2009. Moving outside the box. *Bulletin of the British Ecological Society* 40(3): 35–37. Reproduced with permission of the British Ecological Society.

Ecological Challenges and Conservation Conundrums: Essays and Reflections for a Changing World, First Edition. John A. Wiens.

the climate changes. Endangered species caught in these webs will require ongoing management attention as they encounter new combinations of competitors and predators.

Our inclination will be to approach these new situations using the knowledge and tools derived from our past experience—our understanding of "Nature's rules." But we must consider the possibility that the new conditions may change Nature's rules. We will need to think, and manage, outside the box.

This isn't as easy as it sounds. Despite the positive connotations associated with the phrase, thinking outside the box is not part of mainstream science. Over four decades ago, the historian/philosopher Thomas Kuhn described the role that paradigms play in the development of scientific thought.[1] The prevailing paradigm of a science (the box) embodies what we know (or think we know) and the body of theory and methods that guide and set the boundaries for work in a discipline. It defines the rules and establishes the scientific comfort zone. For example, physics has its paradigm of relativity and biology its paradigm of evolution. Both provide foundations that establish a set of assumptions (the "givens") and stimulate research. The work of science is to elaborate on the rules, to fill in their texture, to expand the realm of their applicability. This is what Kuhn called "normal science." It can be enormously productive, but it is still business-as-usual, pursuing variations on a theme.

Kuhn suggested that, as science pushes the boundaries of the paradigm, observations that do not align comfortably become more frequent. Eventually, the overwhelming weight of these anomalies (the "no-analog observations") reaches a breaking point, paving the way for a shift to a new paradigm that reconciles the anomalies with the previous observations. The "box" is changed. No more business-as-usual.

We hear every day of new evidence that climate change is upon us and that species and ecosystems are changing. Anomalous observations, such as rapid shifts in distribution or phenology or the occurrence of extreme rainfall or temperature, are becoming more frequent. Yet the anomalous observations that would provide sufficient weight to lead us to question the prevailing paradigm of conservation and management and to contemplate a paradigm shift are mostly yet to come. The most dramatic changes, those that suggest a no-analog future (perhaps with different rules), have not yet been observed. They are only predicted, by models that carry with them the baggage of assumptions and uncertainties.

I suggest that we can't wait for more observed anomalies to accumulate. We know already that the future is likely to differ dramatically from what we have seen with the past. We should begin now to rethink our approach to conservation and management to catalyze a paradigm shift, to move outside the current box. There are at least three components to this process.

[1]Kuhn (1970).

First, we should mine and synthesize information about the near and distant past, focusing on the variance rather than on patterns that align with our current thinking. How have environments changed in the past, what have been the thresholds and nonlinearities, and how have species and ecosystems responded? Have the current bioclimatic envelopes we use to model species' distribution changes always been so? What does history tell us about the mechanisms and limitations of species and ecosystem resilience to environmental change? We need to take a fresh look at the past.

Second, we should think about how to broaden the box of conservation and management. This process could begin by articulating the current conceptual framework of these disciplines (e.g., fragmentation, corridors, minimum viable population, succession, habitat selection) and then assessing the sensitivity of the framework to the changes of the magnitudes and directions that are being predicted with climate change. Is it simply a matter of making adjustments around the edges of our concepts, or of developing new ways of thinking—a new paradigm—altogether? There could be great benefits from engaging other disciplines to bring their own perspectives and approaches into the process, not just to add some new sociological, economic, or behavioral variables to the existing paradigm of conservation and management, but also to broaden or move it in a truly integrated way. We need more interdisciplinary or transdisciplinary thinking.

Third, broadening or shifting the scientific paradigm to encompass the novel conditions created by climate change will have little effect on conservation and management unless the new ways of thinking and new approaches can be translated into new ways of action. There is much talk about strengthening communication between scientists and managers, but it will take more than talk. Science and management must be inseparably joined. This requires that the creation of a new paradigm be a collaborative undertaking from the outset, rather than the all-too-frequent pathway of scientists developing concepts and approaches by themselves and then hoping that these insights will be used by the world at large.

On a recent trip to Argentina, I visited the library of the Facultad de Ciencias Exactas, Fiscas y Naturales of the Universidad de Córdoba, where I was shown a richly inscribed atlas of the world from the late 16th century. Large areas of the Americas were blank, labeled "Terra Incognita"—unknown land. This may be an apt metaphor for the future of conservation and management. And just as bold explorers moved beyond the box of what was familiar to fill in those blank spaces on the map, we need to break the bonds of traditional thinking to anticipate how to conserve and manage biodiversity in an uncertain future. The sooner we begin thinking about moving outside the box of conventional approaches, the more likely we'll be able to manage what climate change brings us and do the best job we can of conserving Nature.

CHAPTER 16

Uncertainty: a boon or a bane?

The answer to this question depends on who you are and what you are doing. To a scientist undertaking a research project, uncertainty is essential. Without it, there would be nothing to investigate—all would be known and certain. On the other hand, if one ends an investigation with greater uncertainty than when one started, it would be a failure, or at least unsatisfying. It might take some artful prose to get such a study published.

Uncertainty is the raw material of science.[1] Science takes things that are uncertain and makes them more certain, and thus more understandable. The uncertainty isn't totally raw and unrefined, however. Science builds on the theories, evidence, and understanding of previous work. These provide an acceptably comfortable level of certainty, while leaving many unanswered questions—dangling loose ends of uncertainty.[2] So, science thrives on just the right amount of uncertainty—not too little, which might make a study unrewarding or uninteresting, and not too much, which might overwhelm one with uncertainties everywhere.

To those charged with conserving or managing biodiversity or natural resources, however, uncertainty is a bane. The imperative is on undertaking actions that will predictably have the desired result. Managers look to science for the certainty they need to move ahead. Science provides cover; if things don't work out as planned, managers can still claim to have based their actions on science. In fact, the use of "best available science" is enshrined in numerous laws and environmental regulations.[3] Managers and conservationists also want certainty because the public—and people in general—are uncomfortable with uncertainty. This is where uncertainty can become a bane to scientists. People who don't understand the nature of science

[1]Henry Pollack provides a nice assessment of scientific uncertainty in his book, *Uncertain Science … Uncertain World* (Pollack 2003).
[2]A scientific paradigm provides an umbrella of certainty, while revealing the uncertainties that are the fuel for normal science (see Chapter 2).
[3]Because "best available science" figures so prominently in environmental conservation and policy in the United States, the American Fisheries Society and the Estuarine Research Federation convened a committee to consider what determines the best available science and how it might be used to formulate natural resource policies and procedures. Predictably, there was no single answer—what is "best" or "available" depends on the issues and circumstances. See Sullivan et al. (2006) for the report.

Ecological Challenges and Conservation Conundrums: Essays and Reflections for a Changing World, First Edition. John A. Wiens.

expect it to provide certainty; when scientific statements admit to any uncertainty, it can be used to deny the science, especially if one doesn't like the message. For examples, one need only look to reactions to the reports of the Intergovernmental Panel on Climate Change, which was quite explicit in documenting levels of uncertainty (or certainty) about their conclusions. Casting doubt can be a powerful catalyst for inaction or the rejection of science.[4]

Lying beneath all of this is our old nemesis, equilibrium thinking. The conviction that the world is an orderly and stable place means that certainty will emerge if only we try hard enough—it is ultimately there for the taking. This kind of thinking is what led a generation of ecologists to accept Frederic Clements' ideas about succession, and why published papers ooze certainty, supported by statements of statistical confidence.[5] As I've been saying throughout this book, however, the world is a dynamic place. Stability is elusive and order is ephemeral. Any certainty we have progressively erodes as we delve more deeply into the past.[6] And although history can provide some tangible insights into the past, the future remains cloaked behind a veil of uncertainty. The no-analog assemblages and novel ecosystems described in Chapter 15 differ qualitatively as well as quantitatively from what is familiar, especially if they involve threshold dynamics. Uncertainty is mushrooming.

These changes indicate how uncertainty is related to scales of time—the deeper one goes into the past or farther into the future, the greater the uncertainty (Figure 16.1). However, uncertainty also scales in space. Although they are all based on models (and therefore inherently uncertain), projections of future climate change are most certain at a global scale, somewhat less certain at a regional scale, and annoyingly uncertain at a local scale of hectares to a few square kilometers. In contrast, most ecological studies are conducted at that local scale, where the findings are likely to have the greatest certainty. As the scale broadens, extrapolation of the local results becomes more problematic and studies

---

[4]Some corporations and politicians are skilled at sowing the seeds of doubt; see Oreskes and Conway (2010).

[5]Increasingly, scientific publications in ecology require statistical assessments of probability or certainty and are unlikely to be published if there is too much (statistical) uncertainty in the results. Not so long ago, the statistical convention of $P < 0.05$ set the standard. When we submitted the findings of our assessments of the ecological impacts of the *Exxon Valdez* oil spill on marine birds, we used a probability level of 0.20 to screen our results, preferring not to overlook even weak evidence of an effect (i.e., a Type II error). We had considerable difficulty convincing reviewers and editors of the wisdom of this approach. We eventually prevailed (see Day et al. 1997). More recently, Bayesian statistics have been used to specify probabilities or "degrees of belief" more explicitly.

[6]This is one of several reasons why historical targets for conservation or restoration are problematic; even if we were able to return a system to its former state, we would be less certain about what that former state was as we sought a time less affected by humans.

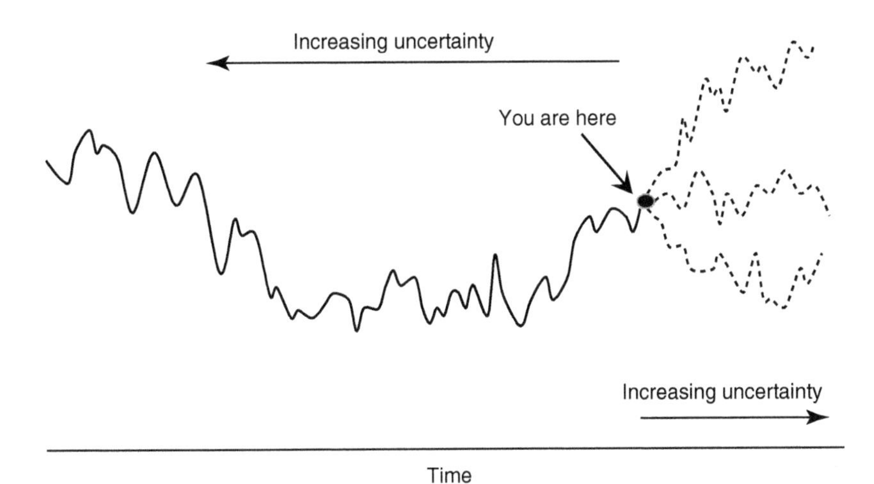

Figure 16.1 In a variable environment, uncertainty increases as one looks farther back into the past or attempts to look into the future. Uncertainty about the past condition of a system may be reduced by historical or paleoecological data, although this becomes more difficult with increasing variation and greater distance into the past. Future conditions may follow one of multiple trajectories, which can be explored using models. The models themselves, however, introduce additional sources of uncertainty.

suffer from a lack of replication and an expanding array of uncontrolled variables; both factors increase uncertainty.[7]

What should ecologists, conservationists, restoration ecologists, or resource managers do? The one thing we can't do is to ignore uncertainty. The overriding message of environmental change is that uncertainty will be with us for a long time, so we'd better learn to live with it.

That doesn't mean, however, that it's hopeless and we're helpless. There are things we can do to deal with uncertainty more effectively. We might begin, for example, by identifying the sources of uncertainty. The scientific process that underpins conservation involves observation, measurement, sampling, data collection, analysis, modeling, interpretation, and other things I have probably forgotten to mention. All of these are plagued by inherent uncertainties, and uncertainties in any one step can propagate to expand uncertainties in other steps.

As I read papers, reports, and policy documents dealing with conservation, restoration, or environmental management, I'm struck by how often adaptive management is proposed as the remedy to such uncertainty. Unsure whether

[7]There may be a "sweet spot" where the scales of uncertainty of climate modeling and ecology coincide rather than overwhelm one another, but I can't think of where it might be.

a restoration effort will work? Propose to do adaptive management to find out. Uncertain about the effects of reducing water releases from a dam on downstream fish populations? Adaptive management will resolve the uncertainties.

In practice, however, adaptive management is a complex and often expensive process. It's easy to talk about, and it's frequently included in plans, but the full adaptive management cycle—plan, implement, assess, and decide—is rarely actually accomplished.[8] And it's easy to see why. Comprehensive adaptive management demands that the goals of a project be clearly stated at the outset and be converted into an operational plan with a design that will enable rigorous analysis of the results. Observations and data must be collected using appropriate monitoring protocols over a long enough period to determine whether everything is working as planned. And the results must then be analyzed, interpreted, and translated from science-talk into language that managers and policy-makers can understand and use to decide whether to continue or modify the project (the "adaptive" part).

All of this takes time. The process can be ponderously slow. Completing a single iteration of the cycle for a large restoration project, for example, can easily take more than a decade. It is also a retrospective process, as it involves looking at the results of actions that may have been undertaken some time ago. As the pace of environmental change quickens and the sources of uncertainty multiply, the need for quick, nimble management grows. Conventional adaptive management may become increasingly irrelevant. When change occurs rapidly, it may be more appropriate to think of adaptive management as a proactive, anticipatory process. Models could be used to project likely outcomes of management actions at some future time as well as to assess and correct current actions in relation to future rather than present or past conditions.

Any conservation or management action entails costs, of course, and there are risks in undertaking an action under the known uncertainties, much less the unknown uncertainties that lurk in the future.[9] Not all uncertainties are equally treacherous, however. In order not to be overwhelmed by uncertainties, it's important to consider which are likely to pose problems and which may be relatively benign. Uncertainty is reduced by gaining knowledge, but the most uncertain things are not necessarily where the greatest efforts should be made. Things may be uncertain because we have only recently come to think about them, or because they are inherently difficult to study and gain knowledge about, or perhaps because people don't care about them or deny their importance.

[8] Allen and Gunderson (2011).

[9] As Donald Rumsfeld, the U.S. Secretary of Defense in the George W. Bush administration, famously said in 2002 in referring to the lack of evidence of weapons of mass destruction in Iraq, "There are known knowns. These are things we know that we know. There are known unknowns. That is to say, there are things that we know we don't know. But there are also unknown unknowns. There are things we don't know we don't know." The same could be said about known and unknown uncertainties in conservation, restoration, or management.

One approach is to assess the costs and benefits of acquiring the knowledge to reduce a given uncertainty, balanced against the risks of not addressing the uncertainty. Tools for formal ecological risk assessment are well developed,[10] but more informal, subjective approaches may also be useful in assessing the tradeoffs between uncertainty and risks, threat levels, management costs, or other factors.[11]

The two essays that follow address different aspects of this uncertainty conundrum. Essay 20, *Taking risks with the environment (2012)* (page 172), develops the notion of risks, costs, and benefits of conservation actions. The so-called "precautionary principle" figures importantly in such considerations. Should one err on the side of caution when undertaking actions that affect species or the environment, minimizing the risk of doing something that might do irreparable harm? Or should one accept greater risk in order to take more immediate and decisive action? The second essay, Essay 21, *Uncertainty and the relevance of ecology (2008)* (page 176), deals more directly with the issue of uncertainty that I've been considering here. In conservation or management situations that entail considerable uncertainty (as most do, if we're honest about it), I suggested that actions might best be based on knowledge that is "good enough" to avoid doing something stupid, rather than pouring more time, effort, and money into achieving a greater degree of certainty (e.g., $P < 0.05$). What is "good enough" is ultimately a matter of costs, benefits, and risks. Decision-support tools and expert knowledge approaches[12] may be helpful in weaving these factors together.

In the end, ecologists, conservationists, restorationists—really, anyone concerned with shepherding, managing, or using the environment—must ask whether the approaches that have worked well in the past will remain appropriate in a future characterized by rampant change, recurrent extremes, and burgeoning uncertainties. Rather than targeting actions toward a single desired outcome, for example, it may be more appropriate to recognize that a variety of future states are possible, some of which may also be desirable (or at least not so bad). If no-analog assemblages and novel ecosystems become more prevalent, as I suggested in Chapter 15, we will have to deal with them, despite the uncertainty about what they may contain or how they may function.

The model of ecological science we have used in the past was based on rigorous empirical or theoretical investigations designed to test well-formulated hypotheses using statistical tools to establish acceptable probabilities and models to address the things we can't measure directly. All of this will still be useful, but there may also be a greater role for observational and subjective ecology—what used to be (and still is) called natural history.

---

[10] See, for example, Burgman (2005).
[11] Game et al. (2013) and Wiens and Gardali (2013) provide examples.
[12] See, for example, Gregory et al. (2012) or Perera et al. (2012).

ESSAY 20

Taking risks with the environment (2012)*

The question I asked in this essay, about how much risk can be accepted when the environment is at stake, has no easy answer. Risks are intertwined with values, so what is an acceptable risk to one person may be totally unacceptable to another. Values also underlie the recurrent conflict between the needs of people and the needs of the environment; as any good conservationist knows, these are not yes-or-no alternatives, but broadly overlapping needs. Still, this conflict emerges in the different approaches to assessing environmental risks that I highlighted in this essay. Should the threshold for acceptable risk be set high, so that actions that might harm the environment are precluded only if there is a substantial risk of damage, or should it be set low, to avoid taking any actions that might possibly harm the environment? The correct answer is probably "it all depends." It depends on the consequences or costs associated with the risks, whether they are irreparable or not, whether being cautious until we know more wastes precious time or whether taking action with incomplete knowledge is dangerously rash, and of course on values. The problem as we move into a more uncertain future is that we'll often be flying blind, making decisions about risks in the absence or near absence of the knowledge needed to assess risks intelligently. Ways of thinking about risks knowledgably yet quickly would be welcome.

Life is full of risks. Normally we make decisions about what to do (or not to do) without giving the risks much thought. When the risks involve the environment, however, the consequences of the decisions go beyond ourselves to affect the larger society and a multitude of organisms. Careful assessment of risks and the balance between costs and benefits is essential.

If ecology has taught us anything over the past century, it is that environments and ecosystems are devilishly complex, full of tortuous webs of interactions, feedbacks, and nonlinearities that create confounding uncertainty. While ecologists may revel in this complexity and view the uncertainties as research challenges, most people prefer the simplicity of either-or choices: water export from streams

*Wiens, J. A. 2012. Taking risks with the environment. *Bulletin of the British Ecological Society* 43(3): 39–40. Reproduced with permission of the British Ecological Society.

Ecological Challenges and Conservation Conundrums: Essays and Reflections for a Changing World, First Edition. John A. Wiens.

versus protection of an endangered fish, jobs in the timber industry versus preservation of old-growth forests. More generally, the choice is frequently portrayed as one between the economy *or* the environment, with the environment regarded as an extravagance that can be afforded only when the economy is strong.

There are risks in shifting the balance too far in one direction or the other. A preoccupation with economic factors can easily lead to a degradation of the environment and erosion of biodiversity, as those of us who care about nature know all too well. A defense of the environment at all costs, however, can mean a loss of jobs and profits and contribute to economic stagnation (and a diminishment of the financial support needed to support conservation). Those in business or industry usually consider risks in terms of profits and growth, while those concerned about the environment evaluate risks in terms of the loss of habitat or biodiversity or a diminishment of ecosystem health.

Faced with this polarization of perspectives and the complexity and uncertainty of the science, how should one go about factoring risk into decisions about the environment? How, for example, should the risks of releasing contaminants into streams as a result of industrial activities, or of introducing a non-native species for biological control of an agricultural pest, or of tightening regulations controlling emissions from power-generating facilities be evaluated? Each of these decisions carries potential economic and environmental costs or benefits. What is an acceptable level of risk?

One way of answering this question, favored for decades in the United States, relies on "risk analysis" to calculate the likelihood that an action will harm the public or the environment. An acceptable level of risk is determined (or specified in regulations). Unless it can be shown that the risk will exceed that level, the action can move forward. The contaminants are released, the biological control agents are introduced, or the emission controls remain lax.

A second approach, embodied in regulations in several European countries and the European Union and, more recently, in several states and municipalities in the United States, rests on the "precautionary principle." When there are risks of harm to public health or the environment or the consequences of a decision may be irreversible, one should err on the side of caution. The release of contaminants is restricted, the biological control agents are not released, or the emission controls are tightened.

Both approaches rely on understanding the risks of taking (or not taking) a particular action.[1] The more fully these risks are known and quantified, the more informed the decisions will be (at least in theory). Ecological risk assessment, for example, provides a formalized framework for modeling cause-effect pathways in complex ecosystems, calculating the probabilities associated with each step of

[1]Burgman (2005) provides an excellent review of these and other approaches to environmental risk assessment.

a pathway, and determining the overall probability (risk) of a desired or undesirable outcome. All risk calculations, however, involve uncertainties, and these uncertainties multiply as system complexity increases. The precautionary principle deals with uncertainty by avoiding potentially harmful actions until the uncertainties are reduced or the risks more fully known. The risk analysis approach instead argues for moving ahead until additional research shows that the risk of harmful effects are unacceptably greater than first thought.

Ultimately, the differences between the approaches are a matter of values—how much value is placed on the environment versus economic returns, and how one is balanced against the other. Those who focus on economic factors such as jobs, growth, or profits may accept a greater level of environmental risk if there are economic benefits, whereas those who value nature and biodiversity may find some risks unacceptable despite the economic benefits. At the extremes, these positions may fuel intense disagreements and produce seemingly endless lawsuits and political paralysis.

It is all too easy to place business and industry at one end of the spectrum and environmentalists at the other. This ignores the reality that most decisions about the environment or economics involve tradeoffs, either between the environment and economics or between different environmental (or economic) costs and benefits. In the southwestern United States, for example, an exotic tree, saltcedar (*Tamarix* spp.), has replaced native vegetation along streams, altering the natural communities and (arguably) reducing water availability to people. Various efforts have been mounted to remove the invasive saltcedar, ranging from bulldozing to introduction of a herbivorous beetle (*Diorhabda carinulata*) as a biological control agent. Although field tests showed that the beetle was effective in defoliating saltcedar, the program was halted because the beetle's success threatened to remove habitat of the Federally endangered southwestern willow flycatcher (*Empidonax traillii extimus*), which adapted to breeding in stands of saltcedar when the saltcedar replaced the native riparian vegetation. In this case, the environmental (and economic) benefits of biological control of saltcedar were outweighed by the risks to the flycatcher.

Balancing environmental risks with economic factors has never been easy. The adoption of either a risk analysis or a precautionary principle approach, however, sows the seeds for confrontation and controversy. As we move into a future increasingly dominated by climate change and its attendant uncertainties, the positions are becoming even more polarized. Some in industry have resisted cap-and-trade policies to reduce carbon emissions because they believe that the climate-change science is too uncertain to risk imposing regulations that would carry substantial economic costs. Others in the environmental community use the precautionary principle to urge immediate actions to reduce carbon emissions, no matter what the economic costs. The lines have been drawn. There is a danger of suffering the consequences of the Nero effect—fiddling while Rome burns. What is needed is a rigorous, quantitative assessment of risks, costs, and benefits, both environmental and economic, followed by a reasoned, dispassionate discussion of options in which hidden values and agendas are laid bare. Neither a risk analysis approach nor one based on the precautionary principle is likely to achieve this by itself, but both can contribute to it.

ESSAY 21

Uncertainty and the relevance of ecology (2008)*

Voltaire's observation that the best is the enemy of the good[1] could be the tag line for this essay. My intent was to temper the endless quest for perfection and certainty that characterizes a lot of science (reinforced, no doubt, by the perception that the peer-review system expects no less) with a bit of realism. Sometimes "good enough" may be just fine, especially when we are faced with a choice between conducting more research to increase certainty or responding to an urgent need with the information at hand. The challenges of resource management and conservation—guiding timber harvests, rescuing an imperiled species, ensuring adequate water for fish and people, and so on—often require taking immediate action even with incomplete knowledge. And as we recognize the complexity of these challenges, attempts at simplification run into the "Yes, but what about this?" problem. The responses of any single target to management or conservation actions will always be contingent on what else is going on (or went on at some time in the past).

Ecologists will always strive for greater certainty—that is their job; but, there will always be lingering uncertainty to be resolved and more to be learned—that is what makes the job so interesting. Unfortunately, the public and politicians sometimes grasp upon this uncertainty to avoid making hard decisions or doing something they'd rather not do. For example, the federal guidelines for assessing the potential impacts of actions that might affect the environment, such as draining a wetland or building a highway, require evaluations to be based on the best available information and tools while avoiding speculation. I was recently part of a team of scientists tasked with reviewing the environmental impact statement for a huge water-conveyance project in California. Although the supporting documents comprised over 30,000 pages, several important issues were not addressed, dismissed because the science was "too uncertain" and any consideration would therefore be "too speculative." Similar arguments have been offered to justify delaying action on climate change.

No one wants to make decisions based on complete ignorance, especially if the risks of being wrong are great (this is "doing something stupid"). However, neither should some degree of uncertainty lead to temerity about doing anything. Figuring out what is "good enough" may be useful, even if it is not scientifically quantifiable.

*Wiens, J. A. 2008. Uncertainty and the relevance of ecology. *Bulletin of the British Ecological Society* 39(2): 31–32. Reproduced with permission of the British Ecological Society.
[1]Voltaire, *La Bégueule* (1772).

Ecological Challenges and Conservation Conundrums: Essays and Reflections for a Changing World, First Edition. John A. Wiens.

Ecology is about figuring out how nature works. A large part of ecology, as of any science, is involved with reducing uncertainty, with deriving ever more secure, robust, rigorous, and certain answers to questions about nature. As ecologists, we do this by following the recipe that has been so successful in other sciences: gathering observations, posing questions, conducting experiments, using models, and developing theory. All of these approaches involve simplifications of reality, stripping away extraneous sources of variation and uncertainty to isolate the variables and relationships of interest. We filter out the background noise so we may hear the symphony of nature, understanding and appreciating the interplays among the instruments and the resulting harmonies.

Yet, as we have learned more about nature, more and more of its complexities have emerged, compounding uncertainties. Feedbacks, indirect effects, nonequilibrium dynamics, nonlinearities, and scale dependencies have made ecology a more exciting but less certain science. Attempts at further simplification in the face of this complexity are likely to reduce the symphony to a solo part, shorn of any accompaniment and lacking the beauty that comes with variety.

This is not a new or novel insight, of course. Books and papers have been written about uncertainty and the methods for dealing with it. Marc Mangel addressed some technical and analytical aspects in the most recent issue of the *Bulletin*.[2] I'd like instead to consider a different aspect of uncertainty: namely, how much uncertainty can be tolerated when making decisions or formulating policies about the environment—what is "good enough"?

In ecology, there are well-established standards for answering this question. The statistical threshold of $P < 0.05$ still holds sway in most ecological work. This level of uncertainty may be adjusted to reduce Type II errors, or Bayesian nonparametric methods may be used to assess probability distributions (as Marc suggested), but the urge to minimize uncertainty remains overpoweringly strong. At its worst, this quest to reduce uncertainty constrains us to ask questions and design studies that can meet such high standards, often at the cost of even greater simplification. Questions about messy systems (so-called "wicked problems"; see http://www.swemorph.com/wp.html) lead to answers riddled with uncertainties. Because the answers don't measure up to the criteria for publication in most scientific journals, the questions may go unasked. Yet these are the very questions that are often most critical and relevant to those charged with managing natural resources or conserving nature. The complexities of nature are the reality they must manage.

There is a tension, then, between the desire of scientists to conduct more research (to reduce uncertainty to acceptable levels) and the imperative of managers to take action ("just do it"). Figure 1 illustrates the challenge. Scientists typically strive to achieve a level of certainty that is sufficient to reduce the probability of making errors while satisfying the rigors of peer review. Of course, managers and conservationists would like that same degree of certainty, but

[2] Mangel (2008).

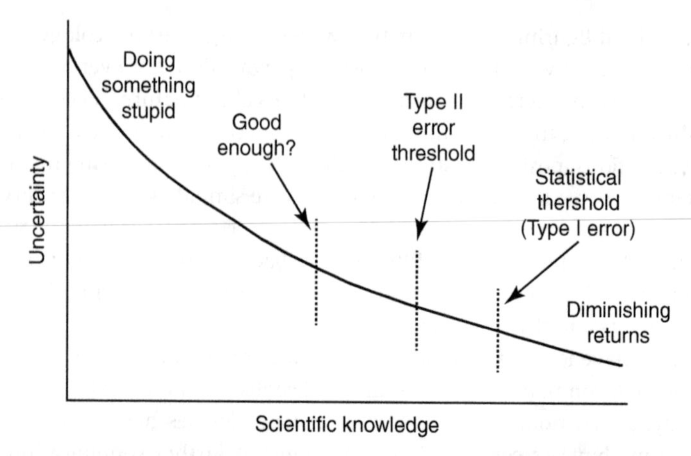

Figure 1 Uncertainty decreases as some function of increasing scientific knowledge. The statistical thresholds that define Type I errors (the likelihood of incorrectly inferring a relationship between variables when none exists) and Type II errors (the likelihood of incorrectly concluding no relationship when in fact one exists) are generally well established. The location of the "good enough" threshold is more nebulous, and shifts toward the right as the costs of making a mistake become greater.

they are often unable (or unwilling) to spend the additional time and money to achieve it; they must be satisfied with answers that are less certain but "good enough." What is "good enough" lies somewhere between attaining a desired level of statistical confidence and having an unacceptable probability of making mistakes because of incomplete knowledge; in other words, doing something stupid.

The location of this tradeoff threshold on the knowledge-uncertainty curve will depend on the costs of making the wrong decisions. Consider, for example, the decision to declare a threatened species extinct. An incorrect decision may be costly, for the removal of legal protection may expose previously protected habitat to exploitation. If the species is later found to be not quite so extinct after all, the habitat critical to its continuing survival may have disappeared in the interim, sealing its fate. Such a situation demands a low level of uncertainty.

There is, of course, a danger in advocating the use of some ill-defined "good enough" standard as a basis for management or policy decisions. Science has extraordinary standing in such decisions because of its credibility. Scientists have the expectation that they can in fact achieve a high degree of certainty, and the public has the expectation that science will deliver such certainty. Relaxing the standards runs the risk of eroding that hard-won credibility. More importantly, it may open the door to pseudoscience, beliefs, faith, or advocacy masquerading as science, further eroding the credibility of science.

As the pressure on managers and conservationists to address complex environmental problems becomes more immediate and more intense, the need to work with greater levels of uncertainty escalates. How can ecologists ensure their relevance—how can we navigate between the rigor and certainty we demand of science and the need for action, which will often entail living with levels of uncertainty ("good enough") that make scientists uncomfortable (and that don't generate scientific publications)? There is no easy answer. We should recognize, however, that clinging tenaciously to rigid standards of certainty might marginalize science in situations where decisions must be made quickly. Instead, rigor must be tempered with realism. We must recognize that reality is really complex, and that the certainty achieved by simplifying it is illusory and potentially misleading. We must change the perception of the public and decision makers (and scientists) about what uncertainty in science really means. And we should find out how to balance what is "good enough" with the costs of being wrong.

CHAPTER 17

Prioritization and triage

California condors are majestic birds. They are the largest terrestrial birds in North America, tipping the scales at 12 kg (26 lb), with a wingspan of 3 m (9.8 ft). Although few would describe them as pretty, they are stunning and, in their own way, charismatic (Figure 17.1). They are critically endangered, and they have become an icon for the conservation movement in North America. Over the past 40 years they have also been a flashpoint for debates that go to the heart of what conservation is all about.

California condors are a relic of the past. They were widespread during the Pleistocene, feeding on carcasses of the Pleistocene megafauna. Most of these species disappeared during the Quaternary, depriving condors of an abundant source of food, which started them on the pathway toward extinction. By the 1960s, only 50–60 individuals remained. The species was listed under the Endangered Species Act in 1967, but by 1982 only 22 birds remained in the wild. A decision point had been reached: what, if anything, should be done to save the condor?

This is when the debates intensified. Rich Stallcup pleaded eloquently that condors should be allowed to "stay as long as you can, and then die with the dignity that has always been yours."[1] Frank Pitelka lamented that "a great deal of energy is being put into an effort to save a species which is almost certainly doomed anyway."[2] Dillon Ripley warned against such rash judgments, urging taking "an ultimate risk in the preservation of this species rather than to continue the generations of benign neglect."[3] The lines were drawn.

The "ultimate risk" was a proposal to capture the remaining condors and undertake an aggressive captive-breeding program. The decision was made, and in 1987 all remaining wild birds were captured. Captive breeding was eventually successful, and reintroduction of condors into the wild was initiated in 1991. As of June 2014, 225 condors were flying in the wild (elegantly outfitted with satellite trackers, leg bands, and plastic wing tags), with another 214 in captive-breeding facilities.

The condor saga is often held up as a shining example of conservation success, an illustration of what can be done when we set our minds and resources to it, an affirmation of hope. And that it is. However, condors are not yet home

[1]Stallcup (1981).
[2]Pitelka (1981).
[3]Ripley (1981).

Ecological Challenges and Conservation Conundrums: Essays and Reflections for a Changing World, First Edition. John A. Wiens.

Figure 17.1 An endangered California condor flies over the Bitter Creek National Wildlife Refuge, California. This individual was hatched and raised in captivity and then released, fully outfitted with plastic wing tags and GPS electronics. Source: Photograph by Gene Nieminen, U.S. Fish and Wildlife Service.

free. The free-ranging birds risk lead poisoning, primarily by scavenging carcasses contaminated by lead ammunition. Despite efforts to eliminate this source of lead, condors remain chronically exposed to harmful levels.[4] Many birds must be recaptured for repeated clinical treatments. And all of this is expensive, averaging over $5 million/year. Funding comes from both government and private sources, which are sensitive to economic and political winds. Yet the needs of condors are unrelenting. Were the birds not treated, mortality would erode recent gains and set the species once again on the pathway toward extinction.[5]

But there is a darker side to the condor story. Resources to support the conservation of endangered species are limited, so funds allocated to condors restrict expenditures on other endangered species. During the time that many millions have been spent to bring condors back, 10 species of Hawai'ian birds have disappeared, most of them now declared extinct. Perhaps additional funds wouldn't have helped, but we'll never know. These species received little attention and

[4]Finkelstein et al. (2012).

[5]Walters et al. (2010) provide a comprehensive review of the recent status of the California condor and recovery efforts.

meager funding, perhaps because they had strange names: akialoa, ō'ō'ā'ā, or kākāwahie. They lived in isolated forest patches in Hawai'i and simply weren't charismatic enough—they didn't have a constituency.

The condor story highlights another conservation conundrum: How do we decide what is worth saving? Should we be idealistic and try to save everything, or be realistic and save what we can, most efficiently as we can? There are those in the conservation community who forcefully argue the former position on both philosophical and practical grounds. Because humans created the conditions that imperiled so many species, they say, we are morally obligated to do our best to save all species. Valuing some species or conservation efforts more than others would be to sacrifice our moral imperative on the altar of crass economics. Even if some valuation criteria could be agreed upon, so little is known about most at-risk species that we would be unable to judge their importance—which species might hold the key to new pharmaceuticals or provide unknown ecosystem services? And since we usually don't know the answers, we should make every effort to avoid *any* loss of species. Better to err on the side of caution.[6] We should never, ever, think of a species as expendable. This would be to sanction extinction, and extinction is forever.[7]

This is one line of argument. While stopping far short of justifying extinction, the other position recognizes the reality that resources and support for conservation are limited. Consider, for example, federal and state expenditures for endangered species in the United States. In 2013, nearly $2 billion was budgeted for conservation and management of listed species.[8] This is a lot of money. However, there are huge disparities in funding levels among the listed species; different runs of salmon, for example, cluster at the top of the list, while two mussels and a thistle bring up the rear. The disparities may reflect differences in population status, recovery costs, or likelihood of success. It helps to be a better-known, charismatic, vertebrate species. And economics and politics are important. Salmon are valuable resources with a large constituency of commercial and recreational anglers and native tribes. Listed species that occur in districts with Congressional representatives on important committees are likely to receive more funding than those lacking such advocates.[9] And budgets for endangered species management are at the mercy of partisan politics, where changes in the balance of power can have profound effects.[10]

---

[6] See Essay 20, *Taking risks with the environment (2012)* (page XX).
[7] Proposals to resurrect extinct species, so-called "de-extinction," notwithstanding. I consider this further in Chapter 23.
[8] Federal and State Endangered and Threatened Expenditures, FY 2013. http://www.fws.gov/endangered/esa-library/pdf/2013.EXP.FINAL.pdf.
[9] DeShazo and Freeman (2006).
[10] As I write this, in spring 2015, the Republican Party has assumed the majority in the U.S. Congress. Senator Jim Inhofe, an Oklahoma Republican who has labeled climate change

The Endangered Species Act presumes that once the recovery goals for a listed species are met, it can be removed from the list and the funds used for other species. The reality, however, is that most of the listed species will require continuing conservation management even after their recovery goals are met, so the expenditure will not end. For example, management of the factors threatening the endangered Kirtland's warbler—loss of young jack-pine habitat and brood parasitism by brown-headed cowbirds—has enabled it to increase beyond the recovery target. Without ongoing habitat management and cowbird control, however, numbers would decrease, sliding the species back into endangerment.[11] The warblers are what Mike Scott and others have called "conservation-reliant species"—species that have come to depend on continuing conservation management to ensure their long-term persistence.[12] Scott et al. suggested that fully 80% of all federally listed species may be conservation-reliant—the threats that led to their predicament cannot be eliminated, only managed. And this may be only the tip of the iceberg. Tom Gardali and I extended the analysis to consider a broader set of some 60 California bird species of special concern and found a similarly high degree of conservation-reliance.[13] It seems likely that recovery of endangered species will be elusive and conserving at-risk species will require a much deeper investment than we once thought.

This is the realistic view. Even though global conservation efforts mobilize nearly $20 billion annually,[14] this falls far short of what would be required to meet the targets of the Convention on Biological Diversity (which do not include costs associated with conservation reliance).[15] The mismatch between needs and resources will become greater as the ranks of endangered and conservation-reliant species swell and environmental changes place greater demands on resources to protect people and infrastructure, leaving even less for the environment. Perhaps societal values will change, politicians will become environmentalists, and funds will flow to conservation. And perhaps, as the saying goes, pigs will fly. Until that happens, however, many conservationists are arguing that resources should be allocated to needs following an honest and transparent appraisal of what might be saved, at what cost, with what likelihood of long-term success. Prioritization is not an option, but a necessity.

"a hoax" and compared the Environmental Protection Agency to the Gestapo, is now chairman of the Environment and Public Works Committee. It remains to be seen how this will affect environmental policy and funding.

[11] Bocetti et al. (2012).

[12] Scott et al. (2005, 2010).

[13] Wiens and Gardali (2013).

[14] The cash flow to non-profit conservation organizations can be impressive. In 2014, for example, The Nature Conservancy had revenue of over $1.1 billion and net assets of over $6.5 billion (http://www.nature.org/media/annualreport/2014-annual-report.pdf).

[15] Waldron et al. (2013).

There are multiple approaches for assigning priorities, ranging from back-of-the-envelope exercises to formal protocols involving much more than envelopes. Depending on the particular conservation goals, priorities may be set among species, threats to species, places to be protected, or management actions to be undertaken. The official designation of species as endangered under the Endangered Species Act, for example, prioritizes them above other unlisted species, and disparities in funding among listed species constitute an additional layer of prioritization.

Prioritization approaches use a variety of criteria, including such things as complementarity among protected areas (what a place adds to those already protected), representativeness (how well a variety of environments is captured by a reserve system), connectivity (the landscape setting), uniqueness and irreplaceability, phylogenetic distinctiveness, the social and political feasibility of conservation actions, or cost-effectiveness ("bang for the buck"). Designations of "hotspots" of biodiversity or areas of endemism can be used to prioritize conservation actions at a global scale.[16] National or regional shortfalls in conservation funding relative to biodiversity richness may indicate where returns on investments may be most promising.[17] The array of prioritization protocols in conservation is varied and rapidly expanding[18]; rather than attempt a summary, I'll develop one example to illustrate the approach.

The Lake Eyre Basin in central Australia's is one of the world's largest internal drainage systems.[19] It is a "boom and bust" system, with infrequent rains creating wetlands that attract thousands of waterbirds interspersed among long dry periods when the "lake" is a barren salt flat seemingly extending forever.[20] It has immense conservation value, supporting numerous endangered and endemic plant and animal species. Like many of the world's distinctive ecosystems, it is threatened by an influx of invasive plants; more than 240 exotic plant species have been recorded in the Basin.

Faced with inadequate resources, scientists and managers with the Queensland University of Technology and the Commonwealth Scientific and Industrial Research Organisation (CSIRO) developed a protocol for prioritizing control strategies that would prevent invasive plant species from dominating native

---

[16]Mittermeier et al. (2005); see http://www.conservation.org/How/Pages/Hotspots.aspx. Going beyond the identification of "hotspots" of biodiversity, Ricketts et al. (2005) identified 595 "centers of imminent extinction" in which highly threatened species were restricted to a single location. More recently, Jenkins et al. (2013) used updated data sets to map global priority areas for conserving threatened species of vertebrates and areas of high species richness at a relatively fine spatial scale (100 km^2).

[17]Waldron et al. (2013).

[18]Reviewed by Moilanen et al. (2009), Arponen (2012), and Cullen (2012).

[19]The Basin includes some 1,200,000 km^2, nearly one-sixth of the Australian continent.

[20]See Essay 8, *Boom and bust: lessons from the outback (2014)* (page 51).

ecosystems and transforming them into novel ecosystems.[21] Prioritization was based on cost effectiveness, calculated as the biodiversity benefits of a strategy divided by the costs of implementing the strategy. Even this simple approach, however, encountered problems. Control of buffel grass, which was introduced to the Basin for pasture production, was the least cost-effective strategy. Buffel grass is still valued by many pastoralists as livestock forage, but it is the most widely distributed exotic species in the Basin, threatening biodiversity and places of cultural importance. These contradicting values create challenges in assessing benefits and costs. Prioritization to meet conservation objectives almost invariably intersects with social forces.

The Lake Eyre project also highlights other factors that affect many prioritization efforts. It is important, for example, that goals, criteria, and metrics be clearly stated at the outset. If the goal in the Basin is eradication rather than reduction in dominance of exotic plants, the costs will be considerably greater and the priority rankings different. Biodiversity benefits, which are often of greatest interest to conservationists, are difficult to measure directly, so surrogate measures must be used. The Lake Eyre analysis used the area of the Basin covered by an invasive as an inverse measure of biodiversity benefits; other measures might have produced different results. Each of the control strategies was evaluated independently of the others, ignoring potential benefits (or costs) of implementing multiple strategies at the same places and times. Calculations of cost-effectiveness could also vary depending on who bears the costs. And because the information needed to conduct a quantitatively rigorous prioritization for the Lake Eyre Basin was limited, expert knowledge came into play. Expert knowledge can be invaluable in framing a prioritization, whether the experts are indigenous people, managers, pastoralists, or scientists. However, experts are people, who carry with them opinions, beliefs, and biases, whether derived from scientific paradigms or cultural mores, and these also influence how a prioritization is done.

The Lake Eyre Basin prioritization exercise was also static—in the absence of information on potential changes, benefits and costs were simply extrapolated into the future. Maps of global biodiversity distribution or areas of extraordinary conservation value likewise ignore temporal dynamics, as do many prioritization exercises. Obviously, however, virtually all of the variables included in prioritization calculations will change, not the least because of climate change. Reserve-selection algorithms, for example, may require adjustment to account for shifting distributions of targeted species in response to climate changes.[22] Identification of global priorities for biodiversity protection may need to incorporate measures of the long-term vulnerability of regions to climate change.[23] And it's not

[21] Firn et al. (2013). Elements of this prioritization approach have been extended by Carwardine et al. (2012).
[22] Reece and Noss (2014) have assessed the related issue of how vulnerability to sea-level rise can be incorporated into prioritizations.
[23] Iwamura et al. (2010).

just climate change that creates a dynamic setting for prioritizations. As places are targeted or purchased for protection or restoration, the value of remaining lands may change, increasing the costs of adding subsequent places to a network.[24]

In the end, prioritization only provides guidance. The Lake Eyre exercise, for example, was intended to help guide planning and investment in the management of invasive plants in the Basin. The decisions about what to do where, when, and how are not made by scientists but by managers, stakeholders, policy makers—and ultimately by society. This is where triage comes in. The combination of priorities with limited resources inevitably means that some species will fall short, some places will not measure up, some threats will remain unaddressed, or some actions will not be taken. This is the reality of doing conservation in a global society full of multiple agendas and priorities, all contesting for limited resources, all staunchly defended and (to their adherents at least) thoroughly justified, and many conflicting with one another.

So, what is to be done? One solution is the "save everything" approach—simply insist that all species merit attention and prioritization and triage are signs of weakness, of giving in to an inexorable march toward extinction. At the other extreme lurks triage. Medical triage on a battlefield, in a temporary treatment facility for the Ebola virus, or in a high-end medical facility in a developed country always involves wrenching decisions. Lives of individual people are involved—people like you and me. To those who care about nature and the environment, however, conservation triage has even higher stakes, since the decision may doom an entire species to extinction. The decision may seem easier if the species is a beetle or a slug or even a tiny fish,[25] but what if prioritization were to say that pandas or koalas or polar bears or condors didn't make the cut?

Triage in conservation is a difficult issue, one that merits probing discussions. I offer some thoughts in the accompanying essay (Essay 22, *Talking about triage in conservation (2015)* (page 188)); here I'll just express my ambivalence. I firmly believe that conservation must confront the reality that we can't save everything, and that we're better off and make better use of the available resources if we adopt a prioritization approach, warts and all. But I also pause when I think about the implications of letting go, of actually making end-of-life decisions for species.

Which brings me back to condors. Along with Frank Pitelka and Rich Stallcup, I argued decades ago that California condors should be allowed extinction with

---

[24]A variation, perhaps, on the Red Queen Hypothesis, derived from a statement by the Red Queen in Lewis Carroll's *Through the Looking Glass*: " ... it takes all the running you can do, to stay in the same place." Originally applied in biology to evolutionary theory, it seems apropos to the conservation challenge of having to spend more and more just to achieve the same results. See Van Valen (1973) for the original evolutionary application.

[25]There are people in California who want to reduce allocations of precious water to the endangered Delta smelt in favor of farmers and irrigation districts, even if this leads to the extinction of the smelt (what one California congressman called a "stupid little fish").

dignity. I'd never seen a condor in the wild, however, or even one suffering the indignity of growing up in captivity and being released into the world burdened with electronics and plastics, only to be captured again for blood treatment. I had difficulty thinking of such an animal as "natural."

But then I saw an Andean condor. It was an experience I won't forget. I was in Chile to collaborate with colleagues, and we spent a day in the páramo in the high elevation foothills of the Andes. I had wandered off by myself to satisfy my fixation with grassland birds when I felt a prickling on the back of my neck (literally!). I turned around and there, approaching like the low-flying B-52 in *Dr. Strangelove*, was a condor. I stood transfixed as the bird soared by, only a few meters above me, giving me only a passing glance ("No food there"). I felt the emotional side of nature, and of conservation.

Talking about triage is awkward. Making the decision, or even conducting the analyses that will lead to the decision, requires an objectivity and detachment that I lack.

ESSAY 22

Talking about triage in conservation (2015)*

Triage in conservation—consciously allowing a species to wink out of existence—is not an easy thing to write about or talk about. It's a topic that most conservationists and environmentalists would rather avoid. However, the stark reality is that some species are doomed to disappear because we don't have the resources or the will to save everything. Having the will is something we can do something about, however, and greater will can generate more resources to do more and save more species. By some accounts, this is the primary mission of conservation.

Yet, even the most heroic efforts will inevitably fall short. The challenge, I believe, is to use the resources we have, or those we are able to generate through increasing societal will, to preserve as much of the natural world as we can. To do this, priorities must be established, and along with priorities comes triage. It's not something we can deal with intelligently unless we think about it and talk about it. Thus, this essay.

To a conservationist, the environment can seem like a battlefield, with habitat loss, invasive species, disease, climate change, and a host of other factors forcing more and more species to the brink of extinction. To deal with this crisis, conservationists have developed ways to prioritize species for action or management. Because there are never sufficient resources to address all the conservation challenges, prioritization inevitably leads to triage.

The practice of triage developed during World War I, when medical staff faced with mass battlefield casualties needed to make quick decisions about allocating limited medical resources. Medical attention was given first to the wounded for whom immediate attention could increase their chances of survival; those likely to live or those likely to die, regardless of care, were dealt with later, if time and resources permitted. Medicine has evolved, with protocols and software now available to make the decisions more informed. The consequences are the same, however: in a catastrophic disaster, when resources are limited, patients who are "doomed anyway" may not receive attention.

*Wiens, J. A. 2015. Talking about triage in conservation. *Bulletin of the British Ecological Society* 46(1): 59–60. Reproduced with permission of the British Ecological Society.

Ecological Challenges and Conservation Conundrums: Essays and Reflections for a Changing World, First Edition. John A. Wiens.

In some respects, conservationists are faced with decisions like those facing battlefield or emergency-response medical staff. As in medicine, triage decisions are not easy.[1] Letting a species go is difficult to talk about. If it is discussed at all, it is often in hushed tones. Yet it is time to open up the conversation.

Many conservationists totally reject the notion of triage in conservation, mincing no words in calling triage "ethically pernicious and politically defeatist when applied to biological conservation" and warning that it offers "an easy way out, a convenient escape from our moral duties to other creatures."[2] To accept the possibility of triage is to conclude that some species are beyond saving, doomed no matter what we do. Thinking this way opens the door to walking away from anything that seems difficult. What if we had given up on the peregrine falcon (*Falco peregrinus*), whooping crane (*Grus americana*), or California condor (*Gymnogyps californianus*) when their numbers were perilously low, the threats seemingly insurmountable, and the challenges hopeless? Massive conservation efforts have brought these species back from the brink—although the crane and the condor have yet to reach recovery goals and are still considered endangered. What other species might be written off prematurely if triage were to become an accepted practice in conservation? Does triage have the potential "to mimic carnage of the 19th century battlefields," as some have suggested?[3]

Other, equally dedicated conservationists argue that the reality of inadequate resources to save everything compels us to do the best job in saving what we can. They suggest that it may be possible to conserve more species by judiciously prioritizing how resources are allocated, even though it may mean that some species get nothing.[4] One could go down the priority list, doling out resources until the funds are exhausted. Lower ranked species would then be left to go it alone. Alternatively, resources might be allocated to ensure that conservation efforts are "good enough" so that high-priority species make progress.[5] This could free resources to halt the decline of some lower-priority species, maintaining them in a holding pattern in the hope that sufficient funding might someday appear.

Critics point out that prioritization is constrained by a lack of information for many species. Moreover, the results are sensitive to which factors are included in a prioritization and how they are weighted or measured. Because the outcome can mean life or death for a species, it is important that the prioritization process be open, honest, and overwhelmingly transparent.

---

[1]In its original medical context, triage referred to the overall process of assigning priorities to patients; in conservation, it has come to mean the process of identifying which species will not receive conservation or management support but will be left alone, increasing the likelihood of extinction.

[2]Noss (1996: 921).

[3]Jachowski and Kesler (2009: 180).

[4]See Arponen (2012) and Cullen (2012) for reviews of species prioritization approaches and Moilanen et al. (2009) for prioritizing places.

[5]See my essay, *Uncertainty and the relevance of ecology*, in the *Bulletin*, Spring 2009 (40(1): 50–51).

The harsh reality, however, is that funds are often allocated on the basis of factors other than need or priorities. In the United States, federal agencies responsible for managing species under the Endangered Species Act are subject to increasing scrutiny from politicians who regard the Act as a constraint on the economy and a failure because so few species have been removed from the list. To justify their activities and the act, managers may choose to focus on species with the greatest likelihood of success, rather than those with the most urgent and difficult challenges. It's also possible that recovery and delisting of species are unrealistic measures of success. Many endangered and threatened species are conservation-reliant: they will require ongoing conservation management, even after recovery goals are met, because the factors that threaten them cannot be eliminated.[6] Some of these species could probably persist without the legal protections of the Act with appropriate, long-term management—rather like a patient who can continue to live a satisfying life provided there is ongoing care.

The concept of conservation triage has led to the suggestion that seemingly doomed species should be allowed to fade into the sunset, to suffer "extinction with dignity." But here's where things get sticky. "Dignity" is a human construct, used in this instance to rationalize or soften the impact of extinction. It's doubtful that the last few individuals of a dwindling species feel very dignified—if anything, they may feel lost and helpless, crying out for companions that are no longer there.[7]

But I'm anthropomorphizing, and I was taught that there's no place for that in science. There's more to conservation than science, however. Science can evaluate what threatens a species, assess the value of conservation targets, calculate the cost effectiveness of actions, and do all those things that go into an objective prioritization. But science cannot decide what is worth doing. Other social, cultural, ethical, and subjective forces influence the decision about what to conserve, at what cost, and whether triage should even be considered.

This is why having a conversation about triage is so difficult, even among scientists. It digs down into our philosophical and moral perspectives, so the discussions inevitably run into personal values. As well they should. But there's much at stake. The conversation must be joined.

One parting thought. All of this is about the threat of extinction hanging over species and what we can or should do about it. However, there's a push underway to resurrect extinct species, "de-extinction"—it was even featured as a cover story on a recent issue of *National Geographic*.[8] Should we really be trying to bring back extinct species, when we can't take care of the living ones that are teetering on the brink? Just something to contemplate.

---

[6] Scott et al. (2005, 2010).

[7] This was the fate of the last surviving Kaua 'i 'ō'ō (*Moho braccatus*), a male.

[8] April 2013.

CHAPTER 18

Protected areas: where the wild things are*

When I was a child, my parents took me to see the Disney documentary, *Nature's Half Acre*. The movie chronicled the changes that went on through the year in a half-acre (0.2 ha) meadow that, in the Disney tradition, seemed to be teeming with all manner of life—flowers, insects, birds, mice, snakes. It was a compelling argument for the value of a small, somewhat ordinary area in harboring biodiversity. That I still remember it after so many years speaks of the impact it must have had on a young boy who would go on to become an ecologist and conservationist.

Nature's Half Acre also helped to stimulate the rapid growth of the conservation movement following the Second World War and to reinforce the value of protected areas, even half-acre ones. Protected areas are widely regarded as the last line of defense against the relentless onslaught of habitat loss and degradation that threatens nature and biodiversity.[1] Worldwide, over 160,000 places have been set aside for the conservation and sustainable use of biological diversity. Protected areas cover 19.7 million km^2, equivalent to some 14.6% of the terrestrial area of the globe.[2] This seems like a lot (and it is), but it would require nearly twice that much to meet global protected-area targets.

"Protected area," however, means different things to different people. What are they for, who should they benefit, and, in particular, what is the appropriate relationship between people and parks? Is the goal to protect places *for* people or *from* people? History provides some perspective. Over two millennia ago, places in India were set aside by Royal decree to protect natural resources, and indigenous cultures in many parts of the world set aside sacred sites that, in addition to their religious functions, protected wildlife. Kings and noblemen in Europe established private hunting reserves early in the Renaissance. The Białowieża Forest in eastern Poland, for example, was set aside as a hunting reserve in the 16th century by King Sigismund I to protect the European bison. As royal powers ebbed, many of these reserves became parks open to the public, particularly as the concept of setting

*With apologies to Sendak (1963).

[1]The role of protected areas in conservation is explored by Terborgh et al. (2002), Chape et al. (2008), Cole and Yung (2010), and Mora and Sale (2011).

[2]Butchart et al. (2015).

Ecological Challenges and Conservation Conundrums: Essays and Reflections for a Changing World,
First Edition. John A. Wiens.

aside areas as "a public park or pleasuring ground for the benefit and enjoyment of the people"[3] gained force in the mid-19th century. Many parks were established where the land was already well settled, so people and their activities were part of the picture from the beginning; benefits to nature were often by-products.

As the conservation movement gained force, however, protecting nature often emerged as the primary objective of parks and protected areas. The forerunners of the National Wildlife Refuge System were intended to protect populations of species threatened by human exploitation, especially for fur or feathers. As more nature reserves were established, a belief took hold that the best way to protect nature was to keep people out. For example, when Białowieża National Park was established in 1932, the purpose was to protect the best-preserved fragment of temperate primeval forest in Europe, and access by people was restricted. In 1963, an Advisory Board to the U.S. Secretary of the Interior[4] recommended that management of wildlife in U.S. National Parks should maintain as nearly as possible "the condition that prevailed when the area was first visited by the white man."[5] The 1964 Act that established wilderness areas in the United States explicitly defined wilderness as "an area where the earth and its community of life are untrammeled by man, where man himself is a visitor who does not remain."[6] Writing four decades later, Carel van Schaik and his colleagues argued that "human impacts must be reduced or, better, eliminated completely, if parks are not to suffer progressive degradation."[7] The emphasis was on protecting places that would preserve "naturalness," the way nature was long ago and was supposed to be before people mucked it up.[8]

When parks and protected areas were first established in America and elsewhere in the late 19th century, many indigenous people were displaced, often forcibly.[9] As the protected-area movement expanded into less developed regions of the world in the latter part of the 20th century, the belief that nature protection and people were incompatible was sometimes carried along, and the pattern was repeated: local people were evicted from places that had been their homeland for generations. This time, however, there was substantial backlash from human rights groups and donors against the conservation organizations and governments that were establishing the parks. Now, the definition of protected areas adopted by

[3]The dedicated purpose of Yellowstone National Park in the United States when it was established in 1872.

[4]At the time, Stewart Udall was Secretary of the Interior; see Essay 38, *Stewart Udall and the future of conservation (2010)* (page 298).

[5]The Leopold Report; Leopold et al. (1963: 4).

[6]These quotes are from the early 1960s, before men (some of them, at least) had become sensitive to the sexist implications of equating people with "man."

[7]In Terborgh et al. (2002: 473).

[8]"Naturalness," is of course an expression of human values and perceptions rather than a scientific assessment, so what is "natural" varies depending on the cultural and social context. Like beauty, naturalness is in the eye of the beholder.

[9]Poirier and Ostergren (2002) and West et al. (2006).

the International Union for Conservation of Nature (IUCN)[10] emphasizes not only the conservation of nature, but also benefits to people and their cultural values. Nonetheless, the tension between parks for people and parks for nature persists, and local involvement in the design and management of protected areas is often still inadequate.[11]

Underlying this tension are varied views about what it is that is to be protected. Most ecologists and conservationists probably think of protected areas in terms of some aspect of biodiversity, with a particular emphasis on places of extraordinary species' richness and uniqueness. Yet protected areas are often established in low productivity areas of little use for anything else—the "rocks and ice" syndrome. The two largest terrestrial protected areas in the world, Greenland National Park and the Ar-Rub'al-Khali Wildlife Management Area in Saudi Arabia, have very low biodiversity. Such areas may do a splendid job of protecting unique ecosystem types, but their contribution to protecting the earth's biological diversity as a whole is disproportionately low.

This raises the issue of how effective protected areas really are in preserving nature. Without a doubt, many places would be developed or suffer degradation were they not protected. Many other worthy places lack protection, however, most often because funding is not available from governments, nongovernmental organizations, or wealthy individuals to buy them or to implement agreements to set them aside for conservation. Placing land or water under some form of protection, however, is only the first step. Whether the intent is to protect a species or ecosystem, stem the loss of biodiversity, or something else, suitable habitats must be available and the factors that threaten nature must be ameliorated. Frequently this will entail restoration, ongoing conservation management, and long-term monitoring, so the initial costs are just the beginning. Effective management also depends on effective governance. In some nations, parks exist only on paper, with little enforcement of boundaries or protection of wildlife within the boundaries. Even if all the pieces are in place to create and manage a protected area, there is no assurance that the desired species or ecological systems will actually occur there. I address this issue in Essay 23, *Build it and they will come (2013)* (page 197).

In some instances, however, places set aside for other purposes may provide some of the most effective protected areas. For example, the U.S. Department of Defense operates nearly 200 military installations, most of which were

---

[10]"A clearly defined geographical space, recognized, dedicated and managed, through legal or other effective means, to achieve the long-term conservation of nature with associated ecosystem services and cultural values"; Dudley (2008). Recognizing that "protected area" covers a broad range of situations, IUCN defines eight categories of protected areas based on the primary objectives of management and involvement of people, ranging from strict nature reserves and wilderness areas to areas managed mainly for the sustainable use of natural ecosystems (i.e., ecosystem services).

[11]Adams and Hutton (2007) provide a nice review of the issues.

established during or soon after the Second World War. Many are large areas that were originally far from population centers. For obvious reasons, public access is restricted. Some of these installations now harbor considerable biodiversity, including species listed as threatened or endangered. Although the primary mission is military training (which can involve tanks rumbling across the landscape or simulated battles), federal law restricts human activities in proximity to listed species. In response, the military has developed comprehensive plans to manage critical habitat for these species in concert with training activities. At Fort Bragg in North Carolina, for example, management of habitat for the critically endangered red-cockaded woodpecker has been integrated with the training mission; under this protection, the birds have thrived. The military takes environmental stewardship seriously.

If protected areas are large, they may be effective in providing sufficient area to harbor substantial biodiversity and viable populations of many species. Yet even the largest reserves may be inadequate to support wide-ranging carnivores or ecological processes that encompass entire watersheds. And the majority of protected areas are small; some (such as Nature's half-acre) are tiny. As the ecologist Dan Janzen sagely observed decades ago, "no park is an island."[12] Protected areas are embedded in broader landscapes. Janzen drew attention to the effects of incursions of natural predators and competitors into a protected area from the adjacent landscape, but people living in the surrounding area can also affect population dynamics and ecosystem processes within a park. Poaching of wildlife in parks is a mounting problem; poachers killed more than 1,000 rhinos in South Africa in both 2013 and 2014. Elephants that wander from parks to raid nearby crops in Africa and Asia are often killed, and bison that leave Yellowstone National Park may be killed by ranchers to protect their livestock from disease (brucellosis). A protected area is a sanctuary only if one stays put, and sometimes not even then.

Landscapes that include protected areas are mosaics of places with different land cover and human uses. Although protected areas receive most of the attention, places that are used by people may have considerable conservation value. To ignore the contributions of other places in a landscape is shortsighted and fails to take advantage of conservation opportunities. Places for nature may or may not be legally protected, but the importance of places in nurturing biodiversity is not so black-and-white, an issue I address in Essay 24, *The dangers of black-and-white conservation (2007)* (page 203). It is only when conservation value is equated with "naturalness" that it may seem so.

Protected areas are fixed in space. There is often an underlying presumption that they are also fixed in time—they will remain the same and retain the desired conservation values. But change happens. Comparisons of historical

---

[12] Janzen (1983).

photographs with matched images from a century or more later[13] vividly illustrate how landscapes have changed, whether protected or managed or not. National Park managers in the United States struggle to explain to the public how disturbances and vegetation succession are a normal part of how nature works, causing the scenery that brings people to the parks to change.

Now, however, both protected areas and the landscapes that contain them are, like everything else, being buffeted by the onrushing tide of climate change. Some effects of climate change on protected areas may be direct, as climatic conditions within the areas change. The indirect effects of climate change on protected areas, however, may be more profound. Species are already shifting distributions in response to climate change, and such realignments are likely to become more frequent and widespread in the future. Some species will move out of protected areas while others will move in, creating novel assemblages. If those leaving are the species for which the protection was designed and those entering are undesired "invasive" species, the effectiveness and value of a protected area may be diminished. In some situations, the climate in places set aside to protect the last remaining remnants of an endangered species may disappear, leaving it stranded.[14] The critically endangered Araripe manakin, for example, exists only in a narrow strip of habitat in northeastern Brazil; 68 ha have been designated specifically to protect the species, but much of the remaining habitat is vulnerable to development. If the habitat within the protected area changes, the manakin will have nowhere to go.[15]

It is clear that the species composition within protected areas *will* change with climate change. Conservation and management will need to change as well. Restricting the focus to areas within the boundaries of protected areas will no longer work.[16] The framework will need to be expanded—both spatially, to include a broader array of protected areas that can be available to absorb dispersing species should they come, and qualitatively, to include the spectrum of land uses in landscape mosaics. Doing this will entail greater costs, which will place a premium on using well-designed prioritization protocols that include climate change,[17] as well as providing incentives to incorporate more private land into the conservation estate.

What, then, is the outlook for protected areas? Many are pretty places with spectacular scenery. They are often managed carefully to achieve the conservation goals. They also provide evidence that conservation can work. Visiting

---

[13]For example, Veblen and Lorenz (1991) or the Rocky Mountain Repeat Photography Project (http://bridgland.sunsite.ualberta.ca/main/index.html).

[14]This is one argument offered in support of moving species to new areas ("assisted migration," or "managed relocation") to adapt to climate change; see Chapter 23.

[15]The Alliance for Zero Extinction lists 587 sites that contain the only known location for 920 endangered and critically endangered species; see http://www.zeroextinction.org/.

[16]See Essay 19, *Moving outside the box (2009)* (page 164).

[17]For example, Hannah et al. (2007).

a nature preserve almost anywhere in the world gives one a feeling of hope, revitalizes the spirit, and renews a commitment to continue the fight. No one can argue that protected areas have not been successful in helping to maintain some semblance of "naturalness" in a developing and changing world, or that they should not continue to be at the core of conservation efforts. However, the swelling tide of extinctions and imperilments and the magnitude of coming environmental changes say quite clearly that this, by itself, will not be enough. Unless efforts are extended to broader landscape mosaics, protected areas may become isolated islands in an ocean of development that leaves little space for nature.

ESSAY 23

Build it and they will come (2013)*

Inspiration for essays comes suddenly and from strange sources. Watching the movie Field of Dreams *one evening, I was struck by how the catch phrase, "Build it and they will come," captured the underlying premise of protection and restoration of places for nature. If we create spaces that include suitable habitats and protect those places from development or other threats, the desired species will come, live long, and prosper.*[1] *At least that's the hope. Unlike the story line of the movie, however, we cannot return to times past, an idealized version of how nature used to be, by building protected areas. Things have changed too much.*

And they will continue to change, probably more rapidly and dramatically than they have in the past that we remember. In the essay, I considered how change might affect our perception of "it" (the places) and "they" (the desired species), and what might affect whether they "will come" (dispersal barriers and the like). I neglected, however, to consider the "build" part of the phrase. Anything that is built requires maintenance and occasional renovation (as any home-owner knows all too well). The expectation that the job will be finished when something is built applies no more to a place that has been protected or restored than it does to a building or a highway. Protected areas need continuing attention if they are to retain their value to nature as the environment changes and species come and go.

In the 1989 film, *Field of Dreams*, an Iowa farmer hears a voice in his cornfield. Following instructions, he plows under his crop, builds a baseball field, and players from the past emerge to play ball. It's a far-fetched plot, but the film was good enough to be nominated for an Academy Award.

More memorably, the film gave us the catchphrase "Build it and they will come," which has become something of a mantra for conservation and restoration ecology. Increasingly, "building it"—protecting or restoring natural areas—is how we counter habitat loss. Placing areas under some form of protection is central to the missions of a host of land trusts and other governmental and nongovernmental programs (Figure 1). Restoration ecology aims to foster ecological processes by repairing places that have suffered environmental degradation.

*Wiens, J. A. 2013. Build it and they will come. *Bulletin of the British Ecological Society* 44(2): 48–51. Reproduced with permission of the British Ecological Society.

[1] Another catch phrase, from Mr. Spock in *Star Trek*.

Ecological Challenges and Conservation Conundrums: Essays and Reflections for a Changing World, First Edition. John A. Wiens.

Figure 1 Sandhill cranes (foreground) and snow geese (background) wintering at Bosque del Apache National Wildlife Refuge, New Mexico. Photograph by John Wiens.

And it has paid off. Worldwide, more than 12% of the land is under some form of nature protection, and habitat restoration is often mandated to repair damages from mining or logging. There is little doubt that biodiversity and people have benefitted. As we venture into an increasingly uncertain future, however, we should ask whether the approaches that have served well in the past should remain the core strategies going forward. In particular, what do we really mean by "Build it and they will come"?

Begin with "it." "It" is normally considered to be a place that contains the resources needed for the persistence of desired species and communities—in other words, nature. Identifying and documenting the resource needs of multiple species is a formidable task, however, so habitats are usually used as surrogate targets for protection or restoration. The assumption is that simply providing a habitat—old-growth forest, tidal marsh, shrub desert, and the like—will allow species and communities to prosper.

But protecting or restoring a place does not guarantee that the habitat it contains will remain unchanged. The viability of a tidal marsh, for example, may be threatened by sea-level rise and salt-water intrusion, and the plant species in a forest or grassland may come and go with climate change. Maintaining a desired habitat in an area against the forces of change will be an uphill battle.

Who, then, is "they"? When places are protected or restored, the expectation is that the places will be populated by the desired species and communities. This has been taken to mean what should be there in a natural community or what was there at some past time (like the ballplayers who came to the field of dreams). These expectations come from theories of ecological succession or community assembly: we think we can project which species will likely occur in an area, given the habitat and the pool of species available to colonize.

It's turning out, however, that nature is not quite so predictable. There is a continuing ebb and flow of species across the landscape. Range expansions, such as the movement of barred owls (*Strix varia*) into forests of the northwestern United States, may displace other species, such as spotted owls (*Strix occidentalis caurina*). The disruptive effects on native species and communities of introduced or invasive species, such as cane toads (*Bufo marinus*) in the Caribbean and Australia or brown tree snakes (*Boiga irregularis*) on Guam, are legion. Shifts in species' distributions will become the norm as the effects of climate change play out, unraveling existing species assemblages to produce novel combinations of species that have not been seen before. There are no assurances that "they" will be the species and assemblages we want or had in mind when undertaking the protection or restoration.

What about "come"? "Build it and they will come" assumes that the desired species will colonize the protected or restored area. Species differ in their dispersal abilities, however, so this may be easier for some species than for others. First arrivals may exclude later arrivals, leading to a community different from what was expected. For species with limited dispersal, human assistance may be necessary to populate the area. This is already an accepted management practice for some critically endangered species. Captive-bred California condors (*Gymnogyps californianus*), for example, have been reintroduced in the Grand Canyon of Arizona, where they had not occurred for centuries. Populations of the saddleback (a bird, *Philesturnus carunculatus*) have been established on several small, rat-free islands off New Zealand following their eradication from the mainland over a century ago. As climate change shifts the locations of species' habitats, there is increasing discussion and controversy about "managed relocation" or "assisted migration"—moving species into protected areas beyond their current range. In this case, the mantra might better be "build it and we will bring them."

None of this is to say that protecting or restoring areas for nature is wasted effort. Without such areas, biodiversity loss would have been considerably greater and the prospects for biotic adaptation to future climate change substantially diminished. Yet, although protected areas may be set aside in perpetuity, what is in them and who will come are by no means certain. One can no longer expect the desired species to appear and thrive simply because a suitable habitat has been created or protected.

This means that we should rethink and expand the goals and expectations for protected areas. Although the aim may be to conserve a particular set of desirable species, this may not always be possible. It may sometimes be more appropriate to protect or restore places in which ecosystem processes can continue to function, regardless of which species are present. We may not know who will come, stay, or leave, and some of the guests may be uninvited and unwanted. But by protecting and restoring places in which external threats are alleviated and essential physical and environmental conditions are maintained, natural processes can continue.

To cope with the rapid changes now underway, we will also need to deal with uncertainty well beyond our comfort levels. Because the outcomes of protecting an area or restoring a habitat cannot be certain, multiple approaches are necessary. A place doesn't have to be protected or restored, for example, to have conservation value (Figures 2 and 3). Some places that are too small or disturbed to merit attention by themselves may in aggregate provide conservation benefits. Many of the places where people live and work—farmlands, pastures, woodlots, hedgerows, even backyards in suburbia—harbor biodiversity and enhance the ecological interconnectedness of landscapes. The importance of such areas has long been recognized in rural landscapes, particularly in Europe, the United Kingdom, and parts of North America. Devices such as conservation easements help to formalize some degree of nature protection while permitting continuing land uses. Conservation will be more effective if it is inclusive of a variety of places over entire landscapes.

Figure 2 Cormorants breeding on the foundations of former buildings at Alcatraz Island in San Francisco Bay, a protected area administered by the National Park Service as part of the Golden Gate National Recreation Area, California. Photograph by John Wiens.

Figure 3 Western gulls breeding on the foundation of an abandoned building on Alcatraz Island. Photograph by John Wiens.

Figure 4 Sevilleta Bosque landscape: this is part of the Bosque del Apache National Wildlife Refuge in New Mexico. Photograph by John Wiens.

Ultimately, how we regard protected and restored areas relates to how we value nature and what we mean by "nature" (Figure 4). If we adhere to a narrow, preservationist approach, we will continue to build it in hopes that "they," the desired species, will come. The approach should be broadened to include other species, ecosystem processes, or areas not intended for nature conservation. This need not mean giving up or accepting weed patches or manicured parks as "nature." As Emma Marris has argued in her fine book *Rambunctious Garden*, even highly altered places may have conservation value. It would be nice indeed if we could protect or restore nature as it used to be (or as we so imagine), but the changing world ensures that this will elude our grasp. Better to recognize that "nature" encompasses a spectrum of conditions, all of which is important but only part of which is served by "Build it and they will come."

ESSAY 24

The dangers of black-and-white conservation (2007)*

This essay was my attack on simplistic, dualistic thinking in conservation. Distilling things to polar opposites—yes or no, good or bad, mind or body—seems part of human nature. It probably has an origin in our evolutionary past, when there may have been real advantages to making quick, black-and-white distinctions (is it something I can eat or something that can eat me?). Such thinking has been richly reinforced by the binary coding that lies at the foundation of our digital age. I argued, however, that it rarely applies in conservation. The world that conservationists or ecologists deal with is riddled with gradients and shades of gray.[1] Thinking of protected areas as separate from their surroundings, or of natural values as either economic or esthetic, only polarizes debates and gets us nowhere. Dealing with the complexities of heterogeneous variation in space and time isn't easy, and as Robert Pirsig observed, a traditional, hypothesis-testing approach to science may be ill-suited to such conditions. However, the real world that we're trying to understand, manage, and conserve is messy, not something that can be arbitrarily partitioned into black or white bins.

The world is a complex place. To simplify this complexity, people often reduce it to simple either-or choices—black or white, do or don't, yes or no, winners or losers, nature or nuture, and so on. Even our computer systems are based on binary logic. Conservationists are people, so we tend to do this too.

Case in point: protected areas. Protected areas have been the cornerstone of conservation, both nationally and internationally. Many of the areas we aim to protect exist as remnants of natural habitat in highly fragmented landscapes or are carved out of larger areas that are rapidly being eroded by human actions. The protected areas—parks, nature reserves, wildlife refuges, wilderness areas—are often viewed as "islands" of nature surrounded by an inimical matrix with little conservation value. Decades ago, ecologists provided a scientific foundation for this view in the theory of island biogeography, which likened nature reserves to

*Wiens, J. A. 2007. The dangers of black-and-white conservation. *Conservation Biology* 21: 1371–1372. Reproduced with permission of John Wiley & Sons.

[1] Less so, perhaps, for environmentalists, who tend to see things as favoring or not favoring particular environmental interests or agendas. Litigation bolsters this way of thinking, by separating parties into plaintiffs and defendants and, usually, winners and losers.

Ecological Challenges and Conservation Conundrums: Essays and Reflections for a Changing World, First Edition. John A. Wiens.

oceanic islands and explained their species richness and rates of biodiversity loss in terms of the size and isolation of the reserves. Even though we know that reserves and protected areas are not really islands and the surrounding landscape is not really the same as an ocean, this binary view—that areas are either protected and have conservation value or they are not and they do not—continues to hold sway over a good deal of thinking in conservation and resource management.

Another case in point: the benefits of nature. Conservation is ultimately founded on the values that people and societies place on nature. These values, too, have tended to be viewed simplistically, contrasting the esthetic, spiritual, and ethical benefits of conserving nature and biodiversity with the more pragmatic and economic benefits of protection. This distinction is not new—many Nature Conservancy preserves, for example, were created based on their beauty or an ethical responsibility to protect remnant populations of rare plants and animals, and the U.S. National Wildlife Refuge System was established largely to manage populations of game species for recreational hunting. The current fad of "ecosystem services" in conservation has the potential to perpetuate or exacerbate this dichotomy. In attempting to create ways of valuing the goods and services that natural ecosystems provide to people, conservation may be drawn into an economic accounting in which nature either has economic value or it does not. A recent exchange of views in *Nature* (Costanza 2006, Marvier et al. 2006, McCauley 2006, Reid 2006) has sharpened this debate and shows that conservation is not quite so cut and dried.

But thinking about protected areas or the benefits of conservation in such simple, black-and-white terms is shortsighted and wrong. Protected areas do not exist in isolation from their surroundings, and the surroundings are not without conservation value of their own. If landscape ecology tells us anything, it is that protected areas and their surroundings are usually richly textured mosaics of different habitats and human uses with differing degrees of naturalness. What goes on in any particular place in a landscape is affected by things elsewhere in the landscape—the predator that lurks in an adjacent woodland, the nutrients that flow from an upslope field. To ensure conservation in one place requires consideration of the broader landscape and the threats, resources, and biodiversity it contains. And people are part of that landscape as well. The ways in which people relate to a conservation landscape are many and varied, some with clear economic benefits, some with quasi-economic benefits, and some that come only through the knowledge that we are protecting biodiversity because it is there. There are multiple constituencies for conservation, and to succeed we must aim to be relevant to them all. Doing conservation requires that we consider nature as a multicolored palate, with varied hues and shades. We must find ways to measure where we are on these spectra and adjust our conservation practices and expectations accordingly.

In one of many thoughtful digressions in his book *Zen and the Art of Motorcycle Maintenance*, Robert Pirsig (1974) reflects on the prevalence of dualistic, yes-no thinking in society and, especially, science. He observes that we actually learn more from experiments in which the result is neither "yes" nor "no" but something in between, which forces us to reformulate and expand the scope of the original question. Being properly trained in dualistic logic, scientists often distain or discard such results. But nature, and the ways in which it benefits people, is not black-and-white. As we wrestle with how to expand our vision beyond protected areas alone, or how to express the value of natural systems in multiple ways, we should be wary of the alluring trap of simplistic, dualistic thinking.

How might this change how we do conservation? We would make greater use of land-cover and land-use information to place the areas we protect in their landscape context, and we would use projections of future land-use changes to evaluate the long-term effectiveness of this protection. We would recognize that it should not be a foregone conclusion that human uses of lands and waters are incompatible with or inimical to biodiversity, and we would develop ways of measuring the trade-offs between different human activities and biodiversity protection. We would recognize that there are many dimensions to the value of ecosystem services to people, dimensions that depend as much on the people and their cultures as on the properties of the natural (or managed) ecosystems. And from this we would fashion conservation that preserves our future as well as protects the present.

All of these things can be done. But none of them will happen if we portray the world as black-and-white.

CHAPTER 19

Ecosystem services and the value of nature

What is nature worth? Philosophers have pondered this question over the ages as also, more recently, have ecologists, conservationists, policy makers, and the public at large. E.O. Wilson considered it with his characteristic insight.[1] It all goes back to the distinction between intrinsic and instrumental values that I discussed in Chapter 1. Some believe that nature has value in and of itself and that it is wrong to impose on it any human-defined values. Others believe that it only makes sense to place nature in a framework to which most people can relate, so that the values of nature can be properly compared with those of competing demands for attention and resources.

The issue is another manifestation of the tension between people and nature that I considered in the previous chapter, but it's not new. In Essay 25, *What's so new about ecosystem services? (2007)* (page 211), I describe one attempt from the 1920s to assign dollar amounts to the benefits that birds provide to people. Such benefits are now labeled *ecosystem services*—the goods and services that people derive from natural ecosystems.[2] They include such things as fresh water, carbon sequestration, soil formation, flood control, firewood, fish and other foods, recreation—the list goes on. Many agricultural crops depend on honeybees and other insects for pollination; the value of this service has been estimated in tens of billions of dollars, and beekeepers keep busy transporting hives to where the pollinators are needed.

Interest in ecosystem services jelled in 1997 with two publications: a book by Gretchen Daily and a paper by Robert Costanza and his colleagues.[3] Daily's book developed a broad perspective on ecosystem services, while Costanza et al. came up with a worldwide estimate of the yearly value of the goods and services nature provides to people: $33 trillion ($52 trillion in 2015 dollars).[4] The unexpected magnitude of this number and the subsequent publication of the

[1] Wilson (2002).
[2] More technically, "the conditions and processes through which natural ecosystems, and the species that make them up, sustain and fulfill human life"; Daily (1997: 455).
[3] Daily (1997) and Costanza et al. (1997).
[4] An updated analysis pegged the number at $140–163 trillion (2015 US$); Costanza et al. (2014).

Ecological Challenges and Conservation Conundrums: Essays and Reflections for a Changing World, First Edition. John A. Wiens.

massive Millennium Ecosystem Assessment,[5] which featured ecosystem services as the organizing framework, put biodiversity and nature's services on the radar screens of economists and governments as well as conservation organizations. Ecosystem services have now gone far beyond being an "eye-opening metaphor" that alerted people to what nature was providing for free,[6] to a framework that engages Goldman Sachs, the World Bank, the United Nations Environmental Programme, and national governments—it has entered the socioeconomic mainstream. Ecosystem services have even been proposed as a structure for assessing damages to people and resources from the *Deepwater Horizon* oil spill.[7]

It is not difficult to see the appeal of an ecosystem services approach. By outfitting nature in the finery of "natural capital," it can be visualized as a stock from which services flow—a conceptualization familiar to any economist. Drawing from economic theory, ecological production functions that relate the chemical, physical, and biological processes of ecosystems to the output of services valued by people can be conceptualized and modeled. Many goods and services can be expressed in financial units (as Costanza and his colleagues did), which allows them to be entered into balance sheets or cost-benefit analyses along with human-produced commodities using a common currency (literally!). From there the extension to paying for ecosystem services is a logical step, so what was free before now carries a price tag.[8] This, in turn, means that ecosystem services can be managed using market mechanisms.[9] Proponents argue that this would create a way to manage the natural capital and thereby prevent further diminishment of environmental resources. Because markets for carbon (e.g., cap-and-trade programs) are at the forefront of efforts to mitigate increases in atmospheric CO_2 and reduce the rate of global warming, they have received the greatest attention, but markets are proliferating for all sorts of ecosystem services.[10]

---

[5]Millennium Ecosystem Assessment (2005). See http://www.millenniumassessment.org/en/index.html.
[6]Norgaard (2010).
[7]National Research Council (2013).
[8]Although, as Costanza et al. (2014) point out, many ecosystem services are best considered as public goods—part of the commons—so dealing with them as conventional market commodities may not make a great deal of sense. Also, it may not always be clear who "owns" an ecosystem service. Thus, payments for maintaining a forested watershed to provide clean drinking water may go to the landowner, but who owns the water that falls as rain on the forest? Ultimately, Mother Nature is the owner who should receive the payments. In this case, the agencies, organizations, or individuals who manage or protect her bank account—the natural capital—should perhaps be the ones receiving the payments.
[9]My colleague Dick Norgaard suggests that the global reach of an almost religion-like belief in market-based economics argues for labeling the current phase of the Earth's history the "Econocene" rather than the Anthropocene. It's not just that humans have come to dominate the Earth, but we have done so (in the past century at least) through the direct and indirect effects of market forces.
[10]The parallel between financial markets and those for ecosystem services only goes so far, however. When banks and important industries failed in the United States during the "Great

Ecosystem services have also become a centerpiece of several large conservation organizations. By casting nature in economic terms, the hope is that public perceptions of the value of nature may be changed and the notion of conservation may resonate with a broader constituency, particularly in the business community. For example, the suggestion that tropical forests could harbor yet-undiscovered medicinal plants or animals that might hold the key to treating cancer or other human diseases has generated interest among pharmaceutical companies in protecting these potential ecosystem services. Such support can have a variety of benefits, not the least by attracting new donors and funding sources to conservation. Most importantly, however, it can shake people out of their complacency in believing that nature will always be there to supply the goods and services we have come to depend on and that we can continue to use them at no cost. There are and always have been costs, of course, some of which have been buried in the supply chain between nature and people, and some of which are the costs of diminished natural resources that have been over-exploited[11]—withdrawals from natural capital that have not been replenished. Humans have been doing this for millennia, especially since the onset of the Industrial Revolution.

One way or another, economics is a main driver of factors such as land-use change, development, habitat loss, air and water pollution, and resource extraction, all of which threaten biodiversity. It stands to reason, then, that forging a closer link between conservation and economics has the potential to improve nature's lot. This prospect concerns many ecologists and conservationists, who see pernicious threats to the conservation agenda lurking behind the façade of market-driven ecosystem services. They fear that, by focusing on the production functions of natural ecosystems and the features of nature that are useful to people, the complex webs of environmental interactions may be artificially simplified.[12] The richness of biodiversity might end up being defined only in terms

Recession" of 2007–2009, the federal government bailed them out with infusions of dollars to retain institutional stability and ensure that the goods and services provided (e.g., jobs, loans, housing starts) did not decline further. It is not clear what should be done to bail out failing environments to rescue the goods and services they provide—cash alone won't work (although, to be a bit self-serving, increased funding for conservation, restoration, and environmental management wouldn't hurt).

[11]For example, many of the world's ocean fisheries—cod off the coast of Newfoundland, anchovetta off the coast of Peru, and many others.

[12]The National Research Council committee that developed an ecosystem services approach to the assessment of damages resulting from the *Deepwater Horizon* oil spill in the Gulf of Mexico (National Research Council 2013) argued that it provides a more holistic view of ecosystem interactions than does the more traditional resource-by-resource approach. The claims that the approach follows these interactions "through all relevant trophic levels and spatial connections to their ultimate impact on human well-being" (p. 15) seems an overstatement, however, especially as (following the Millennium Ecosystem Assessment) provisioning, regulating, cultural, and supporting services are then considered separately.

of money and markets, leaving out things such as spiritual or esthetic values that cannot be directly translated into dollars. If pieces of nature—species, communities, ecosystems—are converted into monetary units, even for the purpose of making comparisons or assessing tradeoffs, then proposals to offset environmental damage or habitat loss in one place by restoring habitats in another place might end up being based solely on equivalent dollar values of ecosystem services, rather than ecological functioning or other noneconomic values.

A full-blown development of the ecosystem services approach is hampered by the difficulty of assigning values to goods and services.[13] Forestry has long valued timber in terms of board feet, and fisheries by the profit margin of catch—products that have clear dollar values determined by market forces. Other services can only be valued indirectly. The value of coastal wetlands, for example, could be tallied by the avoided financial costs of damages from storm surges or the value in increased commercial fish or shellfish production provided by sheltered nursery areas. It's more challenging to come up with monetary values for nonmarketable services. One suggestion is to use contingent valuation—survey-based determinations of the public's willingness to pay for some aspect of the environment, such as preventing the loss of a wetland or maintaining condors in the wild. Following the *Exxon Valdez* oil spill, for example, a contingent-evaluation exercise determined that the public at large would be willing to pay in excess of $3 billion to ensure the existence of an unspoiled Prince William Sound.[14] Although not usually labeled as such, decisions to protect areas such as national parks for conservation (Chapter 18) are often a form of contingent valuation.

More confounding than the challenges of valuation, however, are the effects of change. If ecology has historically been guided by concepts of equilibrium and steady-state, economics has instead followed the core principle of growth. Growth in human and physical capital, combined with increased productivity, will increase the market value of goods and services—economic growth. The explicit assumption, of course, is that technological advances will enable capital and production to continue to grow, avoiding the economic stagnation that would occur if human population growth and resource demands exceed production capacity—the "Malthusian Trap" of classical economics. Because economic growth generally acts to decrease natural capital (just look around), there must be limits.[15]

Nonetheless, the National Research Council report provides a useful and thoughtful in-depth assessment of how the ecosystem services approach might be applied to a complex, real-world situation.

[13]Fenichel and Abbott (2014) describe a detailed but theoretically dense way of assessing the value of the natural capital from which ecosystem services flow.

[14]Carson et al. (2003) and Portney (1994) provide understandable and useful synopses of contingent valuation.

[15]This is the focus of *The Limits to Growth* (Meadows et al. 1972) and its update, *Limits to Growth* (Meadows et al. 2004) and, before that, *The Limits of the Earth* (Osborn 1953). Whether we

Even as economic growth has continued for nearly two centuries, the trajectory has been anything but smooth. Economies undergo booms[16] and busts,[17] such as the Great Depression of the late 1930s or the Great Recession of the early 2000s. The fluctuations in markets inevitably affect the values of ecosystem services. There are also more subtle market forces at play. By valuing an ecosystem service and setting payments for that service, we affect the value of other marketable commodities, changing the economic equation and, in turn, affecting the value of that service—a perverse Heisenberg Effect.[18] Social and cultural forces come into play as well, because the value of a service depends on who is doing the valuing—who is selling and who is buying. The simplifying assumption of some economic models that individual needs and preferences are stable is clearly false; societal demands and cultural norms change, and with them the values placed on marketable and non-marketable things, including ecosystem services.

To reprise the theme running throughout this book, however, environmental changes may have the greatest impacts on ecosystem services and how they are valued. Changing land use associated with agriculture, for example, can modify the production functions for ecosystem goods and services at other places in a landscape, through adjacency effects, altered connectivity, changes in how disturbances arise and spread, and the like. In addition to the direct effects that climate change can have on economics at global to local scales, it can also alter production functions. The services provided by a coastal wetland, for example, will be transformed with sea-level rise or changes in estuarine salinity, and the values placed on those services will change as well. One need only think of the effects of the Dust Bowl of the 1930s or the recent drought in California to begin to grasp the sensitivity of the economic landscape to environmental changes.

Ecosystem services have become part of this landscape. Whether this is good or bad is a matter of perspective and philosophy. The ecosystem services approach reframes the relationship between people and the natural world; it provides "a lens that allows us to view natural systems in a different way: as sources of services and goods for people."[19] Ecosystem services may be selling out nature only if the approach comes to dominate conservation, superseding other approaches or ways of valuing nature. Considered as one of many tools, ecosystem services can have great utility in bolstering financial support for conservation as well as communicating the importance of nature to a broad audience. Like it or not, the ecosystem services approach is here to stay.

---

have already passed these limits or believe that technological fixes will continue to push back them back is at the heart of the issue of sustainability.

[16]What the economist Alan Greenspan referred to as "irrational exuberance" during the dot-com economic bubble of the 1990s.

[17]Like arid environments; see Essay 8, *Boom and bust: lessons from the outback (2014)* (page 51).

[18]Named after German physicist Werner Karl Heisenberg, whose uncertainty principle posited that the act of observing something alters what is observed, confounding predictions.

[19]National Research Council (2013: 37).

ESSAY 25

What's so new about ecosystem services? (2007)*

My premise in this essay was that new ideas are often old ideas in new clothing. Thus, while ecosystem services—the benefits that people derive from natural ecosystems—are commanding a great deal of attention now, the idea itself, and even attempts to calculate the monetary value of such services, is not new. The interesting point is not that "new" ideas have been thought of and dug into before, but why they didn't catch on then but have now. I considered several contributing factors, but one merits particular attention. For much of its history, conservation has been associated with protecting places where nature can flourish.[1] Recently, the approach has been extended to consider broader landscapes, mosaics that include places where people live and work. This shift has coincided with the recognition that benefits to people can derive not just from natural areas, such as a forested watershed that provides clean drinking water, but also from more intensively used places as well. A landscape is a mosaic of ecosystem services, all part of the conservation estate. At the same time, however, there is more to a landscape than the services it provides.

New ideas, *really* new ideas, are hard to come by. More often than not, they are old ideas, polished and dressed up with new labels and jargon to fit the styles of the day. George Salt had the idea of aggregating species into functional groups more than a decade before Richard Root labeled such assemblages "ecological guilds," and Charles Elton had the same idea well before that, and probably Charles Darwin before that, and .... Every new idea that sets off a flurry of activity in ecology or conservation seems to have been thought of before. We periodically reinvent the wheel, albeit with a flashier design, more chrome, and seemingly greater brilliance. But it's still a wheel.

"Ecosystem services" is a case in point. The idea is all the rage now. Financiers envision new markets that will pay for nature's values, agencies see a new way to justify their mandates, and conservationists see an avenue toward broadening the constituency for protecting nature. But the idea itself, of course, isn't really that new.

*Wiens, J. A. 2007. What's so new about ecosystem services? *Bulletin of the British Ecological Society* 38(2): 39–40. Reproduced with permission of the British Ecological Society.
[1]See Chapter 18.

Ecological Challenges and Conservation Conundrums: Essays and Reflections for a Changing World, First Edition. John A. Wiens.

One of my pastimes is browsing through old neglected books. A few evenings ago I picked up one published in 1922, *The Importance of Bird Life*, by one C. Inness Hartley. Hartley prefaced his book with a paragraph extolling how the bounties of America fueled its industry, but what caught my eye were the subsequent paragraphs:

> Coal- and iron-mines are largely responsible for rapid development of the United States. From California, Nevada, Alaska, and elsewhere vast deposits of gold, silver, copper, lead, and a multitude of metals, some precious and other base though valuable, have presented enormous wealth to our country. Our great subterranean lakes of oil have made possible the expansion of the gas-engine and the automobile to their present state of efficiency. For centuries the banks of Newfoundland have filled our markets with fish. To the forests of Maine, Michigan, Wisconsin, and Oregon we owe a debt for timber which can never be repaid. From Alaska come sealskins, fertilizer from the phosphatic accumulations of the Carolinas and Florida, wild hay from the prairies, and so on through the mile-long list. The resources of America are immeasurable.
>
> But, while we prick up our ears upon being informed that the fishery produces so many tons of fish worth so much, or that from the oyster-bed may annually be taken ten million oysters, or that so-and-so's manganese-mine accounted for a hundred car-loads of ore last year, we show little interest when we are told that a sparrow hawk captures ten score field-mice a year and innumerable grasshoppers. Yet these very sparrow hawks save the American farmer considerably more than the combined worth of the fishery, the oyster-bed, and the manganese-mine together.

Leaving aside the realization of how much times have changed, or the somewhat tenuous nature of his calculations, it's clear that Hartley understood the economic value of wild nature, as did Thoreau and Leopold and many others. Why do we consider this idea something new? Partly because several books, the emergence of a new discipline of ecological economics, a report from the United States National Research Council, and the Millennium Ecosystem Assessment have given the idea visibility, quantitative substance, momentum, and coherence that it previously lacked. But there's more to it than that. Initial coarse calculations of the value of ecosystem services (Hartley 1922) have given way to more detailed economic analyses. A recent assessment by Anielski and Wilson (2005), for example, concluded that the annual value of Canada's boreal ecosystems in 2002 was on the order of CAN$131 billion, of which the noneconomic values are roughly 2.5 times the economic values.

As the astounding value of ecosystem services becomes more apparent (Hartley, no economist, thought the cash value of bird life to humanity was "staggering in its magnitude"), dreams of new markets and payments for ecosystem services seem much less fanciful and are mobilizing action in both the financial and conservation sectors. Conservation is growing beyond nature protection to encompass the places where people live and work. These are the places where the values of ecosystem services are especially relevant. The links between ecosystem services, conservation, and human poverty are now made explicit. So, while the idea of ecosystem services may not be new, it is being applied in new ways that are broadening the scope and relevance of conservation.

Sometimes, indeed, some buffing and polishing can give new life and vigor to old ideas. Perhaps the early ideas of Hartley and others didn't have a broad influence because the time just wasn't right, or because the ideas lacked some of the features that give ideas impetus, that make them stick (Heath & Heath 2007). In our rush to embrace ecosystem services as a central theme of conservation, let's not forget that people have been talking about the value of nature for quite some time. We might do well to pursue this "new" idea with a bit less hubris and a bit more humility, albeit with no less enthusiasm.

PART IV
Doing conservation

In the previous parts, I've considered how ecology and conservation are challenged by a changing world, which directly or indirectly has led to various confounding challenges, wicked problems, and conundrums. To move ahead in dealing with the accelerating erosion of habitats and loss of biodiversity, conservation and ecology must jointly confront these challenges. There are many approaches to doing conservation, but I'd like to think that, ultimately, they all depend on good science—the information, analyses, and knowledge that increase the likelihood that our efforts will be successful.

This part includes six chapters. In the first, I consider the issue of what conservation should focus on—species, ecosystems, or something else. This leads into comments on several issues related to data—the numbers on which conservation science is based. However, numbers by themselves don't tell a story, even after they have been crunched and interpreted. What the numbers tell us must be communicated to the wide array of people on whose shoulders the future of conservation rests. This is the focus of the third chapter. In the fourth chapter, I consider several debates that have emerged as conservation has grappled with how to deal with the mounting threat of biodiversity loss. At their core, these debates hinge on differing philosophies and ways of valuing nature, which I touch on in the fifth chapter. Finally, because conservation occurs as much in a social and political as in an ecological or environmental setting, the issue of scientific advocacy (or, more precisely, scientists as advocates) rears its head in the sixth chapter.

Ecological Challenges and Conservation Conundrums: Essays and Reflections for a Changing World, First Edition. John A. Wiens.

CHAPTER 20

What is it we are trying to conserve?

Conservation means different things to different people, who have different views about what is important about nature that should be conserved. The most basic distinction is between the utilitarian and preservationist views I discussed in Chapter 1. The former emphasizes conservation of ecosystem services and its benefits to people, while the latter stresses the intrinsic values of Earth's creatures—biodiversity. I considered ecosystem services in the previous chapter, but what about biodiversity?

Although naturalists and ecologists have long recognized the diversity of various groups of organisms, it was only in the mid-1980s that the term "biodiversity" came into wide use.[1] Now the word rolls effortlessly off the tongues of everyone from Nobel laureates to schoolchildren, to mean anything having to do with nature and the environment. Yet ecologists and conservationists, wanting greater precision, have grappled with what it means. E.O. Wilson offered this definition a decade after the initial birth of the term:

> *Biodiversity is all hereditarily based variation at all levels of organization, from the genes within a single local population or species, to the species composing all or part of a local community, and finally to the communities themselves that compose the living parts of the multifarious ecosystems of the world.*[2]

In other words, pretty much everything about the living world.[3] "Biodiversity" therefore incorporates many things, any or all of which can be a focus of conservation.

[1]Prompted by the National Forum on BioDiversity in Washington, DC, in 1986 and the publication of papers from the Forum in *BioDiversity* (Wilson 1988).
[2]Wilson (1997: 1). This and the *BioDiversity* volume provide a broad sampling of the diversity of things that fall under the umbrella of "biodiversity." See also Norton (2006), who discusses the multiple meanings of the term.
[3]One could quibble, however, about the "hereditarily based variation" part of Wilson's definition. This excludes nonheritable behavioral variation, which (as any ethologist or behavioral ecologist would be quick to point out) can play an important role in enabling individuals to adjust to environmental changes and exploit new ecological opportunities. Think of culturally transmitted behaviors in great ape societies (McGrew et al. 1996) or the spread among blue tits in England of the behavior of opening caps on milk bottles delivered to doorsteps in order to get the cream on top (Aplin et al. 2013).

Ecological Challenges and Conservation Conundrums: Essays and Reflections for a Changing World, First Edition. John A. Wiens.

Most often, however, conservation efforts are directed at some level of biological organization. The four that Wilson mentions—genes, species, communities, and ecosystems—cover a range of possibilities. At the finest level, conservationists often emphasize the importance of maintaining genetic diversity, both within and among populations. Decades of research in population biology and genetics have told us that populations with too little genetic diversity are subject to inbreeding depression, which can allow deleterious genes to be expressed and reduce the capacity of a population to cope adaptively with changing environmental conditions. Reduced genetic diversity is generally associated with small population size, which typifies most endangered species. Even if a species recovers numerically from a small population, it is likely to bear the consequences of passing through a genetic bottleneck. This is what happened to northern elephant seals, whose population hit a low of perhaps 20 in the 1890s, or the cheetah, which experienced a severe bottleneck at the end of the ice age; both continue to have low genetic diversity. Maintaining or enhancing genetic diversity is also a concern of captive-breeding programs, which usually begin with just a few individuals.

Usually, species rather than genes are the units of conservation action. Of all the levels of organization, species seem to have the greatest tangibility and reality. Bird watchers and butterfly collectors deal with species all the time. Yet, gather a group of systematists, taxonomists, and cladists together and the debates will begin. Each will answer the question "What is a species?" in a different way, from different philosophical and methodological perspectives. This is not the place to review the issues,[4] but, because it is so central to conservation, it is important to see how "species" plays out in practice.

One place to look is in the enabling legislation for the U.S. Endangered Species Act. The Act, as amended in 1978, defined "species" to include "any subspecies of fish or wildlife or plants, and any distinct population segment of any species or vertebrate fish or wildlife which interbreeds when mature."[5] Thus, although many full species are listed as threatened or endangered, others are partitioned into subunits, only some of which may fall under the Act's protection.[6] Of the three recognized subspecies of spotted owl, for example, two (the northern spotted owl and Mexican spotted owl) are listed as threatened, while the other (the California spotted owl) is not.[7]

[4]See Williams and Knapp (2010).
[5]ESA Sec. 3(15).
[6]Ruckleshaus and Darm (2006) explore some of the issues related to listable units under the Endangered Species Act.
[7]Some years ago, S. Dillon Ripley, then Secretary of the Smithsonian Institution, argued against such partitioning, noting that "to assume that an endangered sub-species is of equivalent value to an endangered species is a reductio ad absurdum. What really is important is the species unit, and not the sub-species unit" (Ripley 1981: 2).

Subspecies are taxonomically defined units, so they are subject to the classification practices that are in vogue for a particular group or a particular time. A distinct population segment, in contrast, is a demographic and geographic unit, which should have greater biological meaning. However, faced with the difficulty of applying the distinct population segment criterion to species of Pacific salmon that exhibit an array of life histories and migratory pathways, the National Marine Fisheries Service adopted "evolutionarily significant unit" to recognize both the degree of isolation and the evolutionary history and potential of a population segment. Using this categorization, the seven species of Pacific salmon include 58 evolutionarily significant units, many but not all of which are listed under the Endangered Species Act.[8]

Reducing the targets for protection or conservation to such fine categories presents an enigma.[9] Detailed knowledge of life history and ecology is needed to determine what is a distinct population segment or an evolutionarily significant unit. When such knowledge is not available, as is often the case, using such categorizations may be difficult to justify, opening the door to legal challenges. In such situations, focusing on taxonomically defined units such as species or subspecies may make more sense. Doing so, however, may ignore the fundamental ecological differences among groups within a species that would require different conservation and management approaches. Management practices that are appropriate for winter-run Chinook salmon in the Sacramento River, for example, may not be appropriate for spring/summer-run Chinook salmon in the Snake River.

Although the point is debated, there seems little question that the Endangered Species Act (and similar laws elsewhere) has been successful in forestalling the extinction of species or populations that otherwise would have disappeared. Declines of many species have been halted, numbers of some species have built up to recovery levels, and a few have been removed from the list.[10] It would be a mistake, however, to become overly preoccupied with the definitions and dictates of the Endangered Species Act. In fact, by listing species, the Act may have the pernicious consequence of denying protection to other at-risk species.[11] Some of these species may eventually be listed, but others probably not. More to the point,

[8]Waples (2006).

[9]The enigma is further deepened by advances in genetic technology, which have revealed the details of genetic variation within and among species. This has led to debate about such things as subspecies (Patten 2015), which bears on whether such units merit listing.

[10]I comment on what can happen if conservation efforts under the Act are *too* successful in Essay 26, *Be careful what you wish for (2014)* (page 222).

[11]Compilations of "candidate species" under the Act or, less formally, species of special concern (e.g., Shuford and Gardali 2008) recognize a second tier of species meriting conservation attention. All of these approaches lead to prioritizations of species for conservation and management attention and funding: of the top 25 endangered or threatened species receiving government funding in 2013, 14 were populations of salmon species; six of these populations were different evolutionarily significant units of Chinook salmon. Collectively, these 14 salmon units received almost 35% of the funding for all 1,304 listed units.

species are listed when their distribution and abundance diminish to a point that raises concerns about imminent extinction. Distributions and abundances wax and wane, however, and some species that are common now may decline in the future. Better to think about conserving them now rather than later, even though the Act does not apply.

If there is ambiguity, uncertainty, and debate about what it means to conserve a "species," defining what it means to protect a "community" (the third element of Wilson's definition of biodiversity) is even more problematic. To an ecologist, a community is an assemblage of species living in the same place that actually or potentially interact with one another.[12] Apart from the issue of scale (what is "the same place"?), communities tend to be restrictive (e.g., just birds) or inclusive (e.g., all the plants) depending on the interests of who is talking about them. Unlike species, which have a certain degree of reality as taxonomic units, communities do not exist as discrete entities, although plant communities are often categorized as such.[13] Nonetheless, anyone can see that multiple species do occur together, even though their interactions are not always obvious and the specific associations may change in time and space. Focusing on single species, even ones in danger of extinction, may leave out potentially important linkages with other species—predator–prey interactions, pollination, competition, and the like.[14] It may also fail to realize the economies of scale that derive from protecting or managing multiple species at the same time and place. When people talk about conserving communities, however, it is more often in the context of protecting habitats, such as an oak-hickory forest (a community type), or places of extraordinary species richness, such as hotspots.

Despite their ambiguity and subjectivity, communities still have species as their building blocks. The focus shifts from species to processes, however, when the emphasis is upon ecosystems (the fourth of Wilson's levels). This brings a fundamentally different perspective to conservation. The things to be conserved are food chains and trophic webs; the processing of solar energy by plants; the storage and movement of nitrogen, phosphorus, and other chemicals and nutrients through the environment; the sequestering of carbon; the processing of detritus by soil microbes; the uptake and release of water by plants; and so on—in short, the things that make nature go.

---

[12]Robert MacArthur's definition of an ecological community as "any set of organisms currently living near each other and about which it is interesting to talk" (1971: 190) may more accurately capture how communities are viewed among practicing ecologists.

[13]I consider the reality of bird communities in Essay 27, *Concluding comments: are bird communities real? (1980)* (page 227).

[14]In their book, *The Ecological Web*, H.G. Andrewartha and Charles Birch (1984) developed an approach to single-species ecological analysis based on what they called an *envirogram*—a way of organizing and depicting the hierarchical web of factors that influence a species directly or indirectly. Unfortunately, the approach has not received the attention it deserves.

All of these things happen, in real time in real places involving real organisms. The concept of an all-encompassing ecosystem, however, is itself an abstraction. Conservationists talk of preserving an "old-growth ecosystem" or a "tallgrass prairie ecosystem." This is shorthand for protecting places that include the plants, animals, and processes of an old-growth forest or a tallgrass prairie. When they talk of ecosystem services, it is usually in terms of particular processes (e.g., water filtration) or species (e.g., commercial fish stocks) that should be conserved, rather than the ecosystems of which they are parts. Conserving "functioning ecosystems" or "healthy ecosystems"[15] may be the stated goal, but the actual conservation objectives are generally more specific.

In the end, there are plenty of good arguments for why any of these levels of biodiversity—genes, species (or their subdivisions), communities, or ecosystems—is an appropriate target of conservation efforts. It would be good if we could leave debates about definitions to lawyers and recognize that biodiversity includes all of "the living parts of the multifarious ecosystems of the world," none of which is inherently better than the others. Because conservation can have multiple targets that entail different approaches, it is important to be clear about which target one is talking about. The different levels of biodiversity interact with each other, so it's also important to consider how other targets and levels may affect the target of interest. This isn't easy; ecologists have been grappling with such webs of interactions for a century; conservationists must now do so too.

[15] I comment on the issue of attaching the label "health" to ecosystems in Essay 28, *A metaphor meets an abstraction: the issue of "healthy ecosystems" (2015)* (page 230).

ESSAY 26

Be careful what you wish for (2014)*

Unintended consequences—outcomes of actions other than those intended—are prevalent enough that sociologists and economists often refer to the "law of unintended consequences." Actions of people (or governments) always have effects that are unanticipated or unintended. Although some consequences may be beneficial, detrimental or undesired outcomes receive the most attention. Draining of wetlands to create agricultural fields often increases the risk of flooding; suppression of forest fires leads to fuel buildup and an increased risk of mega-fires; damming of rivers to enhance water storage and recreation alters downstream flows, blocking fish movements and restricting sediment renewal of floodplain farms; the list goes on and on. One could argue (as developers often do) that the Endangered Species Act, the Clean Water Act, and similar environmental laws and regulations have had the unintended consequences of stifling economic growth, costing jobs, or constraining innovation.

All of which brings me to the point of this essay. Conservationists often think of their actions as proceeding linearly to the desired outcomes. Protected areas will protect what they were intended to protect, restored ecosystems will function naturally, and endangered species that are nursed back to health ("recovered") will be there for us to cherish and enjoy. But it doesn't always work out that way. Here, I recounted two examples of recovery that were too successful, creating unintended problems (at least for some people) instead of unambiguous benefits. The underlying message is that conservation benefits or costs are determined socially and culturally rather than scientifically.

Can conservation ever be too successful? The ongoing loss of habitats, the growing lists of imperiled species, and the appeals of conservation organizations suggest not. But what if efforts to bring species back from the brink of extinction are so successful that they create conflicts with people and their interests? Let me tell two stories to illustrate this conundrum.

The first story is about geese. Aleutian cackling geese (*Branta hutchinsii leucopareia*) were once abundant, breeding throughout the Aleutian Archipelago and wintering in the Pacific Northwest (where they were first described by the Lewis

*Wiens, J.A. 2014. Be careful what you wish for. *Bulletin of the British Ecological Society* 45(1): 53–54. Reproduced with permission of the British Ecological Society.

Ecological Challenges and Conservation Conundrums: Essays and Reflections for a Changing World, First Edition. John A. Wiens.

and Clark Expedition). During the 18th and 19th centuries, fur traders introduced Arctic foxes (*Vulpes lagopus*) on many islands in the Aleutians. Geese and their eggs and goslings were easy prey, and numbers plummeted. By the middle of the 20th century the goose was thought to be extinct. A breeding population of a few hundred birds was rediscovered on a remote island in 1962, however, and the species was listed under a precursor to the U.S. Endangered Species Act in 1967. A Recovery Plan was drafted: foxes were removed from potential breeding islands, birds were reintroduced as islands became fox-free, hunting was curtailed, and habitat in the wintering and migration areas was protected and managed. The population exploded, and the recovery goal of 7,500 birds was quickly exceeded. The protection and restrictions afforded by the Endangered Species Act were no longer necessary, and the species was "delisted" in 2001. A resounding conservation success.

But goose numbers have continued to increase. By 2011, the population was estimated at nearly 112,000. Well before that, it became apparent that grazing by the thousands of geese gathering at spring migratory stopover areas in California and Oregon was damaging newly emerging pasture and crop vegetation (Figure 1). Birds roosting overnight on offshore islands were degrading habitat in seabird breeding colonies. To protect their lands, landowners began hazing birds to drive them from their fields. An Agricultural Depredation Plan was prepared to address the goose problem.[1] There is now a hunting season; California hunters are

Figure 1 Aleutian cackling geese grazing in a managed pasture, Humboldt Bay, California. Photograph by Ron LeValley. Reproduced with permission.

[1] Mini and LeValley (2006).

permitted to take up to six geese per day over a 100-day season. These measures have reduced pressures on private lands by shifting the geese onto nearby public lands. The current objective is to maintain a population of 60,000 birds, but even with control and hunting, reducing the population to this level will be difficult. A species once thought extinct and then struggling to survive has, in the space of 50 years, become an agricultural pest. A Recovery Plan has been replaced by a Depredation Plan.[2]

Recovery of the Aleutian cackling goose surpassed all expectations. For most of the 1500 species listed under the Act, however, the bar for what counts as "success" is set pretty low. Fewer than 1% of listed species have gone extinct since the Act was passed 40 years ago, and declining trends of others have been stabilized or reversed. To some, this counts as success. But the aim of the Act is not just to avoid extinction, but to enable species to recover so that they no longer require extra legal protection. Some species, such as brown pelicans (*Pelecanus occidentalis*) or peregrine falcons (*Falco peregrinus*), have exceeded recovery goals and have been delisted. Pelicans are now a fixture on the Gulf of Mexico and California coasts, and peregrines have expanded their habitat to nest on skyscraper ledges in many cities. These successes give conservationists hope and justify continuing support for recovery efforts for other species on the cusp of extinction.

My second story is about wolves. Once upon a time (isn't that how all wolf stories begin?), gray wolves (*Canis lupus*) were widespread across North America (Figure 2). As settlement moved westward, wolves were forced out or killed, initially because they were a threat to livestock and later because they competed with hunters (fewer wolves meant more big game). Bounties were paid for killing wolves. In 1902, Theodore Roosevelt called the wolf "the beast of waste and destruction,"[3] and in 1907 the United States Biological Survey declared the extermination of the wolf to be "the paramount objective of the government." By the 1950s, wolves had been eradicated from the United States, although they remained abundant in Canada and Alaska. Wolves were listed under the Endangered Species Act in 1973, leading to lengthy and contentious debates about recovery planning. Finally, in 1995 wolves were reintroduced into Yellowstone National Park and remote areas of Idaho. Numbers grew dramatically, and dispersal established new wolf packs in other areas. Initial proposals to delist the wolf in Idaho and Montana were overruled by a federal court, whereupon the U.S. Congress, as part of an unrelated budget authorization bill, interceded to remove the Act's protection in these states. Last year, the U.S. Fish and Wildlife Service proposed delisting the wolf in most of the United States and Mexico.[4] Another conservation success story.

---

[2]Mini et al. (2011) provide a useful review of Aleutian cackling goose recovery and management.

[3]Roosevelt (1902).

[4]For a perspective, see Buskotter et al. (2013).

Figure 2 Gray wolf. Photograph by Gary Kramer, U.S. Fish and Wildlife Service.

But consider what has followed. Reprising debates from the previous century, ranchers have protested about increasing losses of livestock to marauding wolves, and hunters have complained about reduced big game populations. The ecological argument that wolves act as keystone predators, voiced so eloquently by Aldo Leopold in his essay *Thinking Like a Mountain* (1949),[5] does not resonate with ranchers and hunters. Several states have now opened hunting seasons for wolves, including organized "wolf hunt" contests. Hundreds of wolves have been killed. Idaho has hired a professional hunter to eliminate two wolf packs from a wilderness area.[6] Some legislators in western states are now calling for the eradication of wolves. An Op-Ed in the *New York Times* (June 7, 2013) wondered "have we brought wolves back for the sole purpose of hunting them down?"

[5]For a recent treatment, see Callan et al. (2013).

[6]January 29, 2014: Reports indicate that the hunter was successful in killing all the wolves in the two packs. The purpose is to allow the elk population to grow, which will provide more big game for hunters but alter the "wilderness" ecosystem.

In both stories, legal protection and intense management efforts were successful in bringing a species back from imperilment, only to encounter economic, social, or political pressures to reduce or eliminate the gains. But there is an important difference between the stories. Geese eat grass and grain. Wolves eat cattle and sheep and elk. People have a deep-rooted fear of wolves (and of sharks and tigers and crocodiles—things that now and then eat people[7]). Childhood fables like *Little Red Riding Hood* or *The Three Little Pigs* instill a fear of wolves; *Mother Goose* does nothing of the sort for geese. Culture as much as science influences what counts for "success" in conservation.

None of this is to say that conservationists should be looking over their shoulders for the culture police. It does suggest, however, that it may be wise to think about the consequences of success and plan accordingly. We are usually so preoccupied with fighting against extinction that even modest gains are victories, and we don't look ahead to consider what might happen if we are too successful. In both stories, the outcomes might have been anticipated. Geese are prolific breeders and effective grazers, so removing the threat of predation would sooner or later lead to problems where large numbers of geese aggregate. We might have expected that the deep-rooted attitudes about wolves that led to their eradication in the last century would reappear as soon as wolf numbers increased.

Understanding the ecology of an imperiled species is essential in charting a course toward recovery, but understanding societal attitudes may be just as important once we get there. Determining what "success" means, and whether it is enduring, depends on much more than science.

---

[7]In his essay in this issue of the *Bulletin*, Richard Hobbs calls attention to a similar story unfolding in Australia, and to the distinctively Australian attitude about dangerous animals.

ESSAY 27

Concluding comments: Are bird communities real? (1980)*

The 1970s and 1980s were times of controversy and ferment in animal community ecology. The paradigm that had dominated the field for two decades was under attack. Many ecologists were no longer willing to assume that resources were always limiting and that the patterns of communities were consequences of niche partitioning driven by interspecific competition. Bird communities, which had provided the initial models for the competition paradigm, were also at the center of debates about the ubiquity of that paradigm.

It was against this backdrop that I organized a symposium in 1978 at the International Ornithological Congress in Berlin, in what was then the Federal Republic of Germany ("West Germany").[1] As the organizer, I gave the concluding remarks. I don't remember what I said, but when the symposium papers were published two years later I used the opportunity to address some issues about bird communities that had been gnawing at me. Such things as how time lags in the responses of species to environmental changes could complicate attempts to infer cause-effect relationships from short-term studies, or the importance of the scales in time or space on which communities were viewed. These concerns still ring true.

The more important point was the one raised in the title: are bird communities real? Do they exist as functional biological entities, or are they only constructs in the minds of ecologists, things "about which it is interesting to talk," as Robert MacArthur said? If bird communities, or, more broadly, communities of anything, are only something we have created so we can study, theorize, and talk about them, should they be a focus of either ecology or conservation?

(continued)

*Wiens, J.A. 1980. Concluding comments: Are bird communities real? *Acta 17th International Ornithological Congress*, Berlin. pp. 1088–1089. Reproduced with permission of Deutsche Ornithologen-Gesellschaft e. V. (DO-G).

[1] An unrelated personal story: While attending the Congress, a colleague from the German Democratic Republic ("East Germany") invited me to come over for a visit. Crossing the border to East Berlin at Checkpoint Charlie, I met him at the Museum für Naturkunde. After admiring several stunning dinosaurs, we descended to a basement room. There on a table, illuminated by a single incandescent bulb, lay the fossil sacred to ornithologists—the Berlin Archaeopteryx—undergoing restoration for display. I was allowed to touch it, and did so, reverently. Now, of course, we recognize Archaeopteryx as one of several transitions that have blurred the lines between theropod dinosaurs and birds, but then it was the original bird, dating back 150 million years. This was for me the most memorable part of the Congress.

Ecological Challenges and Conservation Conundrums: Essays and Reflections for a Changing World, First Edition. John A. Wiens.

(continued)

I didn't answer this question back then,[2] and I can't now. Observing communities (or assemblages, or whatever one wants to call them) is an appropriate way of moving from single-species to multi-species analyses and management, and they continue to be an active area of investigation in ecology. Taxonomically defining communities (e.g., "bird communities" or "damsel fish communities") may also make sense if one is interested in the ecology or evolution of niche partitioning among closely related species. Yet the studies by Jim Brown and his colleagues in the deserts of Arizona that I mentioned in this essay illustrate the perils of arbitrarily truncating functional webs of interrelationships among unrelated taxa.[3] If "communities" are to be a target of conservation efforts, they should exist in something other than our minds and publications.

The papers presented in this symposium profile the diversity that characterizes contemporary studies of avian communities. The focus of these studies ranges from local populations to global overviews, and the conceptual foundation is energetics in some cases, competition in others, overwhelming variability in still others. These divergences reflect differences in the level of approach, in the systems studied, and, perhaps most importantly, differences in the initial conception of what is important about the system, what merits our attention. But it is becoming apparent that the pictures of bird communities that we have created are not as clearly defined as we first thought. Several factors are contributing to the increasing fuzziness of our conception, and these must become new foci of future community studies:

1 The responses of birds to variations in their environments are clouded by various time lags, many of them associated with the remarkable learning capabilities of birds. Site tenacity, for example, may contribute to stability in populations over several years in the face of changing habitat conditions, and learning of localized feeding customs may produce unique patterns of resource utilization that differ among local populations of a species quite independently of their community context.
2 The community patterns that we seek to define vary over a wide array of temporal scales, such as tidal cycles, breeding/non-breeding seasons, years, decades, and so on. How do we select the appropriate scale in this spectrum on which to gauge bird community organization?
3 In a like manner, we may consider bird communities over a variety of spatial scales, from individual territories and space-use foci through regional and continental avifaunas. Which of these scales is proper for seeking understanding of communities?
4 Because we are uncertain about the influences of time lags and the appropriateness of the temporal and spatial scales on which we view communities, we

[2]I've always been better at asking questions than answering them.
[3]Brown et al. (1979).

generally can only draw crude inferences about whether the communities are in equilibrium or not. Do biotic interactions produce the patterns that we think are real, or are the organisms decoupled from direct biotic interactions such as competition, and perhaps responsive instead to various abiotic influences? We have generally presumed the former, but that does not make it correct.

5 Finally, we must consider the prospect that bird communities may not really exist at all, in any functional biological sense. While birds are unique creatures in a variety of respects, there may be few compelling reasons to believe that their assemblages are structured independently of those of consumers belonging to other taxa that use the same resources. Workers in southwestern United States deserts, for example, have evidence that suggests that birds, harvester ants, and heteromyid rodents may collectively interact in relation to the seed resources they utilize, and that consideration of any one of these taxonomic groups by itself would lead to incomplete conclusions. We do not know how widespread such intertaxa interactions may be, but by constraining our studies of bird communities over the past decades with such a strong taxonomic bias, we may well have provided a perception of community organization that bears a closer resemblance to science fiction than science fact. MacArthur's definition of a community as any set of organisms living near each other and about which it is interesting to talk may have provided a comfortable rationale for avian community studies, but it has little inherent biological meaning. We rarely know (or seem to care) whether there are any sorts of natural boundaries about the communities we study, whether there are any discontinuities in biological processes that might act to define a functionally interrelated assemblage of organisms in which we might really expect patterns or organization to have some adaptive significance.

What, then, must we do? I submit that we must turn our attention to more intensive, long-term studies of defined local populations and environments, in relation to defined (and measured) resource bases. If assemblages of organisms are not just haphazard, they must result from the operation of biotic processes, and these are expressed through individuals and populations. Only by first looking there, and then by attempting to tease apart the real from the spurious processes with manipulative field experiments, may we determine whether bird communities are biologically real, and if so whether there are any "rules" governing their structure.

ESSAY 28

A metaphor meets an abstraction: the issue of "Healthy Ecosystems" (2015)*,[1]

> *For some time I've been uncomfortable with the notion of "healthy ecosystems" or "ecosystem health." The intent is not hard to grasp: ecosystems that are structurally complete and functionally sound are "healthy," in the same sense that a person who is physically and functionally sound is healthy. The metaphor is effective, and it has been useful in drawing attention to the plight of ecological systems that are missing key parts, such as top predators in food webs, that throw them out of balance.*
>
> *Part of the problem, which perhaps matters only to a scientist, is that "health" of an ecosystem is not something that can be objectively measured—what is healthy depends on the values of whoever is doing the diagnosis. Part of the problem is also that nagging desire for balance that seems to pervade so much of conservation and ecology. Something that is healthy is functioning properly, and by that we generally mean that it is maintaining itself within some relatively narrow bounds. Nature out of balance is usually regarded as sick or diseased in some way—unhealthy.*
>
> *My inner scientist voice tells me that the term and concept are vacuous and useless and should be banished from scientific discourse. However, there's another side, the voice that carries science out beyond the ivory tower to the broader community of people who care about nature and the environment. The term means something to them.*

A Metaphor and an Abstraction walk into a bar. "I'm not feeling so good," says the metaphor. "I feel like a ship adrift in the night."

"That would make you a simile, not a metaphor," says the Abstraction. "But go ahead."

"That's the problem. People think I'm one thing, but I'm actually something else."

*Wiens, J.A. 2015. A metaphor meets an abstraction: the issue of ecosystem health. *Bulletin of the British Ecological Society* 46(2): 44–45. Reproduced with permission of the British Ecological Society.

[1]Thanks to Jerry Franklin, Susan Salafsky, and Tom Spies for comments and photographs.

Ecological Challenges and Conservation Conundrums: Essays and Reflections for a Changing World, First Edition. John A. Wiens.

"Well, how do you think I feel," says the Abstraction. "I don't really exist, except in people's minds."

They pause, staring thoughtfully into their beers. "So, instead of crying on each other's (metaphorical) shoulders, perhaps we can make a go of it if we get together," says the Metaphor.

And so they did. Which is why we talk of "healthy ecosystems" or "ecosystem health."

Although "ecosystem" has a clear meaning to ecologists and much of the public,[2] it is really an abstraction—a conceptual tool that helps to focus thinking, orient research and management, and gather phenomena together so that their interrelationships can be examined and understood. Ecosystems do not actually exist as discrete, bounded entities; they are not "things."

Some readers may dispute my characterization of "ecosystem" as an abstraction. After all, many ecologists study ecosystems, and there is a well-established journal of that name. Instead of debating this point, however, I'd like to consider the "health" part of the expression. "Health" is a metaphor applied to ecosystems to indicate that they meet some standard of well being. By this line of thought, a healthy ecosystem is one that is functioning well, contains all its essential parts, and does not show obvious signs of deterioration or stress. The health metaphor has been used for decades—Aldo Leopold considered land health, "the capacity of the land for self-renewal," to be central to his land ethic.[3] But talk of ecosystem health has now become mainstream,[4] so it is important to understand what it means and what its limits are.

The power of the "health" or "healthy" metaphor is in its allusion to human health, something we all appreciate. We can easily recognize that a lake experiencing an algal bloom or a waterway choked with invasive water hyacinth (*Eichhornia crassipes*) has lost some of its natural functionality—it is no longer a healthy ecosystem. The allure of the healthy ecosystem metaphor is so great, in fact, that it has been enshrined as a goal in laws and policies. Environmental conservation law in New York, for example, establishes the policy "to conserve, maintain and restore coastal ecosystems so that they are healthy, productive and resilient and

[2]To an ecologist, "ecosystem" generally describes a complex system of interactions among biological, physical, and chemical components of the environment; to the general public, "ecosystem" conveys an image of a complex network of interactions that affect something of interest, ranging from the web of life to the elements involved in creating a product and bringing it to market (the "business ecosystem").

[3]Leopold (1949: 221).

[4]For example, the Aquatic Ecosystem Health and Management Society was founded in 1989, and in 2015 the Ecological Society of America and the Ecological Society of China initiated a new online journal, *Ecosystem Health and Sustainability*.

able to deliver the resources people want and need."[5] The agencies administering the Endangered Species Act aim to "promote healthy ecosystems."[6] And the "Healthy Forests Restoration Act of 2003"[7] is intended to reduce the risk of large, catastrophic fires that would degrade forest health by redressing the historic suppression of fires that allowed hazardous fuels to accumulate. Helping damaged or dysfunctional ecosystems regain and maintain their health has become the goal of a good deal of resource management, conservation, and ecological restoration.

It's hard to argue with the common-sense appeal of healthy ecosystems. If we put on a (metaphorical) science hat, however, we must ask, "What is 'health,' and how shall we know it?" There are well-established standards for what makes a person healthy—normal body temperature (measured orally) is 37 °C (98.6 °F); hypothermia sets in at 35 °C (95 °F) and fever at 38 °C (100.4 °F). When I last had a blood test, the results specified ranges of "healthy" values for some 27 measures—the standard range for alkaline phosphatase, for example, is 40–129 U/L. Mine was 83 U/L, so I guess I'm healthy, at least by this measure.

But there's the rub. An individual who is healthy by one measure may be near death by another measure. "Health" is an aggregate property of numerous attributes of an individual. And so it is with ecosystems, except that the measures and the standards are ambiguous and ill-defined.

Consider a forest. To a private forest manager or silviculturist engaged in production forestry, a healthy forest may be one that is thinned to reduce fuel loads, is not physiologically stressed or susceptible to beetle outbreaks or disease, meets the needs of people for products and services, and is easily harvested. To a forest ecologist, a healthy forest would instead include the full array of ecological processes and structural elements that would maintain a diverse and productive ecosystem—heterogeneity rather than homogeneity.[8] More generally, words such as "sustainability," "resilience," "stable," "productive," or "services" figure importantly in assessments of ecosystem health.[9] Conversely, an unhealthy ecosystem might be recognized by signs of stress: erratic population fluctuations, absence of top predators, increased dominance by exotic species, simplification of food webs, declining production and functionality, and so on.

Like people, ecosystems are regarded as "healthy" if they are structurally and functionally "normal." With people, there are well-established benchmarks for normal health, be it body temperature, alkaline phosphatase, or something else. Not so for ecosystems. J. Baird Callicott offered that the "objective condition" of ecosystem health is the "normal occurrence of ecological processes and functions"; that is, "as they have occurred historically."[10] The choice of a historical baseline,

[5]New York Ocean and Great Lakes Ecosystem Conservation Act, Section 14-0103(2).
[6]The Federal Register for Friday, July 1, 1994 (Vol. 59), p. 34274.
[7]Public Law 108–148.
[8]Kolb et al. (1994).
[9]Costanza and Mageau (1999).
[10]Callicott (1995: 345, 348).

or even whether one uses historical conditions at all, is therefore critical. Environments vary, so using different points or windows of historical variation to specify a baseline leads to different designations of what is "normal." To some ecologists and conservationists, for example, the closer a system is to being unaltered by human actions, the healthier it is. A forester or rancher would probably have a different view of what is healthy.

The difficulty of deciding what span of history should be considered "normal" is confounded by thresholds. When a system crosses a threshold, it may enter a new, alternative state. Should its health be assessed in relation to what was previously normal or the "new normal" of the alternative state? And what are the standards for "health" when the environment undergoes rapid, directional changes, as it is doing now? When shifts in species' distributions create novel ecosystems, are these by definition unhealthy because they depart from the historical "normal"?

Such questions strain the appropriateness of the health metaphor for ecosystems. More bothersome, at least to a scientist, is the issue of values. "Health" is a value-laden term, and values are socially, not scientifically, determined. Science deals with determining whether things are true or false (or somewhere in between), but "health" is judged as good or bad, healthy or unhealthy—value judgments. This leads to a conundrum in which a healthy ecosystem is what one chooses it to be, depending on one's values—you'll know it when you see it. A pine plantation may be healthy from the perspective of a production forester or a silviculturalist but may be barren to an ornithologist. Moreover, what is regarded as "healthy" or "unhealthy" from a given perspective may vary depending on the environmental context. To a forest ecologist or wildlife biologist, a healthy coniferous forest in the Pacific Northwest presents a visual image of structural complexity while an unhealthy forest has a simple structure and low diversity. In the drier climate of the Southwest, the positions are reversed: a healthy ponderosa pine forest is open and the unhealthy forest clogged with trees and underbrush (see Figures 1–4). Someone in a forest industry interested in timber production would probably evaluate the health of the forests in a different way.

Differing perceptions of what is "healthy" can lead to miscommunication and conflicts. And when ecosystem health appears as a goal in laws or agency policies, societal values may carry over to twist the debate and affect how scientists pose questions or design research. The science itself may become value-laden or normative—the assumption that some conditions are better than others that is encapsulated in policy infects the science.[11] Values can masquerade as science, especially when the science encounters societal, political, or economic pressures and agendas.

So, is the concept of ecosystem health useful? Scientifically, it is distracting and unnecessary. Rather than attempt to conserve or manage an abstraction for a metaphorical goal, wouldn't it be better to state what is to be managed to

[11] Lackey (2001, 2007) provides especially cogent discourses on ecosystem health in the context of values and policy.

Figure 1 An old-growth Douglas-fir–western hemlock stand with complex structure and high stand diversity; Gifford Pinchot National Forest, Washington. Photograph by Jerry F. Franklin.

Figure 2 A thinned Douglas-fir plantation with simple structure and low stand diversity; H.J. Andrews Experimental Forest, Oregon. Photograph by Jerry F. Franklin.

Figure 3 A mature ponderosa pine stand on the Kaibab Plateau, Arizona. Fire in this forest will cause little or no mortality in the old pines. Photograph by Susan Salafsky.

Figure 4 A ponderosa pine forest that has become dense and loaded with fuel due to an absence of fire over the last century. Old pines are dying due to competition; if a fire occurs, it will be a stand-replacement fire that will kill everything, including the old pines. In trust lands managed by Washington State Department of Natural Resources near Ellensberg, Washington. Photograph by Jerry F. Franklin.

achieve particular results in specific, quantifiable terms? Ecosystem ecologists, for example, talk of ecological stoichiometry, trophic cascades, net carbon balance, or above-ground net primary production. All of these are components of what one might term "health," all of them are measurable, and none carries with it normative assumptions about what is "best." Yet mention of them would dull the senses of anyone other than an ecologist.

The usefulness of "ecosystem health" is in conceptualizing and communicating something about the state of natural systems, particularly to non-scientists. The abstraction—ecosystem—draws us to think about processes and linkages among units other than individuals or species, making it more likely that we will see the mechanisms and interactions. The metaphor—health—draws our attention to the condition of a system and the need for conservation or management. "Ecosystem" is a useful abstraction for things that actually do exist and that can be measured and managed, and "health" is a useful metaphor for conditions that we deem desirable, whether for utilitarian or idealistic reasons.

CHAPTER 21

Numbers, numbers, numbers

Conservation that is based on science entails measurements and data—science is a numbers game. Numbers about almost any aspect of any target—be it genes, species, communities, ecosystems, or something else—can inform and improve conservation. Ecologists and field biologists have gathered numbers about all sorts of things for over a century, massaging them in various ways to reveal the patterns of nature.[1] But the variety of numbers that can be gathered is now overwhelming and the toolkit of methods to analyze them is bursting. The number of numbers threatens to drown us in a deluge of data. There's much too much to consider here, so I'll only comment on two aspects of this data deluge: monitoring and the recent emergence of "Big Data."

Monitoring provides the numbers that tell us how things change through time. Because numbers have been gathered repeatedly, we know, for example, that the abundance of grassland birds in North America has sharply declined over past decades, that invasive cane toads have spread over northern Australia, that the distributions of some species are shifting northward in Europe, and that some plants are blooming earlier in the spring. Monitoring numbers for CO_2 levels atop Mauna Loa in Hawai'i (Figure 5.3) first alerted many people to the buildup of greenhouse gases and climate change. None of these things would have been detected from numbers derived at a single time.

Despite all of this, monitoring is not high on the "to-do" list of most scientists. Because it involves repeated data collection, monitoring is often expensive. It does not generate the quick, exciting results that lead to the publications necessary to keep a scientist's career going. Often there is so much variation among the numbers that it is difficult to detect anything interesting, such as trends or thresholds. And the repetition can be boring. The value of monitoring is often not recognized by scientists or rewarded by funding sources. I well remember from my time working in conservation organizations the difficulty of wringing money for monitoring from donors who asked, "You already have years of data, why do you need more? Shouldn't you know what's going on by now?" I consider this struggle in Essay 29, *Is "monitoring" a dirty word? (2009)* (page 241).

Image is part of the problem. When people think of monitoring, they may visualize someone going out to watch birds on a fine spring day or running through

[1]See Essay 5, *Patterns, paradigms, and preconceptions (2013)* (page 29).

Ecological Challenges and Conservation Conundrums: Essays and Reflections for a Changing World, First Edition. John A. Wiens.

a meadow with a butterfly net. Field biologists tell a different story, of the difficulties of counting frogs in a snake-infested swamp, of seeking the few remaining individuals of an endangered bird in the tangled undergrowth of a tropical forest, or of trying to see and count animals that would rather not be seen and are exceptionally good at it.

Monitoring programs sometimes also suffer from poor design. To provide reliable and relevant numbers *and* be cost-effective, monitoring must be designed for the task at hand. Obviously, the data collected should align with the key attributes of the conservation targets—*what* to monitor. This might include such things as the reproductive output of a listed population segment, species composition and abundances in a community, or nitrogen input into a forest stream. Decisions also need to be made about *where* to monitor. This also depends on objectives. If one's interest is in the effects of a development or environmental accident on a population or community and its subsequent dynamics, for example, sampling in appropriate impact and reference locations, perhaps using a before-after-control-impact (BACI) design,[2] might be appropriate. That much is (relatively) simple. More challenging is determining *when* to monitor—not just the time of year, but how often, and for how long? How often depends on the timing of the phenomena of interest and the questions being asked. Weekly monitoring is obviously inefficient for assessing the reproduction of bison but may be just what's needed to determine the onset of bud break or seed-set in a lupine. If sampling occurs too often it wastes effort and resources; sampling too infrequently increases the likelihood of missing important events. The art is in knowing just when to sample.

This, too, is (relatively) simple if one knows something about what is being assessed. It's more problematic to decide how long is long enough, particularly because there is often pressure to bring things to a close. Maintaining a stream of data, however, may provide unanticipated insights, especially when the environment varies. Prolonged drought, for example, can affect populations in ways that are not obvious unless nondrought conditions are available to provide context. Studies of Townsend's ground squirrels in Idaho shrubsteppe and northern goshawks in forests on the Kaibab Plateau in Arizona have documented shifts in habitat conditions favoring reproduction and/or survival between wet and dry years.[3] If the studies had included only the wet or dry periods, the relationships between climatic variation and demography would have been hidden, leading to different and perhaps ineffective management recommendations.

The longer the monitoring continues, the more likely that unexpected events will be recorded or trends will emerge from the noise of normal variability. The continuing record of annual reproductive output of Cassin's auklets on the Farallon Islands (recounted in Essay 17, *Tipping points in the balance of nature (2010)* (page 144)) began as part of what was expected to be a short study of the species' ecology. As the studies continued, the numbers continued to roll in, eventually showing

[2]Wiens and Parker (1995).
[3]Van Horne et al. (1997) and Salafsky (2015).

pulses that could be related to El Niño oceanic conditions. Only after 33 years did a complete reproductive failure pin down the relationship with prey abundance and distribution. As the monitoring continued, reproduction subsequently rebounded and it became apparent that what had seemed like a threshold was actually not. Unfortunately, such long-term monitoring is still infrequent, even though the benefits of long-term data are widely recognized.[4] It is simply too hard to keep it going. I consider some of the benefits in Essay 30, *The place of long-term studies in ornithology (1984)* (page 244).

Monitoring yields lots of numbers—data. As the ways of gathering data on more and more things have expanded and computational capacity has erupted, new concerns and challenges have emerged. The concerns relate to data quality and the fear that, with so many numbers being harvested and processed, some of the basics of data collection will be neglected and data quality will suffer. Numbers are only as good as the design and methods that are used to get them. Just because there are lots and lots of numbers doesn't mean that accuracy and precision are ensured. Standardization of sampling methods can help, and metadata[5] can provide information on the nuts and bolts of how particular data were collected. However, standardization and metadata don't reveal the underlying assumptions that guided the data collection or the nuances of field sampling, which can jeopardize the efforts taken to analyze or compare numbers from different studies in different systems.

The challenges are tied into the massive sets of data that have been generated on all sorts of things. This has led to the emergence of "Big Data," which I consider in Essay 31, *What use is Small Data in a Big Data world? (2013)* (page 247). Ably aided by the internet, data have been collected and stored in ways that make them accessible to exploration and probing in ways scarcely imagined a few years ago.[6] The mountains of data streaming from biotelemetry sensors or available in newly digitized museum records, for example, are producing new insights about such things as three-dimensional space use by threatened species, the locations of centers of global biodiversity, or the DNA composition of population segments.

The point of my essay was not that Big Data is bad for science (although there are some suggestions that tear at what we currently regard as the scientific process), but that intelligent exploration and probing of the masses of numbers requires

[4]See, for example, Solbé (2005) and Lindenmayer et al. (2012).

[5]Metadata are, literally, data about data: information accompanying a data set that describes the measurements, data-collection procedures, sampling design, and what else one needs to know to place a set of numbers in context.

[6]I learned from Wikipedia, for example, that in 2012, 2.5 exabytes (2.5×10^{18}) of data were created every day; by 2014, 2.3 zettabytes (2.3×10^{21}) of data were created every day. There's a rich lode of numbers just waiting to be mined, but it's way beyond anything I can comprehend.

that one know what one is looking for, or at least have the understanding to recognize meaningful patterns in the numbers and distinguish them from nonsense. This, I argued, is what "Small Data" has to offer—the insights that come from detailed observations of nature.

The data that come from well-designed and tightly targeted investigations are also needed to fill information gaps—the lacunae in the world of Big Data. These gaps become especially apparent when some project that might have adverse impacts on the environment is proposed. Opponents of the project often point to the gaps to support their arguments that the project is ill conceived or not based on science and that "more research" is needed (an argument that resonates with scientists anxious for research funding). Supporters of the project may reply that the lack of information shouldn't hold up the project, that adaptive management will address any uncertainties, and that the mantra of "just do it" should apply. The positions relate to whether, when faced with incomplete knowledge, precautions are needed to avoid possible mistakes or risks should be taken to move ahead anyway.[7] Big Data won't come to the rescue to solve this problem.

So what's the bottom line on numbers? To be science-based, conservation needs numbers. But not just any numbers. The relevant numbers should address some aspect of What?, Where?, When?, or Why?. "What" means that the targets of conservation attention and the important attributes of those units should be clearly identified. "Where" means that the numbers should be obtained from places that have a bearing on the particular conservation challenge, not from some different habitat type in a far-away place. "When" means that the data must apply to a time that is appropriate to the phenomena of interest; decades-old distribution records, for example, may tell one where a target was but not where it is now. And "Why" means that the goals and objectives of the conservation action must be clearly understood and articulated, lest the right numbers be collected for the wrong question (or vice versa).

Numbers are essential, but there is more to science-based conservation than numbers alone, or even the what, where, when, and why that need to accompany the numbers. Words are also necessary, and as with numbers, it's not just any words. So time for another digression, this time about words.

[7] An issue I consider in Essay 20, *Taking risks with the environment (2012)* (page 172).

ESSAY 29

Is "monitoring" a dirty word? (2009)*

There shouldn't be anything controversial about monitoring. It is what anyone does if they do something and want to see what happens, or even if they just observe something for more than a fleeting instant to see how or whether it changes.

But monitoring in sciences such as ecology, restoration ecology, or conservation involves much more than watching or checking up on something. It requires a rigorous sampling design to avoid biases, measurements that are precise enough to detect trends, analyses that can separate changes in the factors of interest from other confounding sources of variation, and a clear understanding of what one expects to learn from the monitoring. All of this can be expensive. And it is, by definition, repetitious and possibly boring. If done properly, it produces the same sort of data on the same things day after day, month after month, or year after year. Consequently, monitoring is often viewed as a waste of time and money that could be better spent on more interesting things—"cutting edge" science.

I'd never really given this much thought until I became involved in conservation, where the focus is explicitly on how species, ecosystems, or habitats are doing. We need to know whether the abundance of a species is declining or its range is shrinking, whether the areas we set aside as preserves are actually protecting what they were intended to protect, or whether the dollars contributed by donors are being well spent. Without monitoring, year after year for more than 40 years, we'd have no idea of what might be causing the dramatic fluctuations in Cassin's auklet reproduction on the Farallon Islands that I describe in Essay 17, Tipping points in the balance of nature (2010) *(page 144).*

The problem with monitoring is that it may reveal something exciting only after long periods of sameness. As the pace of environmental changes quickens and unusual events become more usual, however, the value of monitoring will become more widely recognized. It is incumbent on scientists and practitioners to show how monitoring can be done cost-effectively. How many things need to be monitored, how often, in how many places, with what precision, to tell us what we need to know? This requires that we go back to basics: we need to know what we need to know.

*Wiens, J.A. 2009. Is 'monitoring' a dirty word? *Bulletin of the British Ecological Society* 40(2): 39–40. Reproduced with permission of the British Ecological Society.

Ecological Challenges and Conservation Conundrums: Essays and Reflections for a Changing World, First Edition. John A. Wiens.

mon·i·tor; mon·i·tor·ing *v. trans.* to watch, keep track of, or check.[1]

When I was a university Professor, I rarely thought much about "monitoring." The notion of keeping track of something was scarcely exciting, and certainly not something worthy of a grant proposal. The duration of my studies was largely determined by when the funding ran out, when some new, terribly interesting, questions superseded the previous ones, or (rarely) when the original question was answered. But after I left academia to work in the trenches of conservation, I soon became aware that "monitoring" suffered a stigma beyond merely being unexciting. It is actually viewed with disdain, not just by many academic scientists (who see it as generating answers in search of a question), but also by some of those responsible for providing the funding to manage and conserve natural resources. "A waste of time," I have heard more than once. The perception of monitoring often is of birdwatchers or butterfly collectors going out on a fine spring day, making observations that, if recorded, end up in worn notebooks on dusty shelves. Monitoring, to paraphrase Rodney Dangerfield, "gets no respect."

Yet it is monitoring that tells us about changes in the environment and alerts us to their potential causes. It is from monitoring that we learned about the relationships between eggshell thinning, declines in bird populations, and DDT. It is how we learned that atmospheric levels of CO_2 at Mauna Loa have been steadily increasing since the late 1950s, alerting us to the linkages to climate change. This is how we are learning that songbirds are arriving and breeding earlier in the spring in Britain, or that the distributions of some butterfly species are shifting northward in Europe.

My organization, PRBO Conservation Science, has built its reputation largely on the careful collection and analysis of long-term data on bird populations—keeping track of things. Yet we are increasingly challenged by funding sources to justify why they should support monitoring. "Why gather more data? Surely a few years is enough!" Our continuous monitoring of the annual reproductive success of Cassin's auklets on the Farallon Islands of California over nearly 40 years, however, has revealed fluctuations that can be associated with El Niño episodes and, more recently, complete reproductive failures that may be linked to oceanographic changes and disruptions in marine food webs. There are intriguing indications that salmon recruitment may be affected by the same changes, raising the possibility that monitoring of auklet reproduction might be used to predict salmon stocks in subsequent years.[2] These relationships would not have emerged had the monitoring stopped after 5 or 10, or even 20, years.

So why does the contradiction between the perception of monitoring and its clear value in revealing changes and relationships persist? Monitoring suffers from the sameness that comes with keeping track of something. It strikes some as dull, especially those who do not go into the field to monitor things (and even some

[1]Merriam-Webster Online Dictionary.
[2]Roth et al. (2007).

of those who do). It is often regarded as not "real science," something that is perhaps best left to consultants, technicians, or "citizen scientists." There is often no overarching question, no hypothesis that relates to the issues at the forefront of scientific discussions. Consequently, monitoring is not viewed as research. It is difficult to publish the results of monitoring, so it does not contribute to the reward system of scientific cultures. Monitoring is not the stuff of which scientific careers are made.

Monitoring is also expensive, and its value may not be apparent in a culture that emphasizes short-term returns on investments. Monitoring funds in agencies are especially vulnerable to being raided by managers whenever more immediate needs arise, as they always will. Documenting trends or responses to environmental changes may take years or decades, especially if there are delayed or indirect effects. Enthusiasm for gathering the observations wanes, the patience of funders wears thin, and attention shifts elsewhere.

It would be wrong, however, to conclude that the attitudes about monitoring held by some scientists and managers are pervasive. In his compendium of everything one would want to know about monitoring, Ian Spellerberg[3] lists dozens of national and multinational programs and organizations that explicitly focus on monitoring the environment. Numerous governmental and non-governmental entities publish reports on the "state of the environment"; the Intergovernmental Panel on Climate Change (IPCC) even won a Nobel Prize for their work. Programs such as the Long Term Ecological Research (LTER), the fledgling National Ecological Observatory Network (NEON), or the Breeding Bird Survey (BBS) are dedicated to gathering monitoring data over networks of sites across the United States over many years.

We need to recognize, however, that the value of monitoring extends well beyond the occasional insights that emerge. The environment is changing rapidly, in unanticipated ways. Surprises occur often enough to no longer be so surprising. Management practices of the past may not work in the future. Managing or conserving resources in a changing world requires managing adaptively, and that cannot be done without the information to tell us whether our actions are working as intended, whether the investments are yielding returns.

Our commitment to ecological monitoring needs to be strengthened rather than diminished. Monitoring is how we keep track of how Nature is doing, and the indications right now are that Nature is not doing so well. There is much at stake, and monitoring is too important to be relegated to the backwaters of science. It demands the same attention to design, data quality, analytical rigor, and objective interpretation that are the fabric of mainstream science. And it should merit the same respect.

[3]Spellerberg (2005).

ESSAY 30

The place of long-term studies in ornithology (1984)*

I suppose that I was motivated to write this essay because I was involved in what was turning out to be a long-term study of bird communities in the sagebrush shrubsteppe of eastern Oregon. I may have felt a bit defensive about it. Suffice it to say that the research was creating more questions than answers, questions that related to environmental variation in space and time—the underlying theme of this book. I found it difficult to advise graduate students to frame specific questions that could be answered in a short-term study (rather than taking a decade to complete a degree) when I was providing a role model for long-term studies.

My intent was not to extol the virtues of one or the other, but rather to prompt some careful thinking about where and when either short- or long-term studies are most appropriate. It all goes back to the nature of the system being studied and the questions being asked. The study design and duration should be calibrated based on the phasing of the important system characteristics. If the study duration is too short, it may miss important dynamics; if it is too long, the relationships may be distorted by other intruding factors. What is needed is a Goldilocks Solution—not too short, not too long, but just right.

Over the past several decades, a tradition of short-term research has developed in North American ornithology, in which studies are customarily completed in a few days, weeks, months, or, at most, 1–3 years. Rarely is the same sort of information gathered over a period as long as a decade. Rather, attention is focused on seemingly simple questions that yield quick answers.

There are several reasons for this. Students undertaking dissertation research are required to conduct an original study within a relatively short time. Financial support is generally of limited duration and is not available for gathering the same sort of information for very long. Employment and advancement decisions are increasingly based on publication as an index of scientific progress and productivity, placing a premium on studies that will lead to quick publications. Whatever the causes, an excessive preoccupation with short-term studies can lead to short-term insights. By restricting the duration of investigations, we adopt a

*Wiens, J.A. 1984. The place of long-term studies in ornithology. *The Auk* 101: 202–203. Reproduced with permission of the American Ornithologists' Union.

Ecological Challenges and Conservation Conundrums: Essays and Reflections for a Changing World, First Edition. John A. Wiens.

snapshot approach to studying nature. We can only hope that the glimpses of patterns and processes that we obtain depict reality accurately and that something critical has not been missed because we looked at the system too briefly.

For some situations, this may not be a major problem, and short-term studies are justifiable. Thus, while often time-consuming, descriptive studies of essentially static patterns (e.g., morphology, fossil assemblages, genetic characteristics of species), of processes at an individual level (e.g., growth, physiological responses, behavior), or of evolutionary patterns or systematic relationships do not necessarily require long-term investigation. These are phenomena whose dynamics occur on time frames that are either very short or very long relative to the normal duration of a short-term study. However, for many phenomena, such as social structure, the operation of mating systems, the demographic composition and dynamics of populations, or the patterns of interactions in communities, change occurs often enough so that we cannot consider the system to be static. If we wish to study how the age-structure of a population influences social systems, for example, or whether or not a community is in equilibrium, a short-term approach is likely to produce incomplete or incorrect perceptions of a complex reality. For such studies, a long-term approach that spans the periodicity of the normal dynamics of the system is essential. Moreover, such an approach is more likely than a short-term focus to reveal unexpected patterns or to show the impact of rare but important events.

Conducting long-term studies is not without problems. One must first determine how long is long enough—the termination of such a study should be determined by logic and analysis, not by an ending of funding or by boredom. Studies involving demography might be delimited by the normal generation length of the population, and, for other phenomena, procedures such as spectral analysis or autocorrelation might be valuable in determining the phasing of normal dynamics of the system. Long-term studies, of necessity, must focus on one or two specific situations and thereby sacrifice the breadth of a series of short-term studies for increased detail and intensity. This reduces the potential for generalizing from such studies (even though the knowledge gained will be more firm). Among birds, long-term investigations will be most profitable if they document the details of individual behavior and survivorship and measure the dynamics of their critical resources. Such information is becoming critical with the growing synthesis of ecological, behavioral, and evolutionary studies. In order to obtain such information, however, use of marked individuals is absolutely essential. Such investigations must usually be restricted to a single species. Moreover, studies of resident populations are likely to generate more insight into the causal factors underlying observed dynamics than are those of migrants, because such individuals spend so much of their lives elsewhere that population dynamics are caused by factors integrated over a spatial scale too large to be studied with sufficient intensity.

Thus, while it is well and good to call for more long-term studies in ornithology, they are not essential to all areas of investigation, and they are likely to be successful only in restricted situations. Such studies, however, can make tremendous contributions to our understanding of nature and must be encouraged. This requires that granting and funding agencies recognize the need for long-term, relatively low-level support for such ongoing investigations and adjust their award structure accordingly (as NSF has recently begun to do). It requires that academic institutions recognize that valuable "original" graduate research need not start from scratch but may build upon a base established by previous investigations, so that a long-term perspective is obtained by a sequence of dissertation investigations. The studies of Acorn Woodpecker populations on the Hastings Reservation in California or of Scrub Jays in Florida provide good examples of the success of this approach. If carefully designed, such studies, once established, may foster explorations along many avenues and may produce a steady stream of significant publications, Unless the current emphasis on rapid publication of incomplete or preliminary findings is reduced, however, the more deliberate pace and ultimate value of long-term studies is not likely to be recognized and rewarded. Without some changes in how scientific progress and professional stature are currently perceived, bright young ornithologists will be compelled to avoid investigations of the intermediate-scale phenomena that require 5 or 10 or more years to understand or they will investigate such phenomena on a short-term basis, perhaps obtaining results that are superficial and quite possibly incorrect.

ESSAY 31

What use is small data in a big data world? (2013)*

> *"**Big data** is an all-encompassing term for any collection of data sets so large and complex that it becomes difficult to process them using traditional data processing applications."—Wikipedia*
>
> *That pretty much captures it. Big Data—massive accumulations of data on all sorts of things—do indeed threaten to transform how data are analyzed and, as I worried in this essay, alter how science is done. There's nothing inherently wrong with that, and resistance to progress is always ultimately futile. But the notion that one could probe mountains of data to derive patterns about which one could develop explanations ("stories") without being concerned about the underlying causes is troubling. Even more troubling is the likelihood that what lies behind the data—the insights and impressions that would have come along with the data as they were collected—would be lost. Big Data doesn't seem to leave room for natural history or intuitions, the things that make ecology and conservation something more than number crunching. These are the things, I argued, that are part and parcel of Small Data. After all, small is beautiful.[1]*

Shortly after writing my previous essay, on how ecologists detect and interpret patterns in data, I happened upon an article in *Foreign Affairs* by Cukier and Mayer-Schoenberger (2013) dealing with the rise of "Big Data." Their arguments led me to wonder how Big Data might change the way we do ecology, and in particular about the role of "Small Data" that have traditionally been the bread and butter of ecology. Let me explain.

"Big Data" is all the rage. In areas as diverse as medicine, marketing, particle physics, bird watching, astronomy, crime prevention, genome sequencing, English Premier League football, transportation, social networking, weather and climate forecasting, or national security, massive amounts of data are being turned every which way to reveal unexpected patterns—the tiny needles in gigantic haystacks. The explosive rise of Big Data has been fueled by advances in computational

*Wiens, J.A. 2013. What use is small data in a big data world? *Bulletin of the British Ecological Society* 44(4): 64–65. Reproduced with permission of the British Ecological Society.

[1] As E.F. Schumacher argued in his influential 1973 book, *Small is Beautiful: Economics as if People Mattered*.

Ecological Challenges and Conservation Conundrums: Essays and Reflections for a Changing World, First Edition. John A. Wiens.

technology, informatics, and cloud computing that have made it possible to assemble, explore, and visualize data at magnitudes previously only imagined in science fiction (recall Hari Seldon's "psychohistory" in Isaac Asimov's *Foundation* series). Search engines like Google can be used to track the preferences of individual consumers and target advertising, and smart phones can feed information about personal locations and movements into data banks that absorb tens of millions of records on a daily basis. Big Data, or "data-intensive science," has been called the "Fourth Paradigm," following the earlier experimental, theoretical, and computational paradigms of how science is done (Hey et al. 2009).

Big Data may not yet be as widely embraced in ecology as in some other fields, but it is growing rapidly. Data from multiple sources are being gathered in digital libraries such as the Entangled Bank,[2] the Australian Ecological Knowledge Observation System (AEKOS),[3] the National Phenological Network in the United States[4] or the Avian Knowledge Network (AKN)[5]; eBird[6] (a component of AKN), for example, contains several hundred million observations of bird occurrences, most submitted by bird-watchers. My colleague Grant Ballard tells me that a time-depth recorder on one of the penguins he studies generates a record of depth, temperature, light intensity, conductivity, and changes in speed every second, producing ~200,000 observations during a single foraging trip. Arrays of wireless sensors are being deployed in both terrestrial[7] and oceanic[8] environments to gather terabytes of environmental data.

So "think big" may be the new mantra of ecology. Given the rising tide of Big Data everywhere, one might ask what value remains for "Small Data." Small Data come from specific studies, usually conducted over a short time or in a few locations. For nearly a century, these studies have been conducted to support (or, less often, to refute) concepts or theories in vogue at the time. Such Small-Data studies are the building blocks of modern ecology: observations and data have enriched theory, generating in turn new questions to be addressed by more Small-Data studies. In fact, many ecologists have distrusted analyses of large data sets; decades ago, some worried about being swept out to sea in a deluge of data. At that time, the concerns were with how to manage, access, and analyze all those data. This is no longer an issue—Bayesian statistics, GIS, spatial modeling, radio-tracking, information-theoretic analyses, satellite imagery, and other tools and methods have eased the data-management challenge and sharpened the resolution and rigor of Small-Data studies. But some reluctance to embrace Big Data in ecology remains, for two reasons.

---

[2]http://www.entangled-bank.org.uk/doc/about.php.
[3]http://www.aekos.org.au/.
[4]http://www.usanpn.org.
[5]http://www.avianknowledge.net/content/about/; see Kelling et al. (2009).
[6]http://ebird.org/content/ebird/about/.
[7]http://www.cens.ucla.edu/; see Borgman et al. (2007).
[8]http://www.oceanobservatories.org/.

First, data gathered for particular studies carry with them the idiosyncrasies of assumptions, methods, study design, time and place, scale of investigation, and other factors, all of which are manifestations of the questions asked and who asks them. Combining many such studies to make Big Data encapsulates and magnifies this heterogeneity, blurring the sources of variation and obscuring the underlying assumptions. It may be, as some Big Data proponents argue, that the inaccuracies of heterogeneous and messy data are overwhelmed by the vast quantity. One can't help but feel, however, that something important has been lost.

Second, many Big Data arrays in ecology rely on the contributions of individuals or Small-Data projects to a shared data pool. Yet many investigators have felt a hesitancy to share their hard-won data, even beyond the point where all options for publication have been exhausted. Nonetheless, making data available through repositories or "data commons," with standardized metadata to guide users, is clearly on the rise, and is now mandated for projects funded by the National Science Foundation in the United States. Hampton et al. (2013) have suggested that those who do not participate in data-sharing may "run the risk of becoming scientifically irrelevant." In this view, data that are not shared might just as well not exist.

Big Data, however, is about more than assembling massive data banks or data sharing. Big Data portends a fundamental change in the way science is done, perhaps especially in ecology. For the past half-century or more, ecology has been in a deductive phase, testing hypotheses or addressing questions prompted by concepts and theories. Big Data instead advocates a largely inductive approach, in which the data collection is not directed by particular questions or theories. Using an expanding array of analytic and visualization tools, massive data sets can be mined for unexpected patterns or anomalies. One does not need to know beforehand how the data are to be used, and sampling—the underpinning of a good deal of modern ecology—becomes unnecessary. It's not quite aimless mining (after all, early gold prospectors didn't dig just anywhere, but had a pretty good idea of where to look), but neither is it constrained by particular questions or theories or by the persuasive power of preconceptions, which can lead one to see what one wants to see in the data.

The real paradigm shift, however, may have to do with causation. The search for patterns in massive data sets emphasizes correlations. Some proponents of Big Data argue that the huge amount of data should inspire sufficient confidence in the correlations to allow conclusions to be drawn and actions to be undertaken without understanding the underlying causes. As Cukier and Mayer-Schoenberger put it in their *Foreign Affairs* article, "we will need to give up on our quest to discover the cause of things, in return for accepting correlations." If this is so, the adage that "correlation does not imply causation" that has been drilled into every student in introductory statistics will become irrelevant. The implications of this shift are profound. In science, the detection of a pattern, whether through a few observations or mining of large amounts of data, has long been just the first step. The underlying processes or mechanisms that cause the pattern must also be determined if we are to have robust knowledge and understanding and a reliable foundation for applications of the science, such as predicting the effects of natural-resource management.

Don't get me wrong. Big Data will enable us to uncover patterns we have scarcely imagined. The development of digital libraries and data commons will foster a welcome openness among scientists. Advances in informatics and computational capacity will transform ecology. It's an exciting time. But Small Data still has an important role to play. Mining massive data may yield bountiful correlations, some of which are spurious, some nonsensical, and some revealing of Nature's well-kept secrets. The ability to distinguish among these—to separate the wheat from the chaff—is honed through immersion in detailed studies and observations of particular systems. So long as we continue to seek causal explanations of what we uncover, asking "why" as well as "what," the patterns that emerge from Big Data will prompt new, fine questions that have not been asked before. Interpreting those patterns will depend on the insight, intuition, and understanding gained from Small Data.

CHAPTER 22

A digression on words

I said in the previous chapter that science is a numbers game. That it is, but that's not all it is. How the numbers are gathered, crunched, and interpreted are important, of course, but if numbers are to have any effect, they must be communicated—this takes words. If "the medium is the message," then numbers and data contain the message and words are the medium, which, as Marshall McLuhan famously claimed,[1] shape how the message is received and interpreted and acted upon. It may be a harsh overstatement to say that science that is not communicated might just as well not have been done. However, if it is not available for others to think about, criticize, praise, or absorb into their actions, it will have little effect.[2]

But who are the "others" that science should affect? Most often, the "others" are other scientists. The culture of science, like that of any social grouping, rests on recognition by one's peers. Assuming that a scientist produces good work (i.e., generates the numbers, analyses, and so on),[3] peer recognition is what enables a scientist to fashion a career. Publication in peer-reviewed scientific journals or (now) online carries the stamp of approval of one's work, while awards, appointments to editorial boards, important committees, and the like are usually based on recommendations from established scientists within a discipline. Scientific peers, of course, share interests, experience, knowledge, and (in the context of paradigms[4]) preconceptions, so the premium is on communicating with them in those terms. Using jargon (something akin to the secret handshake of a fraternal order) and following accepted style and format (e.g., avoiding

[1]Initially in his 1964 book, *Understanding Media: The Extensions of Man* (McLuhan 1964).
[2]There are, of course, exceptions. Potentially path-breaking work that lay buried in notebooks or unpublished manuscripts, only to be discovered much later, may then be widely cited and be featured in textbooks. Because the work was unknown or ignored at the time, however, it broke no new paths and is primarily of historical interest.
[3]Alert readers will have noticed that, by emphasizing numbers and data, I have seemingly ignored those scientists who deal primarily with theory, concepts, or models. Any slight is unintended, for these activities are very much parts of science. Yet theories, concepts, and models usually quickly find expression in numbers, from whence the message is expressed in words (the medium).
[4]See Chapter 2 and Essay 4, *The power of paradigms (2014)* (page 23).

Ecological Challenges and Conservation Conundrums: Essays and Reflections for a Changing World, First Edition. John A. Wiens.

footnotes[5]) enhances such communication.[6] Publications that accord with a prevailing paradigm are more likely to be cited than those that do not, increasing the citation metrics that are considered in career-advancement decisions. Successful scientists grow up talking primarily with one another.

Sometime during my years in academia I came to realize the importance of letting aspiring young scientists in on some tricks of the trade, so I offered a graduate class on how to publish a scientific paper. In an attempt to convey how important the first sentence of a paper can be in grabbing the reader and enticing him or her to read further, I quoted the first sentences from several classics in English literature, to wit:

> *Call me Ishmael.*
>
> Herman Melville, *Moby Dick* (1851),

> *To the red country and part of the gray country of Oklahoma, the last rains came gently, and they did not cut the scarred earth.*
>
> John Steinbeck, *The Grapes of Wrath* (1939),

> *He was an old man who fished alone in a skiff in the Gulf Stream and he had gone eighty-four days now without taking a fish.*
>
> Ernest Hemmingway, *The Old Man and the Sea* (1952),

or

> *Once upon a time when the world was young there was a Martian named Smith.*
>
> Robert A. Heinlein, *Stranger in a Strange Land* (1961).

These sentences set the stage, but do so in ways that pique one's interest. I contrasted them with an all-too familiar beginning of a scientific paper (here fictionalized to protect the guilty):

> *The phylogenetic affinities of the ratites, especially the emu* (Dromaius novaehollandiae), *are a matter of some dispute* (Smith, 1999; Jones, 2001) *meriting further study* (McKenzie, 2003; Hillyard, 2007).

or, to use an actual example from what is surely one of the most important scientific papers of the 20th century:

> *We wish to suggest a structure for the salt of deoxyribose nucleic acid (D.N.A.).*
>
> James D. Watson and Francis Crick (1953).

Although I made my point that scientific writing doesn't need to be sterile and should allow room for style, the students were expecting me to help them hurdle the barriers to scientific publication, not to engage in literature appreciation. So I went on to describe the structure of a scientific paper, how to design graphs, how

---

[5] See Essay 3, *In defense of footnotes (2014)* (page 7).

[6] I well remember (but not well enough to provide the citation) a paper that began with "Let us assume that ... " followed by several pages of equations intelligible only to a mathematically astute reader (certainly not me). Words might have helped.

to deal with negative reviews, and so on. The students found it useful. Looking back, I realize that I was guilty of reinforcing the cultural norms of science.

The point of this digression within a digression is that words matter. Even before I began teaching students how to publish (if not how to write), I wrote an essay on this theme: Essay 32, *Word Processing versus Writing (1983, 2011)* (page 254). Words especially matter when one is trying to communicate with people outside one's cozy group of scientific peers, as those of us interested in ecology and conservation must do. However, to communicate effectively with farmers, politicians, environmentalists, and even artists and poets, scientists must speak their language and use words that mean something to them. Doing this has always been hard—scientists, after all, are carefully trained in the art of communicating with other scientists—but it has become even harder as rapid environmental changes present mounting conservation challenges and leave many people bewildered about what to do. Some don't want to hear the message, while others actively deny it. But it's imperative that the message be listened to.

Communication is a two-way street. A writer requires a reader, and a speaker requires a listener. Scientists can't make people read or listen, but they can do their part by communicating clearly and compellingly. Jargon is out. Dry, impersonal prose is out. But facts and carefully reasoned interpretations and conclusions are not out. They are what scientists have to offer, but doing so requires attention to words.

ESSAY 32

Word processing versus writing (1983, 2011)*

This essay did double duty. I wrote it in 1983 as an editorial in The Auk. *Then, because I felt that the points still needed to be made (and I was faced with a looming deadline with no other ideas about what to write), I resurrected it as an essay in the* Bulletin of the British Ecological Society.

The points remain relevant. Writing is a craft, and not an easy one to master.[1] *It is too much to expect that scientists will become virtuosos of prose (although a few are), but there are no reasons why more cannot write as if they cared about writing. Except, perhaps, that it's not the tradition, and reviewers and editors seemingly frown on clear and concise writing.*

As an aging curmudgeon, I've long been dismayed by the decline of quality writing, especially in scientific literature. Few scientists are known as great masters of prose, of course. But is it asking too much not to have to fight through the writing to understand what's being said? In any case, I was drawn to reflect on this as I reviewed some chapter manuscripts for a book I'm editing. And I had one of those *déjà vu* experiences.

Some years ago I edited an ornithological journal (*The Auk*—ornithologists have a propensity to name their journals after birds, even extinct ones). One evening, in a fit of pique brought on by reading through a stack of unusually sloppy and mind-numbing submissions, I'd had enough. So I wrote an essay, largely decrying the subservience of good writing to the electronic gods of word processing.

This was 1983. Word processing on computers was just getting started. WordStar was the leading software, WordPerfect was catching up, and Microsoft Word had

*Wiens, J.A. 1983, 2011. Word processing versus writing. *The Auk* 100: 758; *Bulletin of the British Ecological Society* 42(2): 39–40. Reproduced with permission of the British Ecological Society.

[1] If you'd like some help, Ursula Le Guin's *Steering the Craft: Exercises and Discussions on Story Writing for the Lone Navigator or the Mutinous Crew* and Stephen King's *On Writing: A Memoir of the Craft* provide delightful guidance.

Ecological Challenges and Conservation Conundrums: Essays and Reflections for a Changing World, First Edition. John A. Wiens.

only just become available. It was the Neanderthal era of word processing, but the stage was set.

So I thought it might be informative, at least from a historical perspective, to resurrect here some excerpts from that essay of 28 years ago. Here (shortened somewhat) is what I wrote.[2]

High technology has invaded the peaceful domain of scientific writing. Rather than laboring over a typewriter (much less pen and paper) to turn their thoughts into sometimes readable prose, writers sit comfortably before the video screen of a word processor. They tap out a few lines. Lights flash, and the lines appear at once before them. If they're not right, the offending words or lines disappear with the touch of a key. More lines are tapped out, guided by the mesmerizing, blinking cursor. When all seems satisfactory, the touch of another key may correct misspellings; another commits the composition on the screen to the certainty of a computer's storage, later to be retrieved and converted to typewritten text, presumably with equal certainty. Writing with a word processor is efficient, easy, and even fun. Word processing is rapidly becoming equivalent to writing. Or so it would seem.

With the ease and efficiency of word-processing writing, however, come some perils. Unless writers are aware of these, their prose may become boring or senseless, no matter how exciting their results or how pure their science. I see three major perils: duplication, verbosity, and laxity.

Of these, duplication is perhaps the most insidious. Word processors encourage one to store entire paragraphs or sections of manuscripts in the computer and then call them forth to be used, verbatim, in subsequent papers. It is much easier, for example, to recall a methods section from a previous paper and insert it as a block into a current manuscript than to rewrite it or make the revisions that would tailor it to the current presentation. In addition, the ease with which entire paragraphs or sections of manuscripts can be moved from one place to another may often destroy the transitions that give well-written prose a smooth flow. These practices promote not only dull but inaccurate prose.

The second peril, verbosity (what I call word-processor diarrhea), is related to the first. Our thoughts naturally tend to ramble, and good writing imposes logical and stylistic constraints on such excesses. Word processors promote an effortless flow of words from our mind to the screen, and the temptation to write now and edit later is strong. The result is an increasing words:meaning ratio.

Laxity imperils the final preparation of a manuscript. Mistakes appearing on a screen may not be as apparent as those typewritten on paper. Inevitably, some errors will be missed. Because the text appears so clean and crisp on the screen and when it is printed out, the final editing and proofreading may be only superficial. If the copy represents corrections of an earlier draft, there is a temptation to avoid

[2] *The Auk*, 100: 758 (1983).

proofreading the printed copy altogether, or to check over only the corrections, not the entire manuscript. Computers may not make errors, but they do have a nasty way of exercising commands that their human masters have forgotten or neglected. The consequence may be unanticipated slips between the screen and the paper.

There is no doubt that word processors represent an advance in writing that has great potential [see how prescient I was!]. Their speed and efficiency offer substantial time savings and thus provide the opportunity *really* to pay close attention to editing and composition and to express thoughts or present results in just the right way, because the tedium of typing and retyping (and retyping) is circumvented. But, in the end, word processors can only process what the writer provides. They produce clear copy, but they do not guarantee clear thinking or clear writing. Those will remain the responsibility of the careful writer.

Re-reading this essay, it's apparent even to me that I was too dismissive of the power and inevitability of word processing. We no longer call it word processing, of course, but simply writing. Revisions are easy and can be made on the fly. A thesaurus function helps one find alternative words (although it doesn't indicate which word is exactly right). Drafts can be circulated for comment quickly, increasing the opportunities for feedback before a paper is submitted. Electronic submission and publication have shortened the lag between the completion of a manuscript and its appearance to weeks instead of months or years, accelerating the pace of science. Preparing a manuscript is much easier now than it used to be.

But that doesn't mean that the problems of the past have disappeared. Although journal editors are more alert for duplicate submissions, the issues of verbosity (or, more precisely, a lack of clarity and focus) and laxity persist. Manuscripts submitted electronically still contain spelling errors, incomplete sentences, and disconnected paragraphs, and revisions often come back with the same errors firmly in place.

The deeper problem lies not with word processing (or its modern equivalent). The speed with which a paper is composed is no longer impeded by the task of typing and retyping—the limiting factor is now the mind, not the machine. Clear writing compels and demands clear thinking, which is the essence of science. This entails more than grammatical correctness or careful proofreading. The point of scientific publication, or of any writing, is to *communicate*, and imprecise writing, riddled with jargon (or, worse, a deluge of acronyms), is an anathema to clear communication. Good writing requires thought and care. It is not easy. Technology helps, tremendously, but Word cannot write by itself.

I guess I was asking, back then, that scientific writing be literature (we do call it the "scientific literature," after all). Why not? Who said that scientific writing could not have style, or even the beauty of an aptly turned phrase? I guess I'm still a curmudgeon. Harrumph!

CHAPTER 23

Debates in ecology and conservation

Debating gives most of us much more psychological satisfaction than thinking does; but it deprives us of whatever chance there is of getting closer to the truth.

C.P.Snow (1963: 56)

To debate is human nature. Raise any point and someone will argue the point. Debate ensues. Science, despite its objectivity and reliance on facts, is often a hotbed for debates. This is because science is as much about ideas as it is about facts, and ideas by their very nature are debatable. Scientists are also skeptics, primed to debate at the drop of an idea (or publication of a paper).[1] As the conservationist Stuart Pimm admitted, "I spend my life in scientific debate: it's what makes science so effective."[2]

The history of ecology is replete with debates.[3] One of the earliest, which I mentioned in Chapters 2 and 15, was between Frederic Clements and Henry Gleason in the 1920s over the nature of ecological succession and plant communities. Clements envisioned ecological succession as a predictable process leading to tightly organized, stable communities, whereas Gleason considered succession to be largely a matter of chance; the outcome was unpredictable. Clements pushed his ideas with an almost religious fervor, and his views dominated community ecology for decades.

The Clements–Gleason debate was followed in the 1950s by a debate over population regulation. Several ecologists, notably David Lack and A.J. Nicholson, argued that population size is governed by negative feedbacks that reduce population growth as population density increases—density-dependent processes. H.G. Andrewartha and Charles Birch championed the alternative, density-independent view that population growth is affected by a welter of environmental factors acting independently of the density of a population at a given time. The debate was contentious at times and continued to dominate animal population ecology well into the 1970s; elements of it persist.

[1]A topic I considered years ago in Essay 33, *On skepticism and criticism in ornithology (1981)* (page 264).
[2]Pimm (2014).
[3]Kingsland (1985, 2005), Worster (1994), Hassell (1998), and Cooper (2003) provide perspectives on some of these debates.

Ecological Challenges and Conservation Conundrums: Essays and Reflections for a Changing World, First Edition. John A. Wiens.

These debates were fueled primarily by empirical observations that morphed into theory. In contrast, the debate that developed in the 1970s over the role of competition between species in structuring animal communities derived largely from the theories of community organization advanced by Robert MacArthur and G. Evelyn Hutchinson.[4] Proponents such as Jared Diamond, John Roughgarden, Tom Schoener, and Martin Cody argued (and presented supporting evidence) that competition drives ecological differentiation among coexisting species, resulting in optimally structured, stable assemblages. Dan Simberloff, Don Strong, and several others (myself included) challenged this view (also with supporting evidence), emphasizing how environmental variation can overwhelm the effects of competition and questioning the evidence offered to support the theory.[5] The debate raged on for more than a decade, in the process generating a side debate on assembly rules, the use of null models, and the role of hypothesis testing in ecology that involved Dan Simberloff, Ed Connor, and Don Strong sparring with Jared Diamond and Michael Gilpin.[6] In the end, there was no clear resolution other than concluding that "it all depends," and most of the protagonists went on to other things.[7]

There have been other debates in ecology and conservation. Most often, the conclusion has been the same, that it all depends. This is not too surprising, given the variable and indeterminate nature of ecological systems. Rather than rehash old debates, however, let's look instead at some ongoing debates on conservation that are generating a good deal of heat and, now and then, some light.

Several of these debates relate to the issue of whether, when, and how to intervene with nature. Of course, people have been mucking around with nature since the onset of the Anthropocene. With the advent of resource management and

---

[4]MacArthur (1972).

[5]I initially accepted the competition perspective, noting in the conclusions of my doctoral dissertation that "differences in habitat occupancy and utilization were seemingly adequate to circumvent direct competition and to explain the co-occupancy of the study area by the seven breeding species [of birds]" (Wiens 1969: 89). It was only later, after continuing my studies, that I came to realize the extent of spatial and temporal variation in environments and communities and the failure of traditional competition-based theory to account for the variations (Wiens 1974, 1977). See Essay 6, *Fat times, lean times, and competition among predators (1993)* (page 33).

[6]Insights into the competition and null model debates can be gained by perusing the chapters in Cody and Diamond (1975), Strong et al. (1984), Diamond and Case (1986), and Gotelli and Graves (1996).

[7]The issue later resurfaced, with an ironic twist. My ecological credentials were established in part through my challenges to competition theory. Then along comes my son, David, who conducted his PhD research on the interactions between spotted owls and barred owls, demonstrating convincingly the central role of competition and the consequences of dominance by barred owls. To extend the irony further, Martin Cody was a consulting editor for his publication (Wiens et al. 2014).

conservation, however, interventions have become more targeted, either toward using nature to our benefit (e.g., ecosystem services) or preserving nature to its benefit (e.g., nature preserves). Conservation is, by its very nature, interventionist.

But intervention comes in a variety of guises. At one extreme, the default (because it's easiest) may be to do nothing, to let nature take its course and only watch what happens. Quite apart from the fact that doing nothing isn't really conservation at all, the very act of observing a system can alter its dynamics—an ecological variant of the Heisenberg Uncertainty Principle.[8] Between doing nothing and the opposite extreme of completely engineering a system and calling it "nature"[9] lies a wide variety of interventionist conservation actions: prescribed burning, habitat restoration, captive breeding and release, predator control, supplemental feeding, and so on. Each of these is accompanied by uncertainties, creating tension between acting now or waiting until we know more and can be more certain of the outcomes of our actions. Climate change is accelerating the pace of environmental change, so we may not be able to wait until we learn more and become more certain. This has prompted intense debates about several approaches that lie toward the more extreme (or innovative, depending on your point of view) end of the intervention spectrum.

One is the suggestion that species be proactively moved to somewhere else to maintain populations or ecosystem functions, keep ahead of a changing climate, or avoid extinctions. Such *managed relocation*[10] has been going on in some form for some time as people have moved species about, sometimes intentionally, and often unintentionally. Polynesians introduced taro, pigs, dogs, and rats to many Pacific islands, and the British took many familiar plants and animals with them to the places they colonized. Conservationists have often reintroduced species into areas of their former range where they have been recently extirpated.[11] Foresters have established highly productive tree species in areas far removed from their native range—Monterey pine, which is native to a limited area of coastal California, is now the dominant introduced tree species in Australia, New Zealand, and Spain and is a favored plantation species in parts of South America and Africa. Now foresters are suggesting that managed relocation be used to establish plantations of common and widespread tree species at locations within their current ranges to facilitate adaptation to changing climate conditions.[12] Managed relocation is also being proposed as a conservation strategy to preserve species that are increasingly

[8]Animal behaviorists are acutely aware of this issue: just try watching a bird without feeling that you're also being watched. This is why nature photographers use blinds.

[9]For example, nature theme parks such as Animal Kingdom in Disney World (see Higgs 2003) or zoos, which despite their good intentions and an important role in preserving and breeding some vanishing species, don't pretend to be natural.

[10]See McLachlan et al. (2007) and Hoegh-Guldberg et al. (2008) for initial development of the idea. The terms "assisted migration" and "assisted colonization" have also been used.

[11]For example, the Aleutian cackling geese discussed in Essay 26, *Be careful what you wish for (2014)* (page 222).

[12]Pedlar et al. (2012).

jeopardized by climate change by moving them to places they haven't been historically.[13] For example, a group of botanists and environmentalists have planted seeds of *Torreya taxifolia*, an endangered conifer restricted to a few locations in the Florida panhandle, into areas in North Carolina well outside its historical range in anticipation of environmental shifts resulting from climate change.[14]

Managed relocation emphasizes the species that is being moved. The concept of *re-wilding* instead focuses on restoring ecological communities and functions to the recipient ecosystems. Variations on this theme span multiple scales in space and time,[15] but most attention has centered on a particularly contentious proposal: *Pleistocene re-wilding*. By the end of the Pleistocene in North America, many large herbivores and carnivores—the "Pleistocene megafauna"—had become extinct, at least partly at the hands of humans. Ever since, the ecosystems that included these species have been incomplete (or at least different), missing basic functions such as dispersal of large seeds by herbivores or the evolutionary pressures of large predators on their prey. What we see now (and have created sometimes elaborate ecological stories to explain) is a fragment of what once was. The aim of Pleistocene re-wilding is to restore the functional ecosystems that existed in North America in the Pleistocene by introducing large herbivores and carnivores from Africa and Asia to replace their missing counterparts. The idea was first broached in 1998 by Michael Soulé and Reed Noss but has since been developed in greater detail by Josh Donlan and his colleagues.[16]

Even farther toward the ecological engineering end of the spectrum lies *de-extinction*. The impetus for this approach comes not from mainstream conservation science but from recent advances in genetic and genomic technology.[17] These tools offer the prospect of resurrecting species that became extinct decades, centuries, or even millennia ago, starting by reconstructing their genomes from bits of DNA recovered from existing specimens. Discussions have focused on several extinct species—woolly mammoths, thylacines, and passenger pigeons are most often mentioned—and the approach has generated considerable media attention.[18]

The ultimate aim of de-extinction is to resurrect extinct species that can then be reintroduced into the wild, restoring ecosystems that are incomplete due to the absence of extinct species. Currently, the emphasis is on the genomic technology. Resurrecting a passenger pigeon or woolly mammoth, however, is only

---

[13] See Essay 34, *The demise of wildness?(2007)* (page 267).
[14] See http://www.torreyaguardians.org/.
[15] For example, the Lynx UK Trust has recently proposed releasing lynx into forests in England and Scotland where they have not occurred for more than 1,300 years.
[16] Soulé and Noss (1998) and Donlan et al. (2005, 2006).
[17] The efforts are spearheaded by Stewart Brand and the Long Now Foundation, through the Revive & Restore program; see http://longnow.org/revive/.
[18] For example, a cover story in *National Geographic* (April 2013), a lengthy piece in *The New York Times* (February 27, 2014), and coverage in *Wired* (March 15, 2013) and *The New Yorker* (March 11, 2014).

the first step. These species had distinctive ecological and behavioral traits that are only partly encoded in their genomes, if at all. Passenger pigeons were loosely migratory, moving in vast flocks in response to masting in trees such as beech. Woolly mammoths disappeared some six millennia ago, so we have only fragmentary knowledge of their ecology and know almost nothing of their behavior. De-extinction advocates recognize these difficulties, but the technological challenges and the very idea of bringing long-vanished species back to life drives them on. And even those who question the conservation value of de-extinction seem resigned to its inevitability, given the developing technology and the "wow" factor.

These three proposals—managed relocation, Pleistocene re-wilding, and de-extinction—represent points along a spectrum of conservation intervention. Each generates debates, the pros and cons of which are generally similar. Advocates emphasize the potential benefits of forestalling extinctions and restoring missing ecosystem functions. Managed relocation, for example, may prevent the extinction of species that are rare, poor dispersers, or hemmed in by habitat fragmentation. The addition of species to contemporary ecosystems, either through re-wilding or de-extinction, could enrich biodiversity and bring back long-missing herbivore-plant or predator-prey relationships. Most directly, however, advocates argue that bold, radical actions are justified given the magnitude and immediacy of the threat of climate change—many species may not survive unless we help them.

Critics worry about unintended consequences. Whatever balance exists in communities and ecosystems could be disrupted by interjecting species from elsewhere, as has happened so often with introductions gone awry. No matter how carefully considered, the potential for elephants, cheetahs, and camels (much less mammoths) to have devastating impacts on contemporary ecosystems is great. There's also the challenge of finding suitable places to put an assortment of alien species, particularly large, wide-ranging, and potentially dangerous ones. The ecosystems that now exist in North America, for example, are not the same as those in the source areas in Africa and Asia, and they bear only a faint resemblance to those of the Pleistocene. What use is it to resurrect an extinct species if there's no place for it to go—the "no country for old species" problem? Critics also worry that these high-profile proposals will divert attention and support from efforts to deal with the root causes of extinctions. Habitat loss, for example, is a major factor in species imperilment, so it might be better to focus efforts instead on habitat restoration. The most basic concern, however, is that we simply know too little to be undertaking such aggressive actions. There are too many uncertainties, and the risks are too great, so we should err on the side of caution.[19] Some have suggested moving cautiously ahead to see what works,

[19]The "precautionary principle"; see Essay 20, *Taking risks with the environment (2012)* (page 172).

following specified guidelines.[20] It seems unlikely, however, that guidelines will do much to mollify the debates.[21]

Lurking over all the talk about moving or resurrecting species is the specter that they may become invasive species, which I considered in Chapter 7. Although no one suggests that invasive species should be ignored, there is ongoing debate about how native versus non-native species are valued. Mark Davis (and 18 other ecologists) suggested that the emphasis should be on the ecological effects of species, not their citizenship. Many non-native species that invade communities, for example, may not have negative impacts or may provide resources for native species—the importance of invasive saltcedar to endangered southwestern willow flycatchers I mentioned in Chapter 7 is an example. Eradication of many invasives is an unattainable goal, they argued, so we should learn to live with the non-natives unless they are clearly harmful. Dan Simberloff (and 140 other ecologists, showing at least strength in numbers) countered that the vast majority of problem species are not native to the invaded ecosystems; that communities dominated by native species tend to be better at resisting invasions; and that the effects of invading species may only emerge long after the invades have gained a foothold.[22] Clearly, the emphasis of Simberloff and others is on the value of maintaining existing native assemblages.

When species move or are moved to some place beyond their historical distribution, they contribute to the formation of novel ecosystems (see Chapter 15). There are also debates about novel ecosystems and whether they should be accepted as inevitable. These debates have a familiar ring. One side argues that the development of novel ecosystems may unravel the fabric of well-established and smoothly functioning ecosystems and that accepting novel ecosystems means condoning invasive species, opening the door to further degradation of natural ecosystems. Others counter that species have been moving around forever so novelty is not new—ecosystems adjust. In any case, they argue, disassembly and reassembly of

[20]For example, Rout et al. (2013) and IUCN/SSC (2013) consider guidelines for relocations, and Seddon et al. (2014) discuss criteria that could be used to select candidate species for de-extinction following the IUCN guidelines for reintroductions.

[21]Some feeling for the pros and cons of these debates may be gained by reading Ricciardi and Simberloff (2009), who mounted a vigorous attack on managed relocation that generated a series of responses in a subsequent issue of the same journal; Hewitt et al. (2011) and Schwartz et al. (2012) provide reasonably balanced reviews of the same issue. Rubenstein et al. (2006) and Caro (2007) consider some aspects of the Pleistocene re-wilding debate, and Stewart Brand and Paul Ehrlich present opposing views about de-extinction in an informative Point/Counterpoint in *Environment 360,* a newsletter of the Yale School of Forestry and Environmental Studies (January 13, 2014); available at http://e360.yale.edu/feature/the_case_for_de-extinction_why_we_should_bring_back_the_woolly_mammoth/2721/.

[22]Davis et al. (2011) and Simberloff et al. (2011, 2012) describe elements of the debate.

communities will increase as the pace of environmental changes accelerates, so it's better to deal with that reality rather than fight it.[23]

If these debates rested solely on science, it might be possible to get closer to the truth that C.P. Snow alluded to in the quotation that opened this chapter. But there's more than science involved. Differences in paradigms, beliefs, and philosophies about nature run through these debates. There are also some substantial ethical and moral concerns. For example, is it ethically and morally acceptable *not* to take action if we can do something to avoid extinctions or restore ecosystem functions? Can we justify taking actions that might jeopardize other species? Is it morally or ethically just to translocate elephants and cheetahs from the wild into fenced parks under the guise of re-wilding? Should we be undertaking efforts to resurrect long-dead species when we are unable to deal with the growing queue of living species teetering on the brink of extinction? If we (or our ancestors) caused the extinction of the Pleistocene megafauna, isn't it our moral responsibility to do everything we can to bring them back, to redress the sins of the past? And if so, how far back into the past does this responsibility extend—a century, to include the passenger pigeon, or 6,000 years, to include the woolly mammoth? Should we automatically regard non-native species that show up in an area as invaders that threaten the sanctity of native ecosystems, or should we reserve judgment and show them the same respect we accord native species (which were themselves, in an earlier time, non-native invaders)?

These questions strike at the heart of how and why we do conservation—whether our aim is to protect and preserve nature or manage it, how we value species, and where our responsibilities lie. I consider these issues in the following chapter.

[23]Hobbs et al. (2013, 2014b), Hobbs (2013), and Murcia et al. (2014) lay out the dimensions of this debate.

ESSAY 33

On skepticism and criticism in ornithology (1981)*

From 1976 to 1984, I edited The Auk, *the journal of the American Ornithologists' Union.*[1] *Along the way, I began to notice how unusual it was for a submitted paper to challenge established ideas, and how scathing the reviews could be if it did. Skepticism and criticism seemed to be the exclusive province of reviewers. There also seemed to be a belief that if a paper did run the gauntlet of reviewers and editors to make it into print, it must be the truth (or at least true within statistical confidence limits). At the same time, graduate students in journal club sessions would mercilessly demolish whatever published paper was selected as that week's victim (even, or perhaps especially, one of my own). So publication clearly did not equate with perfection, at least in the eyes of graduate students.*

Fast forward to the present. Scientific publication in most journals still depends on peer review. Reviewers are still likely to seize on things they would have done differently and to be skeptical of results or interpretations that challenge their established beliefs. Perhaps anomalous or contrary findings should in fact receive extra scrutiny, but it's also clear that what is published is not necessarily true, and certainly not perfect. Skepticism and criticism are still needed. As public policy has come to depend more and more on "best available science" (for which read "published in peer-reviewed journals"), it may be worth remembering that "best" is a relative term and not everything that is published is by definition right.

Skepticism was once the hallmark of the scientist. Recently, however, it seems that uncritical acceptance of prevailing notions has become more fashionable. This is perhaps most clearly evidenced by the seeming reverence in which a good deal of published work is held. We tend to regard published research or ideas as having been certified as "true," or at lease substantially correct. There is, indeed, a certain sense of security in believing that the rigors of peer review and editorial scrutiny assure that only quality papers, in which the data, analyses, and interpretations can be implicitly trusted, will be published. The work that is published can then

*Wiens, J.A. 1981. On skepticism and criticism in ornithology. *The Auk* 98: 848–849. Reproduced with permission of the American Ornithologists' Union.

[1]Ornithologists have a curious predilection for naming their scientific journals after birds, some of them extinct or nearly so: *The Condor, The Ibis, The Emu, The Ostrich, The Passenger Pigeon,* and so on.

Ecological Challenges and Conservation Conundrums: Essays and Reflections for a Changing World, First Edition. John A. Wiens.

be taken to represent the foundation of established facts, verified theories, and logically correct ideas upon which subsequent work can be built. It is infrequently subjected to close critical evaluation, especially if it is consistent with established beliefs, supports neat ideas, or agrees with (and thus certifies as also "true") one's own findings.

But this sense of security is false. Every study has its limitations—honest mistakes in the methods that are used, the observations that are made, or the analyses employed; biases in the interpretations of the data or in the logic that is followed; or deception in the way in which findings are presented. Some studies suffer from such failings more than others, and the more blatant mistakes are presumably detected during peer review of manuscripts that are submitted for publication. But reviewers are not infallible or unbiased, and errors or inconsistencies do pass by their scrutiny, more often than we might wish.

Published work is thus not perfect. Much of the time it is not even close to it. Readers who turn to the scientific journals for knowledge therefore cannot simply accept what is printed as correct. One must read critically, with a skeptical attitude, continually evaluating the methods, the data, the analyses, and the interpretations. Because the reviewing and editorial processes generally (but not always) filter out the absurdities, the errors or misinterpretations that do emerge in print are often subtle or carefully hidden. The need for careful, critical reading and evaluation is thus all the more important.

But it is not sufficient to read critically and, with an inner feeling of satisfaction, note the flaws in a published paper. If the problems are serious, they should be noted in print. This seems not to happen very often. In part, this is a consequence of the structure of our journals, which do not readily offer the opportunity for open dialogue on the points raised by a critical evaluation of previously published work. But it also reflects the general politeness with which we approach science. We often seem to regard published criticism as a personal attack on an author rather than an objective evaluation of the work itself and, as a consequence, tend to avoid such unseemly behavior. Science does not progress much, however, if fair, open discussion of published work is shunned because it might be taken personally. Authors who would view objective criticism in this way are guilty of taking their work too seriously, of investing too much of their egos in it. The purpose of publication is to present our work and ideas in a manner that makes them available for the close, critical scrutiny of our peers, and we should be gratified when our work is so evaluated, even if the outcome is negative. After all, a published paper that induces critical thinking on an issue, even if it is not completely correct, is important. Science progresses as much from the challenge of wrong ideas or incorrect observations as it does by the accumulation of "basic truths," but this requires continual skepticism and criticism.

My points, then, are these. Authors have an obligation to present their findings and interpretations in a clear, careful, nondefensive manner. Despite the best of intentions, however, such work will probably be flawed, and not all of the errors will be detected during the review process. Published work is thus not certified as necessarily correct, and it should therefore be subjected to critical evaluation, undertaken from a stance of skepticism. When basic flaws are detected, they should not pass unchallenged. Journals should be structured in a way that fosters open, responsible, impersonal dialogue on the issues that are thus raised. Only in this way, I believe, can we resist the seductiveness of neat but illogical ideas, or of observations that are consistent with some favored view but are flawed, and approach science as a challenging intellectual activity rather than a system of shared beliefs.

ESSAY 34

The demise of wildness? (2007)*

Conservation has always been torn between idealism, aiming to preserve nature as close to its pristine state as possible, and pragmatism, aiming to save the parts of nature most useful to people. In this essay I ruminated on two expressions of this duality, both of which might be considered extreme. One proposes that conservation should be proactive in returning nature to a long-vanished past by introducing ecological counterparts of extinct animals—"re-wilding." The other argues that, since nature everywhere is now heavily affected by humans, we should put our conservation efforts into protecting the benefits people derive from nature—"ecosystem services"—and abandon any pretext of preserving (much less re-creating) wildness. I favored the middle ground, suggesting that re-wilding is unrealistic but giving up on wildness is premature. Debates about both propositions have continued and intensified.

Conservation is firmly rooted in the notion of wildness, of nature untrammeled by people. Organizations such as The Nature Conservancy or the Wildlife Conservation Society aim to protect "the last great places on earth" or "the last of the wild." Such mottos emphasize the clear distinction between these places and those that bear the human footprint, and that few such places remain. Conservation has largely been about protecting those areas of essential wildness, where nature still rules and people are visitors who come and experience, to go away renewed and enriched.

Two recent publications have brought the issue of "wildness" and its place in the conservation agenda into sharp relief. In a paper in *Nature*, Donlan et al. (2005) have proposed "re-wilding" parts of North America. Proxies for the large animals—camels, elephants, cheetahs, and lions—that once roamed North America but became extinct 13,000 years ago would be introduced, restoring the ecological and evolutionary processes (and renewing a presumed balance of nature) that existed during the Pleistocene. Tim Flannery (1994) suggested a similar re-wilding of Australia, restoring ecological communities to the pre-Aboriginal state of 50,000 (or more) years ago. Such re-wilding is ecological restoration at its most ambitious.

*Wiens, J.A. 2007. The demise of wildness? *Bulletin of the British Ecological Society* 38(4): 78–79. Reproduced with permission of the British Ecological Society.

Ecological Challenges and Conservation Conundrums: Essays and Reflections for a Changing World, First Edition. John A. Wiens.

The idea has generated considerable debate among conservationists, which need not be rehashed here (see Caro 2007). The relevant point is that such proposals aspire to restore wildness to a previous state, not just by protecting threatened populations or mending ecosystems degraded and diminished by decades of abuse, but by aiming at an endpoint that few believe possible (or even desirable). Proponents of re-wilding imagine a much wilder world.

At the polar opposite, Peter Kareiva and his colleagues, writing in *Science* (2007), have suggested that there are no places on earth that remain untainted by people, that "ours is a world of nature domesticated" and further domestication of nature is inescapable. Domesticated nature means nature exploited and controlled for human benefits. Sure, there is a spectrum of degrees of domestication, but the game of preservation of the wild is essentially over, and wildness lost. Better, they say, for conservationists and ecologists to help humanity domesticate nature wisely, to preserve a balanced mix of ecosystem services. Rather than protecting nature from people, conservation stewardship should manage the trade-offs among ecosystem services so that both people and nature thrive.

It is this proposition of domesticated nature that I wish to challenge (mildly, because Peter is a colleague and a friend, and because their argument has much merit). Part of my concern is pragmatic. The shift of conservation focus from biodiversity to ecosystem services might be regarded as little more than an embrace of a new ecological buzzword, but it has a more pernicious side. Once one admits to the domestication of nature, one embarks on a utilitarian pathway. It's fine to talk of people and nature thriving together, but if recent history is any guide that doesn't often happen. If it's a choice between people and nature, nature rarely wins. And the emphasis on ecosystem services opens the door to "designer ecosystems," ecosystems engineered to perform vital ecological functions and provide essential goods and services to people in the most efficient (read cost-effective) manner. Think of the possibilities for genetically modified organisms. Biodiversity is diminished in such a world, leaving little room for wildness.

My other concern is philosophical. Philosophers and environmental ethicists talk of "instrumental" versus "intrinsic" values. Instrumental values are those that directly or indirectly benefit humans—the ecosystem services of domesticated nature. The moral imperative is only to conserve those parts of nature that have utilitarian value. Intrinsic values are those of nature in its own right—wildness, for example. In this case, there are strong moral arguments for conserving all aspects of biodiversity. I think it is too early to give up on wildness. When Thoreau observed that "in wildness is the preservation of the world" or Aldo Leopold spoke of watching the "fierce green fire" die in the eyes of a dying wolf, they understood something of the intrinsic value of nature. That's not something we should easily give up.

It's true that there are probably no places on earth that are not tinged by human actions (global climate change is making sure of that). The Pleistocene megafaunas of North America and Australia are gone forever. No amount of re-wilding, even with proxies, can recreate the communities or ecosystems of those past millennia.

Better to focus on the wildness we still have. And there is still much wildness in the world, whether it be in the Serengeti of Africa, the Great Basin shrubsteppe of North America, or the Bialowieza Forest of Poland. Wilderness Areas in the United States are crown jewels of wildness that should be cherished, and managed, in that context. But wildness also persists in many of the places where people live and work—overgrown fields, hedgerows, urban parks, military training areas, and the like. Instead of creating a dualism of wild nature versus domesticated nature (or, for that matter, between instrumental and intrinsic values), we should recognize that these are simply points along a continuum of "naturalness" and values.

"Where the wild things are" should be more than a product of Sendak's (1963) fertile imagination, yet we should also focus on "where the people are." As we seek that elusive balance between people and nature that Kareiva and his colleagues advocate, we should be sure to leave some room for wildness.

CHAPTER 24

What lies behind the debates? Philosophy, values, and ethics

Why do ecologists and conservationists get into such intense and rancorous debates? Some of it has to do with the way science works. As one moves from data to interpretation to theory (or the reverse), honest differences of opinion can arise. Combine these with strong personalities and the psychological boost of a debate and the stage is set. However, there are also deeper forces at work—philosophy, values, and ethics.

Let me use a debate that has recently erupted in conservation to illustrate my point. The debate has to do with what's been called "new conservation," "new conservation science," or simply "conservation science," and it is shaking the very foundations of conservation.

The initial salvo in the debate was fired by Peter Kareiva, Michelle Marvier, and Robert Lalasz in an article in the journal of the Breakthrough Institute, followed by a peer-reviewed paper in *BioScience*.[1] Kareiva et al. argued that, despite its apparent progress and widely touted achievements, conservation is failing in its mission to protect and preserve biodiversity. Humans are transforming the Earth at an alarming pace and biodiversity continues to decline. Aiming backward to restore species and environments to historical states won't work, and appeals based on the value of nature for its own sake fall on the public's deaf ears. A new approach is needed, quickly.

Enter "new conservation." Kareiva et al. proposed that nature is not as fragile as is often portrayed, that ecosystems have considerable resilience, and that many extinctions of species have had little effect on the overall functioning of the ecosystems. The logical train is as follows. "Nature is so resilient that it can recover rapidly from even the most powerful human disturbances." Consequently, "conservation's continuing focus upon preserving islands of Holocene ecosystems in the age of the Anthropocene is both anachronistic and counterproductive." By moving away from a preservationist tradition, then, conservation can focus on "development by design, done with the importance of nature to thriving economies foremost in mind."[2] A key goal of the newly anointed "conservation

[1] Kareiva et al. (2012) and Kareiva and Marvier (2012).
[2] Kareiva et al. (2012).

Ecological Challenges and Conservation Conundrums: Essays and Reflections for a Changing World, First Edition. John A. Wiens.

science"[3] is the "improvement of human well-being through the management of the environment."[4] Although the traditional focus of conservation on preserving biodiversity is not abandoned, it is considered only when it improves human well-being. To be successful, they suggest that conservation must acknowledge peoples' self-interests and work within economic systems by forming partnerships with large corporations. Conservation must become more about people; as Kareiva puts it, "at the end of the day, it is all about us."[5]

The response from Michael Soulé, one of the founders of conservation biology, was immediate outrage. Soulé called new conservation "a powerful but chimeric movement" that would "hasten ecological collapse globally, eradicating thousands of kinds of plants and animals and causing inestimable harm to humankind in the long run."[6] In Soulé's view, new conservation replaces biological diversity with economic growth as the goal of conservation. Consequently, it is neither new[7] nor conservation.

The battle was joined. Michelle Marvier fanned the flames by calling new conservation "true conservation" and accusing Soulé of attempting to "marginalize a group of dedicated individuals and discredit new approaches to conservation."[8] Others entered the fray, accusing one another of mis-stating positions or facts and warning that conservation is either being sold out to corporate interests or that it is hopelessly mired in the past and needs fresh thinking and approaches to mobilize broad public support.[9] Kareiva and Soulé then wrote about "finding common ground."[10] Any common ground, however, seems limited to a shared commitment to the value of biodiversity; they continue to stand apart on what that value is and who or what should benefit from it. Malcolm Hunter et al. tried to calm the waters by noting that the perspectives are really complementary and there is room in conservation for both.[11] A commentary in *Nature* called on conservationists to

---

[3]Kareiva and Marvier carefully frame the new approach as science, contrasting it with the current "conservation biology": "conservation science that is focused primarily on biology is likely to misdiagnose problems and arrive at ill-conceived solutions" (Kareiva and Marvier 2012: 963). Kareiva and Marvier argue that conservation must include human-related disciplines such as economics, sociology, and health sciences. It's not clear how this makes new conservation any more scientific than conservation biology; after all, biology is itself a science.

[4]Kareiva and Marvier (2012: 962).

[5]Kareiva (2011).

[6]Soulé (2013: 896).

[7]The conservation value of places where people live and work, touted as a centerpiece of new conservation, was an important element of Aldo Leopold's 1933 book on game management (Leopold 1933).

[8]Marvier (2014: 1).

[9]See, for example, Kirby (2014), Doak et al. (2014), and Marvier and Kareiva's response (2014) to Doak et al. (2014).

[10]Kareiva (2014) and Soulé (2014).

[11]Hunter et al. (2014).

stop their "vitriolic, personal battles," recognize the value of diverse perspectives, and get on with the business of protecting the planet's biodiversity.[12] Writers and journalists have been attracted to the debate like moths to a flame, spreading the arguments far beyond the cozy confines of scientific journals and conferences.[13]

As is often the case in such debates, there are valid points on both sides. The "new conservationists" contend that protected areas are not enough—conservation must also occur in places with people—and that it will succeed only with broad public support. It must also be conducted in a way that respects basic human rights. The central thesis, that conservation as practiced in the past will be inadequate to address future challenges, also seems valid. While opponents of new conservation might accept many of these points,[14] their concerns are also valid. Will past approaches be abandoned or minimized as conservation shifts to populated places? Will the economic tail end up wagging the conservation dog? Will conservation be led astray if it gets too close to the corporate world?[15]

I'm not hopeful that this debate will be resolved and put to rest anytime soon. The basic issues aren't new—there have been arguments about the relationship between people and nature for at least as long as people have talked about conservation. John Muir and Gifford Pinchot carried on an intense debate in the early part of the 20th century over a proposal to construct a dam in Hetch Hetchy Valley in Yosemite National Park, in California. Pinchot argued the utilitarian position, that building the dam would provide the greatest good for the greatest number of

---

[12]Tallis et al. (2014).

[13]There is a growing proclivity for participants in such debates to air their views through interviews and articles in the popular media. It has not escaped notice that this is an effective way to garner attention for one's cause, as well as one's self (Pimm 2014). Fresh perspectives may also emerge when journalists rather than scientists provide the accounts. For example, D.T. Max wrote a piece about The Nature Conservancy (TNC) and new conservation for *The New Yorker* (2014) that included insights that would not make it into peer-reviewed scientific papers. Science writers are also involved—Emma Marris' book *Rambunctious Garden* (2011) has become somewhat of a bible for the new conservation movement and is frequently cited in scientific publications.

[14]In fact, by linking conservation to human rights and well-being, advocates of new conservation have made conservation a humanitarian endeavor, placing those who would question new conservation in the awkward position of seeming to disregard the needs of people.

[15]Having worked for 6 years in the worldwide office of TNC, I have seen both the benefits and risks of corporate partnerships. Conservation NGOs like to work with large corporations partly because (to paraphrase John Derringer) that's where the money is. That's what has enabled TNC to conduct large land deals that protect natural areas from development and to influence environmental policies at multiple scales. There is also a hope that, by exposing corporate leaders to a businesslike approach to conservation, corporations will become more attentive to the environmental impacts of their activities. At the same time, the leadership of TNC has increasingly emphasized the human benefits of nature, particularly ecosystem services, and has linked this form of conservation to the humanitarian efforts of other international organizations.

people. Muir was passionate in his opposition: "Dam Hetch Hetchy! As well dam for water-tanks the people's cathedrals and churches, for no holier temple has ever been consecrated by the heart of man."[16] Muir lost the debate and the dam was built (although there is now a battle over removing the dam).

The debates between Muir and Pinchot or between Kareiva and Soulé (or those described in the previous chapter) are not really about science, or even about conservation, but about values and philosophy and how they translate into ethical and moral responsibilities. Pinchot believed that conservation must work within the realities of a socioeconomic world, and advocates of new conservation and of the interventions I discussed in Chapter 23 clearly believe the same. They see little problem with moving species about, condoning all but the most egregious invasives, accepting novel ecosystems, prioritizing species and conservation efforts on the basis of benefits to people, or working closely with corporations to achieve their goals. The end justifies the means, and aggressive and innovative means are needed to keep conservation from failing in its mission. It's an instrumentalist philosophy.

Those who counter these practices hold to a deeply intrinsic philosophy. Nature has standing and value in its own right, independent of any values (good or bad) to people. To them this is a core principle, an axiom, which no amount of pragmatism can rent asunder. Michael Soulé puts it this way: "wild things and places have incalculable intrinsic value, at least as salient as the value of humanity."[17] This does not mean that nature cannot simultaneously have intrinsic and instrumental values, even though in this debate, and in the broader history of conservation, they have often been framed as being in conflict. The environmental philosopher Arne Næss considered this issue some time ago.[18]

These are not scientific positions but beliefs about philosophy and ethics.[19] They determine what is valued about nature, by whom, for whom. Advocates can always find evidence to support their position and to refute opposing positions, all dressed in the finery of science. However, the arguments are always colored by values. Values are the baggage that we carry with us as members of a culture, and they are deeply ingrained. That doesn't mean that values and science don't

---

[16]Muir (1912: 262); see Callicott (1990).

[17]Soulé (2014: 637).

[18]Næss (1986); see also Vucetich et al. (2015).

[19]There's also a belief in a balance of nature that lies buried deeply (sometimes very deeply) beneath these and other debates. Implicitly, those who argue against the more extreme forms of intervention discussed in Chapter 23 do so because of a feeling that undisturbed (or unengineered) ecosystems are more resilient, as natural systems ought to be. Talk of the fragility of nature implies a sensitive balancing act. Putting nature first, as those who cherish the intrinsic values of nature do, often carries with it an assumption that natural systems, if not stable, are at least organized well enough to have internal feedbacks that prevent chaotic variations. It's also possible, however, that I'm overly sensitive to this issue, and see traces of a balance-of-nature view where none exist. That would surprise me.

mix; indeed, values are part of science. It does mean that scientists, like anyone else, should be aware of when and how beliefs and values are infusing their work, affecting the questions they pose, how they interpret results, and what they choose to emphasize in talking (or debating) with other scientists or with the public. Scientists in general, and perhaps conservationists in particular, should be open and transparent about articulating their values.

Values are expressed as moral or ethical positions—what an individual feels is right or wrong and what a society deems to be acceptable. Almost any question in conservation has moral and ethical overtones; I considered several such questions at the end of the previous chapter. Most of these questions express the tension between the ethics of putting people first or nature first. As the role of technology in conservation continues to grow, conservationists will be faced with an even more daunting ethical dilemma: because we *can* do something, does that mean we *should* do it? This question is at the core of the de-extinction debate I described in Chapter 23.

I'm neither a philosopher nor an ethicist, so I shouldn't venture any further into that realm.[20] Suffice it to say, however, that philosophy, values, and ethics affect not just how we think about nature, but how we conduct science to learn about it and how we communicate what we know and think to others. Ethical standards are what hold science together, make it work, and give reason for why it should be granted credibility.[21] This is why scientific facts can generally be regarded as being at least provisionally "true."[22] Ethical standards for how science should be communicated in the mainstream (i.e., peer reviewed) scientific literature are also well established, but the ethics of how it should be communicated more broadly are blurred. This blurring opens the door to science advocacy, which I consider in the following chapter.

---

[20]Much has been written about environmental ethics; I've found these to be particularly enlightening: Charles Birch's *A Purpose for Everything* (1990), Holmes Rolston's *Conserving Natural Value* (1994) and *A New Environmental Ethics* (2012), Eric Higgs' *Nature by Design* (2003), the collection of essays edited by Kathleen Dean Moore and Michael Nelson, *Moral Ground: Ethical Action for a Planet in Peril* (2012), Allen Thompson and Jeremy Bendik-Keymer's *Ethical Adaptation to Climate Change* (2012), and the writings of Paul Errington (*A Question of Values*, 1987), E.O. Wilson (*The Creation*, 2006), and, of course, Aldo Leopold in *A Sand County Almanac* (1949). A different, instrumentalist perspective is developed by Steven Vogel in *Thinking Like a Mall* (2015).

[21]In *Merchants of Doubt* (2010), Naomi Oreskes and Erik Conway describe what can happen when ethical standards are bent or disregarded, and elsewhere I have considered how expectations and agendas may have influenced science in assessing the impacts of the *Exxon Valdez* oil spill (Wiens 2013b).

[22]The history of any science, of course, is replete with examples of what were once regarded as "true facts" that were later questioned and eventually abandoned; this is what Thomas Kuhn wrote about in *The Structure of Scientific Revolutions* (1970). We like to think of "facts" as being immutable, but facts are based on observations by humans, which (the observations) or who (the humans) are fallible.

CHAPTER 25

A digression on advocacy in conservation

It is part of being human and a member of society to have beliefs and values that are deeply held. They define what an individual regards as good and right and true. They affect how people live their lives as well as the opinions they hold.

Of course, scientists are people, so they have beliefs, opinions, and values just like anyone else. But these are not supposed to intrude into their science—the facts, obtained fairly, should speak for themselves. Scientists, then, suffer from an internal tension between their beliefs and their objectivity. This opens the door to advocacy, which becomes problematic when the science is used to push a preferred policy or political agenda.

Such policy advocacy is a particularly contentious issue in ecology and conservation. The subject matter itself—nature and biodiversity—elicits strong feelings. Many ecologists and conservation scientists have been drawn to these disciplines because of a love for nature that develops in childhood, instilling a worldview about what is good and right.[1] Aldo Leopold, who said so many things so well, put it this way:

> *"A thing is right when it tends to preserve the integrity, stability and beauty of the biotic community. It is wrong when it tends otherwise."*[2]

Seeing nature on the run all about them, then, it is hard for ecologists or conservationists not to take offense or be outraged and to want to do something about it. As Leopold also observed:

> *"One of the penalties of an ecological education is that one lives alone in a world of wounds. Much of the damage inflicted on land is quite invisible to laymen. An ecologist must either harden his shell and make believe that the consequences of science are none of his business, or he must be the doctor who sees the marks of death in a community that believes itself well and does not want to be told otherwise."*

The pull to advocate for nature, to speak out, is therefore great.[3]

[1]I wrote about my own experience in the Prologue of this book, and in his autobiography, *Naturalist* (1994), E.O. Wilson described how his boyhood interests in nature set his course for a career in evolutionary and ecological science.
[2]Leopold (1949: 262).
[3]Leopold (1966: 165).

Ecological Challenges and Conservation Conundrums: Essays and Reflections for a Changing World, First Edition. John A. Wiens.

Put simply, the choice is to stay in the ivory tower, producing and publishing science and letting others decide what to do with it, or to promote the incorporation of the science into policy actively, perhaps with a hidden agenda. It is not difficult to distinguish science from advocacy at these extremes, but the reality is more nuanced. For more than a decade, Bob Lackey has been warning about the dangers of "normative science"—science that carries with it personal values or policy preferences[4]—as an insidious form of "stealth advocacy."[5] Talking about a "healthy forest," for instance.[6] Or saying that protected areas provide favorable conditions for wildlife and should be expanded. Or that a clearcut forest is degraded and damaged. Such statements are normative—they contain subtle judgments about values that can influence policy. Lackey would have us say instead that a forest has particular attributes that increase its productivity, or that the abundance of species X is greater in protected areas than elsewhere, or that a clearcut forest has been altered—statements of scientific fact rather than opinion.

It's difficult, however, to avoid normative language in writing or talking about conservation. Virtually all conservationists, for example, believe that biodiversity is good and should be protected. Although this premise may be rationally defended, it's a statement of values nonetheless. In a review of papers published in conservation and natural-resource journals, Mike Scott et al. found evidence of normative statements in nearly all of the papers, and over half expressed a policy preference.[7] One could dismiss this widespread use of normative language as something relatively harmless—it's just the way ecologists and conservationists write and speak, unaware of any normative content. However, to a layperson or decision-maker it may carry the force of science, telling them not just what is, but whether it's good or bad and what should be done about it. That's when it becomes policy advocacy.

The effects of advocacy don't stop when scientists mix their opinions with their science in presenting their findings. Once a value premise is assumed, even subconsciously, it can feed back to affect the science—how questions are asked, studies are designed, or results are interpreted.[8] I consider some of these more subtle effects of advocacy in Essay 35, *Scientific responsibility and responsible ecology (1997)* (page 280). As a hypothetical example, however, consider conserving a species whose numbers have been declining. To understand why, you might conduct a study on one or another human impact—fragmentation of its forest habitat, for example. This choice is not value-free, but follows from the expectation that the decline was due to human actions and the widespread belief that habitat

[4]Lackey (2004). A more specific definition of normative science is "science developed, presented, or interpreted based on an assumed, usually unstated, preference for a particular policy or class of policy choices"; Lackey (2007: 13).
[5]Roger Pielke, Jr. explores the issue of stealth advocacy in his book, *The Honest Broker* (2007).
[6]See Essay 28, *A metaphor meets an abstraction: the issue of "healthy ecosystems (2015)"* (page 230).
[7]Scott et al. (2007).
[8]Thus, advocacy can have much the same power over science as a paradigm; see Chapter 2.

fragmentation is detrimental. If you find a negative relationship between fragment size and abundance, the interpretation and conclusion might have already been planted in your mind. You may then brief forest managers on your findings, free of any interpretations (but probably using normative language), or you might join a movement to change management policies so that harvesting that might further fragment the forest is restricted.

Is there anything wrong with this picture? It depends on how one thinks of advocacy. Values can enter at any point in the process, from the choice of what to study and measure, through the interpretation of results, to the use of results to influence forest management and policy. Most ecologists and conservationists would probably agree that much of this is what we normally do in conducting a scientific study and that only the last step is policy advocacy. However, advocacy in science can be pernicious, starting with a mere choice of words and growing into something that affects how the science is done and how it is communicated, and to whom. There are multiple opportunities for bias and agendas to creep into the process. Those who are concerned about advocacy in science maintain that we must guard against it to ensure that scientists do not become just another policy advocacy group, sacrificing whatever credibility and standing they may have.

There is, of course, a counter view. Michael Nelson and John Vucetich have argued that scientists are, above all, citizens, and like any citizen they have a moral obligation to promote what they think is right to the best of their abilities.[9] It is, in fact, their ethical responsibility as members of society. A scientist's hard-won scientific credibility carries with it an obligation not just to inform policy deliberations (i.e., present the facts), but also to advocate in a justified and transparent way for what their science tells them are preferred policy options (i.e., what they think or believe). Bob Lackey and others agree that scientists have a moral obligation to provide accurate and relevant information to those making policies, but they must do so in a policy-neutral way—just the facts, not what they think or prefer. In this view, it's simply not acceptable to advocate for personal beliefs or political agendas under the guise of science.

So what does one do? Cold-hearted, emotionless, totally objective scientists[10] do not tend to become ecologists or conservationists. All of us have feelings about nature and about what is right or wrong. For example, some passionately believe that it is wrong to condone the extinction of *any* species, while others believe in cutting the losses and doing the best job we can of protecting as many species as possible—this conflict is the essence of the debate about triage.[11] No conservationist is insensitive to extinction. The challenge is to determine how best to balance science with personal values.

---

[9]Nelson and Vucetich (2009).

[10]If such exist, they probably wouldn't be much fun to be around.

[11]See Essay 22, *Talking about triage in conservation (2015)* (page 188).

Some scientists become actively engaged in policy and the political process and can be very effective. They play an important role, using their stature and credibility to bring science into places where it might otherwise be neglected. For example, Albert Einstein sent a series of letters to Franklin Delano Roosevelt during World War II, using his standing as a scientist to draw attention to the potential uses of nuclear reactions and his concern that German scientists were rapidly developing the capacity to build an atomic bomb. Einstein's policy advocacy contributed to the development of the atomic weapons that ended the war with Japan and unleashed the Cold War and subsequent nuclear proliferation.[12] Steve Schneider worked tirelessly to draw attention to the reality and consequences of climate change, using his scientific standing and knowledge in speaking and writing on the coming crisis and helping to guide the efforts of the Intergovernmental Panel on Climate Change (IPCC).[13] Several ecologists and conservationists have adopted a strong advocacy role, using their science to argue persuasively for policies that benefit nature, biodiversity, and conservation.

But not everyone has the scientific credentials, interpersonal skills, courage, or commitment to be on the front lines of scientific advocacy. Regardless of how far one goes down the road to active advocacy, however, it's important to know where you are on the road. It may involve no more than using the normative language that surreptitiously creeps into publications. It may be in the way one frames a research question, designs a study, or analyzes data so that conclusions consistent with your expectations (your beliefs) become almost inevitable.[14] It can occur when you are asked to present the results of your science to managers or politicians who then ask what you would recommend—an invitation that is hard for a scientist to refuse. But making a recommendation about actions or policy is advocating a position. It may be based on "best available science" (e.g., your own), but it is filtered, consciously or not, by your values and beliefs—what you think, not just what your science showed you.[15] If you've decided to step outside

[12]Einstein's letters are worth reading as examples of reasoned and direct science advocacy; see http://hypertextbook.com/eworld/einstein.shtml.

[13]Schneider's experiences in the trenches of both science and advocacy about climate change are recounted in his 2009 book, *Science as a Contact Sport.*

[14]This, of course, would simply be poor science or advocacy based on poor science. Biases and preconceptions do have a nasty way of working their way into one's thinking, however, and their effects are often subtle and hard to detect. Even good science may contain unconscious biases.

[15]Here's an example. I serve on a board to advise the State of California, through its Delta Stewardship Council, on science and how it is being used in the management of the Sacramento-San Joaquin Delta and its water (see Chapter 10 and Wiens and Hobbs 2015). We are also asked to make recommendations that may affect water policy. Recommendations about the amount of tidal wetland targeted for restoration or the timing of release of water from upstream dams, while based on science, are shaped by our thinking about what is important ecologically. While we try to avoid any intrusion of personal beliefs or preferences into our recommendations, we nevertheless feel a scientific and ethical responsibility

your role as a scientist and actively advocate for or against a policy, then say so, but first ask yourself whether it's an appropriate use of your position and whether you're straying too far from your area of expertise into areas about which you're as uninformed as everyone else.

Advocacy by itself is not unethical so long as the advocacy position does not masquerade as a science position. The ethical dilemma is in recognizing the borderline between advocating for what one thinks and believes about nature and the environment, versus letting what one thinks and believes affect how one does science. Easy to say, but much harder to do.

Several decades ago the ecologist Robert Whittaker addressed this dilemma:

> *" ... we should as scholars both keep our roots in our own research and seek understanding of and speak out on problems of human ecology when we may. There may be great need for balancing, if possible, the politics of slogan and doctrine with probing, thoughtful commentary on our problems from an ecological perspective, in terms of underlying forces, complexities of interrelation, and humane implications for the future."*[16]

---

to tell people what we conclude, based on the science. However, we try to be transparent about it.

[16] Whittaker (1969: 196).

ESSAY 35

Scientific responsibility and responsible ecology (1997)*

In 1997 I was asked to comment on an article by the Canadian forest ecologist Gordon Baskerville that appeared in the inaugural issue of the online journal Conservation Ecology.[1] *Baskerville maintained that ecological science provided little useful advice to guide resource management and policy, due largely to a failure among ecologists to appreciate the realities of management and a scientific culture that emphasized numerical precision over useful generality. Baskerville's arguments were thought-provoking, and his paper and the responses it elicited are well worth a read. My comments focused on two aspects of Baskerville's thesis that continue to generate debate among ecologists, conservationists, and resource managers today.*

One was the nagging issue of scale. Baskerville noted that ecologists and those who attempt to apply ecological knowledge operate at different scales of time and space. Such mismatches in scale are not confined to the gap between science and practice—they bedevil comparisons among ecological studies conducted at different scales, attempts to conserve species that function at different scales, implementation of regional policies at local scales, and so on. The problem is real, but Baskerville's solution was for ecologists to adjust their studies and theories to the broader scales of management and policy. I suggested instead that both ecologists and managers should align their work with the scales of the phenomena of interest. Let nature determine the appropriate scales, be it that of a bark beetle infestation, an owl home range, forest fire dynamics, watershed hydrology, or something else. The problem, of course, is that all of these things function on different temporal and spatial scales, so multi-scale approaches are required. There are also very real practical constraints on the scales at which either scientific studies or management actions can be implemented. Scale remains a conundrum.

The other issue was the tension between science and advocacy. Purists argue that, to maintain credibility as an unbiased source of information and insights on which to base actions, science and scientists should shun advocacy. As I pointed out, however, it's difficult to prevent elements of advocacy from surreptitiously creeping into even the best-intentioned science. This can happen, for example, by asking value-laden questions, adopting a study design that foreordains conclusions, or interpreting results with an eye toward what is expected rather

*Wiens, J.A. 1997. Scientific responsibility and responsible ecology *Conservation Ecology* [online] 1(1): 16. Available at http://www.consecol.org/vol1/iss1/art16. The journal is now *Ecology and Society* (http://www.ecologyandsociety.org/).

[1] Baskerville (1997).

Ecological Challenges and Conservation Conundrums: Essays and Reflections for a Changing World, First Edition. John A. Wiens.

than what is actually shown. It's even more difficult to avoid advocacy when science intersects with management or policy, both of which have societal goals and therefore are suffused with advocacy for actions to achieve those goals.

So it's not just a matter of aligning scales to coalesce ecology and management, but a need to recognize goals and agendas and ensure that, in striving to be relevant, ecology does not become biased, and therefore ultimately irrelevant.

Ecology has been called the relevant science. If this relevance is to be anything other than a catchy phrase, two things are necessary. First, ecology must generate reliable information and insights about environmental systems. This means that the information must be gathered in a rigorous and unbiased way and interpreted objectively: the science must be sound. Second, ecological information must be incorporated into management practices and policy decisions, the arenas where this information can make a difference. This means that the information must be gathered in a way that will provide useful insights to management and policy, and it must be communicated in a way that is understandable to people who have not been raised on a diet of ecological jargon.

Baskerville's essay ("Advocacy, science, policy, and life in the real world," *Conservation Ecology*) addresses primarily the second point. Here, I will comment on the issue of scale as it relates to Baskerville's thesis and then offer some thoughts on the first point, which I believe is the more important of the two.

Baskerville notes that the scale on which ecological research is conducted rarely matches the scale of management. In his view, this leads to a preoccupation with describing fine-scale patterns rather than discovering how systems actually function. The latter concern is simply a misreading of current trends and activities in ecology, which are increasingly focused on ecological mechanisms and processes. The concern with mismatched scales, however, is very real, and it permeates all of ecology, whether basic or applied. We know that ecological processes, and the patterns they produce, change as the scale in space or time changes. We also know that these changes are often nonlinear (Wiens 1989, Levin 1992). What we do not know is the nature of the "scaling functions" that describe these relationships for particular phenomena. Thus, although logistical necessity and ecological tradition (e.g., a preoccupation with experiments) usually dictate that ecological investigations be conducted at relatively fine scales of space and time, it is not clear how these findings should be extrapolated to the broader scales on which management is usually practiced. Simple linear extrapolations usually will not work. The issue of extrapolation is one of the most vexing in ecology, but if ecologists wish to contribute to effective resource management and scientifically based policy, it must become a central focus of ecological research. Some progress might be made by implementing carefully designed multiscale investigations (e.g., Koch et al. 1995), by integrating some of the approaches of macroecology (Brown 1995) with fine-scale, mechanistic studies, or by using theories of self-organizing processes in ecosystems (e.g., Holling et al. 1996) as a framework for evaluating scale dependency and scaling functions.

Baskerville argues that, if ecological information is to be relevant to management and policy, ecologists must scale their studies to match the scales used in management. In my view, this is an unrealistic demand, not because ecologists are unlikely to do this, but because it is not likely to advance ecologically based resource management. Management scales have been determined by a variety of factors: some economical, some political, some simply traditional, but all essentially anthropocentric. These scales of management do not necessarily coincide with the scales on which organisms respond to their environments, on which the processes affecting biodiversity or disturbance regimes operate, or on which ecosystems function. Ultimately, the health and profitability of the resources that are being managed depend on these organismal, population, and ecosystem scaling relationships, and to regard the scales of management as fixed and inviolate is a mistake. Rather than imposing a management scale on nature, efforts should be made to adjust the scales of management to those of natural processes, insofar as economic, social, and political constraints permit. This is, in fact, the approach being developed in the "new forestry" practiced in parts of Sweden and elsewhere (Haila 1994, Pastor et al. 1996).

Thus, problems with incorporating ecology into management and conservation stem, at least in part, from problems in translating patterns and mechanisms across scales. Detecting such scale-dependent effects depends, of course, on the scientific rigor of the studies conducted. More importantly, how (or whether) ecological science is applied to broader issues of public concern depends critically on the integrity of the scientific process. Let me turn now to the issue of sound science.

Ecologists are increasingly being drawn into environmental debates, whether about the effects of land uses or management practices (such as grazing of rangelands or clearcutting of forests), conservation issues (such as the design of natural reserves or the management of endangered species), or environmental perturbations (such as oil spills or global change). These are often emotionally charged issues. They attract media attention and, not infrequently, foster litigation. Because they are socially relevant, they are often associated with opportunities for research funding. Collectively, these pressures create an atmosphere in which advocacy for a particular position in a debate may affect the scientific process. At its worst, advocacy may masquerade as science (Wiens 1996) or science may be perceived as advocacy (Westoby 1997). Both erode the credibility of honest science.

Advocacy can influence the scientific process in several ways, beginning with the questions we ask. Most questions in ecology are influenced by our preconceptions about nature or current fashions in the discipline. Questions that relate to environmental or management issues often carry with them values (e.g., oil spills are bad) that can affect the way the questions are framed and the range of answers that can be obtained. Thus, instead of asking, "Did an oil spill have environmental effects, and if so, what?", the question may become "How bad were the effects?" The distinction is important, for the first question leads to an unbiased examination of environmental effects, whereas the second restricts attention only to environmental damages. We often initiate a study because of some environmental debate and the need to bring scientific evidence to bear on the issues, so some element of advocacy in the questions we ask is probably unavoidable. Biased questions, however, do not lead to good science.

Advocacy can also affect the way a study is designed. It can lead to conscious or unconscious bias in the selection of study areas, the way sampling stations are distributed, or the degree to which pseudoreplication is tolerated. Control areas may differ systematically from treatment areas, for example, but these differences may be ignored in analyzing results; as a consequence, all differences are mistakenly attributed to treatment effects (Wiens & Parker 1995). By specifying that certain variables will be measured while others will not, the results of a study may be constrained, enhancing the likelihood that one will find what one expects (or wants) to find. Whether or not values are implicit in the questions we ask, the study design and analysis must be rigorous and unbiased. Weak or biased study designs lead to weak or biased "scientific evidence," which is worse than worthless in environmental debates.

Perhaps the most pernicious and subtle effect of advocacy is on the interpretation of results. Even if a study is objectively framed and conscientiously designed and analyzed, the findings still must be placed in a context. Rosseau (1992) drew attention to what he called "pathological science," in which researchers unknowingly lose their objectivity in interpreting data that are near detection limits when much is riding on the results. Advocacy can reinforce this tendency, particularly because environmental debates are often emotionally charged. We care about the environment; that is why many of us became ecologists in the first place. Faced with the uncertainty that characterizes most findings in ecological research, it is all too easy for these feelings to influence how we view data, which results we choose to emphasize or to disregard, or whether what begins as speculation becomes transformed into "fact" because it is consistent with an advocacy position.

The responsibility of the ecologist, then, is to do science, and to do it as rigorously and objectively as possible. We must accept what our results tell us, not what our emotions might say. This is not to say that ecologists must retreat into the ivory tower and refrain from taking positions in environmental debates. There is an urgent need to bring scientific evidence to bear on environmental and management issues. These issues are so pressing that ecologists have a responsibility not to remain quiet when their findings can contribute to the debate. We should communicate the results of our science clearly and vigorously, in understandable terms, to the public and policy arenas. In so doing, however, there is also the paramount responsibility to recognize our own advocacy and to distinguish clearly between statements that are based on science and those that are based on personal values or viewpoints (Pitelka & Raynal 1989, Murphy & Noon 1991). We might take our lesson from the atomic scientists who, following the development of atomic energy at the end of the Second World War, spoke out frequently and vigorously about the potential abuses of this power, without compromising or distorting the science itself.

Ultimately, of course, ecological science is only one of many inputs to the development of management protocols or environmental policy. Some of these inputs reflect advocacy positions based on economics, religious beliefs, or political agendas. As ecologists, our agenda should be science, and our responsibility is to ensure that scientific findings carry the greatest possible weight in societal decisions about the environment.

PART V
Concluding comments

It's time to pull my thoughts together. Not an easy task, given the variety of changes and the breadth of challenges and conundrums confronting ecology and conservation. And it's made all the more difficult because so much of this fuels feelings of depression and despair about the state of nature. But it would be a mistake to conclude that it's all hopeless. Conservation depends on hope, not just because doing otherwise would be to concede defeat to the forces of environmental degradation and extinction, but because there really are reasons to be hopeful. More and more people are recognizing the multiple values of nature and the environment, and it's not just because there are more and more people. Popularizers such as Carl Sagan, David Suzuki, David Attenborough, and M. Sanjayan have made science and environmental issues understandable to the public, helping to build a broadening platform for conservation.

This part includes a single (hopeful) chapter, along with three essays that, in different ways, reinforce the importance of optimism.

Ecological Challenges and Conservation Conundrums: Essays and Reflections for a Changing World, First Edition. John A. Wiens.

CHAPTER 26

Whither ecology and conservation in a changing world?

Before I went to college and began to get serious about things, I was an avid follower of Pogo. Pogo was a comic strip created by Walt Kelly that dealt with the bumbling antics of various animals in Okefenokee Swamp, in southeastern United States. It blended humor with social commentary, philosophy, puns, and fine artwork—it was much more than your average comic strip. Pogo Possum, the lead character, was given to making sage observations about the environment. Viewing the accumulation of garbage and litter in his swampland home, he concluded, "We have met the enemy and he is us." The phrase caught on, and Kelly made it the centerpiece of a poster for the first Earth Day in 1970 (Figure 26.1).[1]

Pogo's observation might well be the mantra for this book. Humans—"us"—are the enemies of nature. We are behind the forces of change. We are the source of the challenges and conundrums that confront ecology and conservation. We are the cause of the feelings of discouragement, disillusionment, doom, and despair that the previous chapters in this book may have left you with. Yet, we must also be the benefactor and savior of nature—there is simply no one else to do it. Conservation cannot afford to be the dismal science—that's best left for economics—but should assume the mantle of the hopeful science. This was the point of the essay about woodpeckers I used at the beginning of this book.[2]

And there really is cause for hope. More land and water is being protected from development. The conservation values of places where people live and work are acknowledged by conservation organizations and legal agreements.[3] At least some endangered species have recovered. Peregrine falcons now patrol the canyons among the skyscrapers of New York City, London, Chicago, and other cities; packs of gray wolves roam in several western US states; southern white rhinos have come back from the brink of extinction; there are now self-sustaining populations of black-footed ferrets alerting prairie dogs to watch their backs; and

[1]Appropriately, I'm writing this on Earth Day 2015. The global reach of Earth Day is cause for hope, but that leaves 364 days in a year in which the Earth needs the same attention. Every day should be an Earth Day.

[2]Essay 2, *Found! The survivor in the swamps (2005)* (page 4).

[3]For example, conservation easements, or Safe Harbor Agreements under the Endangered Species Act.

Ecological Challenges and Conservation Conundrums: Essays and Reflections for a Changing World, First Edition. John A. Wiens.

Figure 26.1 Walt Kelly's poster for the first Earth Day, April 22, 1970.

Aleutian cackling geese have transitioned from being endangered to agricultural pests. Where I live in the Willamette Valley of western Oregon, habitat restoration efforts (many by private landowners) have enabled numbers of the Oregon chub to increase from less than 1,000 when it was listed under the Endangered Species Act in 1993 to over 140,000 today; it's the first fish to be removed from the list because it recovered fully. Andrew Balmford recites other conservation success stories in his book *Wild Hope*.[4]

Unfortunately, such success stories stand out because there are so few. But it doesn't have to be so. We can balance realism about the magnitude of threats to nature with optimism and hope in our ability to deal with them. But it will require changes. We must, as Lincoln said, "think anew, and act anew."

We can begin by changing how we think about nature. Natural systems have been in flux since well before the onset of the Anthropocene; "pristine nature" hasn't existed in most places for a long time. Because there is no longer a "natural" state of nature, our perceptions of what we aim to conserve should be revised. Novel ecosystems and invasive species, for example, are novel and invasive only with reference to what was there before, and they are bad only if what we value in nature remains rooted in the past. This doesn't mean that we should stand back and meekly accept any novelty or invasive that comes our way, but neither should we fight, however valiantly, to save every species and protect every place. Resistance to change, much as it is part of human nature, will not help the cause of conservation. As the world changes, we should be judicious about which changes we can accept, which we should fight, and which we can manage to the benefit of species, ecosystems, and people.[5] Recalibrating the goals of conservation does not mean that we are giving up on nature.

Adjusting to change, however, entails tradeoffs—any action in a complex system evokes reactions elsewhere in the system. Michael Rosenzweig has described "win-win" situations, in which both people and biodiversity benefit from management or land-use actions.[6] This reconciliation ecology approach extends the conservation agenda beyond protected areas to acknowledge the value of places where people live and work. Yet win-win situations inevitably involve tradeoffs. Somewhere in the system, there will be unintended consequences—losers. "Win-win" is usually accompanied by "Yes, but . . . ."

Tradeoffs mean that we can't do just one thing—a balance must be found among competing priorities. When my wife and I decided to build a new house while also trying to be environmentally conscious ("green"), for example, we encountered a cascading series of decisions that pitted practicality or personal preferences against greenness; I recount our first-hand experience with tradeoffs in Essay 37, *Being green isn't easy* (2010) (page 295). Our experience is not unusual. People

---

[4]Balmford (2012).

[5]I address change and the need to break free of the bonds of the past and past thinking in Essay 36, "*It was the best of times, it was the worst of times* ... (2009)" (page 292).

[6]Rosenzweig (2003).

encounter tradeoffs and set priorities all the time, deciding to do one thing but not something else. How conservation fares in the future will depend to a great extent on how people factor nature and the environment into their personal decisions and priorities. I'm seeing this now in California as the state enters its fourth year of extraordinary drought.[7] Some people are willing to reduce their use of water to leave enough for fish, while others insist on using water as they always have while leaving the fish to make do as best they can. Fortunately, many people care deeply about the natural world, see the tradeoffs, and try to balance their own interests with those of the environment. They are the wellspring of hope in conservation. So, too, are politicians who will speak out and act as advocates for people *and* the environment. Stewart Udall, who I profile in Essay 38, *Stewart Udall and the future of conservation (2010)* (page 298), was one such. Other, like voices, are sorely needed today.

I've devoted most of this book to reflections about how the world is changing and the challenges and conundrums this creates for ecology and conservation. I summarized these thoughts as desiderata for a new approach to conservation—a "new conservation paradigm"—in Chapter 1, so I won't repeat them here. However, there's one more tradeoff that has permeated debates over what conservation is really about for decades. How conservation moves forward to cope with the changing world hinges on the tension and balance between instrumentalist and intrinsic philosophies of nature's values. Is nature, in whatever form and wherever it occurs, something that has value and is therefore worth conserving because it provides things that we can use—the goods and services now commonly associated with ecosystem services? Or does it go beyond that? Do the values of nature in and of itself supersede any benefits that might be provided to people? Does conservation entail a moral obligation, something more than drawing down nature's capital for our own gain?

These are questions for philosophers and ethicists, not scientists. But scientists, or anyone thinking about conservation, cannot ignore them. So where do I stand? I believe that the people who care intensely about biodiversity and the environment will do all they can to ensure that nature persists in some form in some places. But it will not be enough, especially as the "nature" they care about changes before their eyes. It will take more—a deeper commitment by society at large—to maintain nature. And it will be a new nature, perhaps Emma Marris' "rambunctious garden,"[8] full of novelty and alien species. It will be different from what we remember and have idealized. As awareness of the instrumental values of nature

[7] Some say the drought is "unprecedented," which depends on how far back one looks to find precedents. Paleoecological studies record equivalent dry periods lasting several centuries. What is unprecedented depends on how one's window of time shapes perceptions of environmental variation.

[8] Marris (2011).

grows, support for conservation should also grow as people realize that it's in their own best interests.

Is that all there is? Is the future of conservation to tabulate nature's values on a spreadsheet so that they can be included in the mystical machinations of economists? Those who extol the intrinsic values of nature would vehemently say not. However, justifying conservation on the philosophical grounds of morality and ethics may have little sway in an increasingly materialistic society. Placing nature above people when it is people who make the decisions is a hard sell, especially when times are tough (as they will be as climate change plays out). It leads to simplification of the choices—give the water to little fish or to farmers who are the drivers of economies. For most people, it's not a choice at all.

But not so fast. Human societies since the dawn of civilization have appreciated beauty. This is the basis for art, literature, music, and poetry. Science also has beauty—an elegant theory or a cunningly written computer code has beauty, and the stunning images produced by the Hubble Space Telescope reveal the beauty of the galaxies and the universe. And while it's true that beauty is in the eye of the beholder, the common essence of beauty is the feeling that wells up inside when one views a work of art, listens to music or poetry, appreciates the nuances and intellectual power of theory, or experiences the joy of discovery—what the physicist Richard Feynman called the pleasure of finding things out.[9] And although neurologists are probing the mechanistic basis for an appreciation of art or music and cognitive psychologists are digging into how people appreciate literature, something intangible and lovely—beauty—remains.

How is this at all relevant? Nature, I suggest, has beauty, whether it is that of a manakin performing a courtship dance, the graceful spirals of a nautilus, or the smoothly functioning way a forested watershed filters water and processes nutrients. This sort of beauty cannot be digitized, nor, unlike a Picasso or a Jasper Johns, can it be monetized. But it is there. Whether we label it as instrumental or intrinsic is beside the point. These, after all, are ultimately *our* values and *our* philosophies. But beauty transcends these values. Ultimately, nature's beauty, in all its forms, is what we must conserve, for wherever it is lost our lives are diminished.

[9]Feynman (1999).

ESSAY 36

"It was the best of times, it was the worst of times ... "[1] (2009)*

I must have been in a rare optimistic mood when I wrote this essay. Yes, it was easy to list ways in which the environment and the economy were getting worse, but I naïvely suggested that this might provide an opportunity for change and progress—the best of times. Perhaps politicians would begin listening to the public about climate change and actually take actions rather than obstruct them. Perhaps the top scientists that President Obama brought into his administration would help to usher in a new attitude of respect rather than disdain for science.

But this was not to be. The scientists tried, they really did. But the euphoria of Obama's election soon wore off. Congress continued with business as usual, although deepening partisan divides led to increasing gridlock and even less business. Public acceptance of the reality of climate change waned as concerns about the economy, jobs, and various wars grew stronger.[2] It seemed that the worst of times were taking over.

But I remain optimistic, partly because I believe that crises breed innovation and change. My concluding points still hold. Broader, more inclusive discussions of how to deal with environmental issues can lead to more effective management. But we must acknowledge that the ecological systems we have known and loved in the past are likely to change along with everything else, so we need to think anew about how to conserve them.

It's easy, following the news these days, to think that Charles Dickens had it only half right. Surely, these are the worst of times. The environmental crisis deepens every week, it seems. No sooner than the IPCC reports on the accelerating rate of climate change, we learn that the predictions of sea-level rise and global warming are already outmoded underestimates. Changes in land use reduce the quantity and quality of habitats and threaten to place even more species on the pathway to extinction. Fresh water is becoming scarcer, and what remains increasingly carries the exotic contaminants of modern chemistry. It's becoming a familiar litany.

*Wiens, J.A. 2009. It was the best of times, it was the worst of times ... *Bulletin of the British Ecological Society* 40(1): 50–51. Reproduced with permission of the British Ecological Society.
[1] Charles Dickens, *A Tale of Two Cities*.
[2] A Pew poll in 2014 found that 61% of Americans thought there was strong evidence for global warming, down from 70% in 2008.

Ecological Challenges and Conservation Conundrums: Essays and Reflections for a Changing World, First Edition. John A. Wiens.

And then there's the economy. In the United States, thousands of people have lost their homes, respected financial institutions have suddenly dissolved into insolvency, and bankruptcies of businesses both large and small occur daily. The economic turmoil in one place creates waves that surge back and forth across the globe, enveloping both developed and developing countries. Add to this the political turmoil of wars in Iraq and Afghanistan, ongoing conflicts in the Middle East, rebellions in the Congo and elsewhere, and one has the makings of a "perfect storm"—the worst of times.

So how can it also be the best of times? I would argue that the worst of times may in fact be the best time to reconcile conflicting agendas and to reconsider business as usual. There is a "politics of hope" in the United States that presents an opportunity to think anew about how to balance ecology and economics in ways that will benefit the well-being of both biodiversity and people, and to mobilize support to address the environmental changes that are bearing down upon us.

As I write, my home state of California is facing a budgetary crisis that has led to the suspension of all work on some 4,000 publically funded projects that would benefit conservation and the environment. These programs range from the restoration of damaged wetlands, to the purchase of remnant natural habitat, to the construction of trails and interpretative centers. Many of these projects were initiated under a boom of economic growth. Now we have the opportunity—indeed, the imperative—to assess whether they all make equally good sense. We should think of conservation and environmental management in terms of returns on investments, of the overall balance sheet of economic costs and environmental and social benefits.

Even efforts to mitigate or adapt to climate change, which in the past have been hindered because the costs of actions must be borne immediately but the benefits will accrue at some time in the future, are now gaining support among an increasingly aware public. According to a recent survey conducted by the Pew Research Center, more than 70% of Americans now think that there is solid evidence of global warming.[3] Efforts are underway at many levels of government not just to develop alternative energy sources, reduce CO_2 emissions, and bolster carbon sequestration, but also to adapt to the changes in the earth's ecosystems that have already been set in motion and that will play out over the coming century.

Here in the United States, there is now a growing sense of purpose and optimism among those who care about environmental conservation. After 8 years in which the environment has been largely an afterthought and science has been willfully distorted or suppressed, President Obama has selected respected scientists to fill high-level cabinet and agency positions. Stephen Chu, a Nobel laureate in physics for developing methods to cool and trap atoms with laser light, is the incoming Secretary of Energy; John Holdren a physicist who has focused on the causes and consequences of environmental change and on energy technologies and policy,

[3]Research conducted by the Pew Research Center for People and the Press in April 2008 (http://people-press.org/report/417/a-deeper-partisan-divide-over-global-warming).

will be the President's Science Advisor; and Jane Lubchenco, an ecologist who cut her teeth studying plant–herbivore interactions in rocky intertidal communities, will lead the National Oceanographic and Atmospheric Administration (NOAA).

These scientists, and others like them in the new administration, will re-establish scientific thinking and scientific integrity as essential components of the government's approach to environmental issues. But they will need help. What else is needed, and what can ecologists contribute? Of course, our knowledge of the complex interactions between organisms and their environments should be brought to bear on the conservation and management of ecological systems. But simply making that knowledge available to those framing policy or making decisions is not sufficient, for two reasons.

First, the gap between scientists and practitioners, managers, or policy-makers is still deep. Assessments of where and when the effects of climate change or land-use change on species and communities may be greatest, for example, must be framed in ways that can direct appropriate management actions. This will require that scientists, land managers, and decision-makers—as well as the public—work much more closely together, *listening* as well as talking, to understand differing perspectives on shared challenges.

Second, it is becoming apparent that the future will not be a simple extension or extrapolation of the past. The environmental changes that have been set in motion will lead to nonlinear, threshold shifts in the physical and biological settings of ecological systems, producing novel combinations of species and webs of species interactions. Attempting to maintain current species distributions or assemblages in the face of changes in the environment may be futile. The pervasiveness of altered fire regimes, invasive species, disease spread, and a host of cascading effects of climate change will increase the likelihood that management actions will have unintended consequences (see, e.g., http://www3.interscience.wiley.com/journal/121638291/abstract). Traditional ways of conserving or managing ecological systems may no longer work. It is said that necessity is the mother of invention; surely now is the time for ecologists to be inventive.

In the winter of 1862, at a turning point in the history of the United States, Abraham Lincoln addressed Congress with these words:

> "The dogmas of the quiet past, are inadequate to the stormy present. The occasion is piled high with difficulty, and we must rise—with the occasion. As our case is new, so we must think anew and act anew."

The present is stormy once again, but the opportunities are there. To forestall the worst of times, we must once again think and act anew.

ESSAY 37

Being green isn't easy (2010)*

> *"Being green" is a rapidly growing environmental lifestyle. It involves factoring such things as carbon footprints, energy sources and costs of production and transport, sustainability, and a host of other factors into everyday decisions about what foods to eat, what clothes to wear, or how to get to where you're going.*
>
> *This essay grew out of the trials, tribulations, and tradeoffs we experienced in building a new house while also being green. And the experience was mostly about tradeoffs. It seemed that every decision placed the ideal of being* really *green against the practicalities of costs and comfort. Doing the right thing in terms of greenness (minimizing window area to reduce energy loss, for example) often conflicted with things we personally valued (watching the Oregon rains move across the landscape or a hawk vainly trying to catch birds attracted to our feeders). And in that sense, building the house really was a parable for the broader issue of how people relate to the environment, which is all about tradeoffs between environmental values and personal values.*
>
> *We've lived in the house for more than four years now. The house isn't energetically self-sufficient, but it is efficient. The Swedish wood stove makes the entire house warm and cozy on the coldest winter nights (which, in western Oregon, aren't really all that cold), and we've scarcely made a dent in the stacks of firewood from the trees we felled. And Bea's garden is beginning to produce a bounty of vegetables that will carry us through the summer and well into the winter. But we had to buy a freezer to store it all. More tradeoffs.*

Bea and I recently moved to western Oregon. It's intended to be the last of a series of moves over the years. So we've decided to build a house. And because we're both ecologists, it can't be just any house—it must be a *green* house (which isn't the same as a greenhouse, but … oh, never mind).

It's a nice thing to think about. We'll be able to design the house to reflect what's important to us—a kitchen that's fun to cook in while talking with each other, large windows for views and openness, a site landscaped to provide habitat for wildlife, and plenty of space for books. But we also want to do our part to save the Earth. We'll lower our energy consumption and generate power from alternative sources, reducing our carbon footprint. We'll contribute to sustainable forestry. We'll grow our own organic vegetables. And all that. We'll feel good. And while we're sitting

*Wiens, J.A. 2010. Being green isn't easy. *Bulletin of the British Ecological Society* 41(4): 46–47. Reproduced with permission of the British Ecological Society.

Ecological Challenges and Conservation Conundrums: Essays and Reflections for a Changing World, First Edition. John A. Wiens.

on our sustainably forested deck in chairs made of recycled plastic bottles while sipping locally produced organic wine, we may even feel a bit self-righteous.

But it's turning out not to be easy to be green. Of course, we could build something like Eeyore's house of sticks in the woods or perhaps a yurt or a hogan, using all natural materials we gathered ourselves. Now *that* would be green! But it rains a lot in western Oregon. Besides, we've been there—we've done the 60s. Bea wants a place to do her gardening (yes, we'll also have a greenhouse), and I find it difficult to write without classical music blasting in the background. So there's a tradeoff between being really green and a lifestyle we want.

Building a green house is all about tradeoffs. First there were the trees. A dozen Douglas-fir trees were growing on the lower hillside. The trees had been there for half a century, happily sequestering carbon. They provided wildlife habitat. But Douglas-firs grow tall in western Oregon (did I mention that it rains a lot?). They would eventually fall, possibly taking the house with them, and their shade would block the photovoltaic panels we want on the roof, preventing us from using solar energy to reduce our off-site electrical demands. So the tradeoff. Reluctantly, we cut down the offending trees, keeping what we could as firewood and sending the rest to be milled or pulped. We still have trees, but not as many.

So we lost some trees, but we can recycle the firewood in our high-efficiency stove to heat the house. But even an efficient wood stove releases carbon to the atmosphere. We'll also need tight insulation and a ventilation system to circulate that warm air and keep it from accumulating in the upper reaches of the house—but that requires extra space and energy. More tradeoffs.

And then there's the roof. On a recent trip to Norway I was duly impressed by the sod or turf roofs, which have been part of the Scandinavian culture for centuries. These living roofs are becoming an integral part of green construction in many parts of the world—they provide insulation, filter rainwater (yes, good in Oregon), cool the surface through evapotranspiration by the vegetation, and fairly reek of greenness. But plants won't grow under solar panels. So we've decided on a recycled metal roof, which provides fire protection (an increasing concern in forested landscapes, given climate change) and is long-lasting. Another tradeoff.

It doesn't end there. Should we use local wood decking to avoid carbon emissions during transport? Or should we use wood sustainably grown in Brazil, which lasts five times longer? What about composite decking made from recycled materials? Should we use steel framing, which has recycled content but is thermally inefficient, or sustainably harvested wood framing that provides better insulation? Windows are a major avenue for heat loss, even with low-emissive (low-e) glass, argon gas filling, and double-glazing, so it's best to minimize them. But we don't want to live in a dark box where we can't see the habitat.

In the end, each of these decisions (and there are many more) involves balancing tradeoffs. We're both scientists, so we should adopt a rational, even quantitative, approach. We could have done the calculations, for example, to determine how the reductions in energy demands and emissions produced by the solar panels compare to the carbon released by felling the trees (taking into account the processing of the trees into useable products, the energy involved in doing so, and the disparate time scales over which these factors operate). But we are also people. We have likes and dislikes, many determined by our upbringing, our professions, and our culture, so there is a strong emotional element to such decisions. And over it all hangs economics. Being green is not cheap. Every time we explore another green option, the construction costs mushroom. At some point, economic realities restrict how green we can be, but emotions influence when we will reach that point.

I'd like to think there is a broader message here. In a sense, the story of our green house and our agonizing over the multiple tradeoffs we face is a parable for the broader societal debate about how we build a greener world. Some pundits and politicians pose this debate as a choice between the environment and the economy (or jobs). Of course, it's much more than that. No matter what position one takes, decisions will involve assessing tradeoffs and reaching some balance or compromise that is far from ideal. "Win-win" outcomes are elusive. And how these tradeoffs are evaluated is always influenced by cultures and perspectives. Discussions about how best to mitigate greenhouse gas emissions or how to balance protection of endangered species with the need to exploit their habitat involve assessing tradeoffs, and the discussions are often emotionally charged. Bea and I share a common set of values, so it's relatively easy for us. When multiple, conflicting value systems are involved, assessing the tradeoffs can seem intractable.

The parable of our green house may be useful, but only to a point. We'll have our house built in another year. We are committed to building green, so we'll muddle through the tradeoffs and get it done. I doubt that we'll see much progress in that time on the far thornier issues of "the environment versus the economy." This will require breaking through deeply held ecological, economic, spiritual, and esthetic values, what the British polymath Gregory Bateson[1] called the "habits of mind." Muddling through must begin with a shared commitment to building a green world and the recognition that there are no perfect solutions, only some that are better than others. And we had better get moving if we hope to outpace the change that is upon us.

[1]Bateson (1972).

ESSAY 38

Stewart Udall and the future of conservation (2010)*

Politicians who can be considered environmental visionaries are few and far between. Stewart Udall was one. His passing prompted me to re-read his book, 1976: Agenda for Tomorrow[1] *and to realize, once again, how sensible his arguments were and how much they are missed in today's debates. Udall saw clearly the environmental problems that could come from materialism and unbridled growth, and he called for civil, open, and reasoned debate among citizens and politicians about how best to resolve them.*

Udall was obviously an optimist, and in writing this essay I was perhaps also guilty of hopeful optimism. Debates now seem to be dominated by political posturing, mostly designed to appeal to fringe groups and aimed at establishing positions for the next election. And because there is always a next election (terms in the US House of Representatives are for 2 years), it never lets up. But the environmental changes wrought by global warming, consumptive land use, and burgeoning human populations also never let up.

I suggested that we start thinking now about what sort of world we might inhabit in 2076. The most recent projections, however, indicate that rates of climate change and sea-level rise are likely to accelerate markedly come mid-century. If we are going to have much left to plan for in 2076, we'd best start acting, rather than just thinking, now. Unfortunately, there is only weak evidence that either is happening. Not a time for optimists.

Stewart Udall died the other day. He was 90. Readers of the *Bulletin* may not know much about him, but they will surely recognize his accomplishments. Udall was the United States Secretary of the Interior during the Kennedy and Johnson administrations, from 1961 to 1969, when the cornerstones of the current environmental movement in the United States were being laid. Udall used his position and his political skills to help pass the Clean Water Act, the Clean Air Act, the Wilderness Act, the Endangered Species Preservation Act, the Wild and Scenic Rivers Act—and the list goes on. He was instrumental in establishing dozens of National Parks, Seashores, Monuments, and Wildlife Refuges. He probably did

*Wiens, J.A. 2010. Stewart Udall and the future of conservation. *Bulletin of the British Ecological Society* 41(2): 39–40. Reproduced with permission of the British Ecological Society.
[1] Udall (1968).

Ecological Challenges and Conservation Conundrums: Essays and Reflections for a Changing World, First Edition. John A. Wiens.

more to promote conservation and protection of the environment than anyone since Teddy Roosevelt.

Udall was also a visionary. Shortly before he left office in 1968, he gathered his thoughts together in a short book, *1976: Agenda for Tomorrow.*[2] Reading it, one realizes how much progress has been made, but at the same time how little some things have changed. Udall's agenda was an antidote to what he saw as the rise of materialism and mediocrity in the United States during the 1950s and 1960s. He was concerned about the "gospel of growth" that prioritized parking over parks. Udall wanted people to find that elusive balance between economic growth, the livability of cities, the value of open space, and the condition of the environment. Conservation was a central element: "plans to protect air and water, wilderness and wildlife are in fact plans to protect man."

A politician himself, Udall saw politics as the pathway to implementing his agenda. He argued for a fresh political framework, one that recognized local and state governments as sources of innovation that should be encouraged rather than restricted by the federal government. He proposed that Congress conduct an annual national assessment, in which leaders from all segments and levels of society would use town meetings to "think aloud for the country, to try out new ideas, to question possibly outdated assumptions." People would debate and discuss what kind of future they really wanted. There would be civil and intelligent discourses, carried on within a civil society.

Open debates could strengthen the bond between Congress and the people, and shift the focus to shared goals and priorities instead of special interests. Udall bemoaned the failure of Congress to think and speak for the nation as a whole. He saw Congress as too constrained by arcane rules, such as the Senate's "egregiously unconstitutional filibuster rule." The recent debates about health care and the initial skirmishes over legislation to address climate change suggest that Congress still suffers from these constraints. Public support for environmental conservation, which Udall considered central to civil society, has waned as economic conditions have worsened, even though consumers are moving toward new green-energy options. Almost half of the public now believes the seriousness of global warming is exaggerated.[3]

Udall's agenda may seem idealistic, but he framed it as a way of addressing the real problems facing the nation, problems lying at the intersection of society and the environment. Udall might despair at the persistence of special interests and gridlock in the body politic today, but he would no doubt be astounded at the progress that has been made. Health care legislation has been passed and climate change is now being discussed in the halls of Congress and international conferences. President Obama highlighted climate change in his Nobel Peace Prize acceptance speech. Millions of acres have been set aside for conservation, enabled

[2]Harcourt, Brace & World, Inc., New York; the year marked two centuries since the nation's founding in 1776.

[3]Gallup Poll, March 2010; http://www.pollingreport.com/enviro.htm.

by the growing influence of conservation organizations and facilitated by some of the legislation that Udall helped to frame. Although town meetings sometimes disintegrate into yelling and confrontations, more often they promote the civil discourse that Udall envisioned.

As ecologists, we can help to accelerate the implementation of Udall's agenda. Ecologists deal with the couplings between organisms and their environments. We now have a better understanding and better projections of how future environmental changes may affect ecological systems and what adaptations to these changes are likely or feasible. Our work carries messages about how the environments in which people live are likely to change and how the generation and delivery of ecosystem services for people may be altered. Udall's agenda called for a broad, inclusive, and rational discussion of such challenges. Ecologists can help by explaining relentlessly, clearly, and simply, again and again until we are numb, how science and nature work, what the future is likely to hold, and how we can help nature and people adapt to the coming changes.

But we must proceed with caution, with an awareness of our own limitations. At one point, Udall observed that society developed "a new Myth of Scientific Salvation, persuading ourselves that the fouling of our own nest could be quickly cleansed by a sorcerer's apprentice from the house of science." The public at large still expects that, whatever the problem, science and technology will find a solution. But scientists are not sorcerers, and we must partner with the government and the people to address the pressing social and environmental problems that so concerned Udall.

The United States will celebrate its third century in 2076. It is not too early to begin the discussion about that future, what we would like it to be, and what will be needed to achieve it. We've made progress, but if we heed Udall's agenda we can do better.

References

AAAS. 2014. What We Know: The Reality, Risks and Response to Climate Change. Available at: http://whatweknow.aaas.org/wp-content/uploads/2014/07/whatweknow_website.pdf.

Adams, W.M., and J. Hutton. 2007. People, parks and poverty: Political ecology and biodiversity conservation. *Conservation and Society* 5: 147–183.

Ale, S.B., and H.F. Howe. 2010. What do ecological paradigms offer to conservation? *International Journal of Ecology*. doi: 10.1155/2010/250754.

Allee, W.C., A.E. Emerson, O. Park, T. Park, and K.P. Schmidt. 1949. *Principles of Animal Ecology*. W.B. Saunders, Philadelphia, PA.

Allen, C., and L. Gunderson. 2011. Pathology of failure in the design and implementation of adaptive management. *Journal of Environmental Management* 92: 1379–1384.

Allen, C.R., J.J. Fontaine, K.L. Pope, and A.S. Garmestani. 2011. Adaptive management for a turbulent future. *Environmental Management* 92: 1339–1345.

Anchukaitis, K.J., and M.N. Evans. 2010. Tropical cloud forest climate variability and the demise of the Monteverde golden toad. *Proceedings of the National Academy of Sciences* 107: 5036–5040.

Andrewartha, H.G., and L.C. Birch. 1954. *The Distribution and Abundance of Animals*. University of Chicago Press, Chicago, IL

Andrewartha, H.G., and L.C. Birch. 1984. *The Ecological Web. More on the Distribution and Abundance of Animals*. University of Chicago Press, Chicago, IL.

Angelo, M.J. 2008. Stumbling toward success: A story of adaptive law and ecological resilience. *Nebraska Law Review* 87: 950–1007. Available at: http://digitalcommons.unl.edu/nlr/vol87/iss4/3.

Anielski, M., and S. Wilson. 2005. *Counting Canada's Natural Capital: Assessing the Real Value of Canada's Boreal Ecosystems*. The Pembina Institute, Canada. Available at: http://www.pembina.org/pubs/pub.php?id=204.

Aplin, L.M., B.C. Sheldon, and J. Morand-Ferron. 2013. Milk bottles revisited: Social learning and individual variation in the blue tit, *Cyanistes caeruleus*. *Animal Behaviour* 85: 1225–1232.

Arponen, A. 2012. Prioritizing species for conservation planning. *Biodiversity Conservation* 21: 875–893.

Bagchi, S., D.D. Briske, B.T. Bestelmeyer, and X. Wu. 2013. Assessing resilience and state-transition models with historical records of cheatgrass Bromus tectorum invasion in North American sagebrush-steppe. *Journal of Applied Ecology* 50:1131–1141.

Balkin, J.M. 1989. The footnote. *Faculty Scholarship Series*, Paper 287. Available at: http://digitalcommons.lawyale.edu/fss_papers/287.

Ecological Challenges and Conservation Conundrums: Essays and Reflections for a Changing World, First Edition. John A. Wiens.

Balmford, A. 2012. *Wild Hope. On the Front Lines of Conservation Success.* Chicago University Press, Chicago, IL.

Banks-Leite, C., R. Pardini, L.R. Tambosi, W.D. Pearse, A.A. Bueno, R.T. Bruscagin, T.H. Condez, M. Dixo, A.T. Igari, A.C. Martensen, and J.P. Metzger. 2014. Using ecological thresholds to evaluate the costs and benefits of set-asides in a biodiversity hotspot. *Science* 345: 1041–1045.

Baskerville, G.L. 1997. Advocacy, science, policy, and life in the real world. *Conservation Ecology* 1(1): 9. Available at: http://www.consecol.org/vol1/iss1/art9/.

Bateson, G. 1972. *Steps to an Ecology of Mind.* University of Chicago Press, Chicago.

Benson, M.H. 2012. Intelligent tinkering: The Endangered Species Act and resilience. *Ecology and Society* 17(4): 28. Available at: 10.5751/ES-05116-170428.

Bestelmeyer, B.T. 2006. Threshold concepts and their use in rangeland management and restoration: The good, the bad, and the insidious. *Restoration Ecology* 14:325–329.

Bestelmeyer, B.T. 2014. Deforestation of "degraded" rangelands: The Argentine Chaco enters the next stage of the Anthropocene. *Rangelands* 36(4): 36–39.

Bestelmeyer, B.T., and J.A. Wiens. 1996. The effects of land use on the structure of ground-foraging ant communities in the Argentine Chaco. *Ecological Applications* 6:1225–1240.

Bestelmeyer, B.T., M. Duniway, D.K. James, L.M. Burkett, and K.M. Havstad, 2013. A test of critical thresholds and their indicators in a desertification-prone ecosystem: More resilience than we thought. *Ecology Letters* 16: 339–345.

Birch, C. 1990. *A Purpose for Everything. Religion in a Postmodern World.* Twenty-Third Publications, Mystic, CT.

Bocetti, C.I., D.D. Goble, and J.M. Scott. 2012. Using Conservation Management Agreements to secure postrecovery perpetuation of conservation-reliant species: The Kirtland's warbler as a case study. *BioScience* 62: 874–879.

den Boer, P.J. 1968. Spreading of risk and stabilization of animal numbers. *Acta Biotheoretica* 18:165–194.

Borgman, C.L., J.C. Wallis, and N. Enyedy. 2007. Little Science confronts the data deluge: Habitat ecology, embedded sensor networks, and digital libraries. *International Journal on Digital Libraries* 7:17–30.

Botkin, D.B. 1990. *Discordant Harmonies. A New Ecology for the Twenty-first Century.* Oxford University Press, New York, NY.

Botkin, D.B. 2012. *The Moon in the Nautilus Shell. Discordant Harmonies Reconsidered.* Oxford University Press, New York, NY.

Bradbury, R. 1952. A sound of thunder. *Collier's Magazine* 28, 1952.

Brandner, J., A.F. Cerwenka, U.K. Schliewen, and J. Geist. 2013. Bigger is better: Characteristics of round gobies forming an invasion front in the Danube river. *PLoS One*. doi: 10.1371/journal.pone.0073036.

Breshears, D.D., O.B. Myers, C.W. Meyer, F.J Barnes, C.B. Zou, C.D Allen, N.G. McDowell, and W.T. Pockman. 2009. Tree die-off in response to global change-type drought: Mortality insights from a decade of plant water-potential measurements. *Frontiers in Ecology and the Environment* 7: 185–189.

Brodie, J.F., E. Post, and D.F. Doak (eds.). 2013. *Wildlife Conservation in a Changing Climate.* University of Chicago Press, Chicago, IL.

Bronowski, J. 1977. *A Sense of the Future.* The MIT Press, Cambridge, MA.

Brown, J.H. 1995. *Macroecology.* University of Chicago Press, Chicago, Illinois, USA.

Brown, J.H., D.W. Davidson, and O.J. Reichman. 1979. An experimental study of competition between seed-eating desert rodents and ants. *American Zoologist* 19: 1129–1143.

Brown, M., and R. Haworth. 1997. Culturally-embedded sustainability practices among the Walpiri: Conservation and commerce in the Tanami Desert. *Rural Society* 7(2): 3–16.

Burgman, M.A. 2005. *Risks and Decisions for Conservation and Environmental Management*. Cambridge University Press, Cambridge, UK.

Buskotter, J.T., J.A. Vucetich, S. Enzler, A. Treves, and M.P. Nelson. 2013. Removing protections for wolves and the future of the U.S. Endangered Species Act (1973). *Conservation Letters*. (online) doi: 10.1111/conl.12081.

Butchart, S.H.M., et al. 2015. Shortfalls and solutions for meeting national and global conservation area targets. *Conservation Letters* 8: 329–337.

Callan, R., N.P. Nibbelink, T.P. Rooney, J.E. Wiedenhoeft, and A.P. Wydeven. 2013. Recolonizing wolves trigger a trophic cascade in Wisconsin (USA). *Journal of Ecology* 101: 837–845.

Callicott, J.B. 1990. Whither conservation ethics? *Conservation Biology* 4: 15–20.

Callicott, J.B. 1995. The value of ecosystem health. *Environmental Values* 4: 345–361.

Caro, T. 2007. The Pleistocene re-wilding gambit. *Trends in Ecology and Evolution* 22: 281–283.

Carson, R.T., R.C. Mitchell, M. Hanemann, R.J. Kopp, S. Presser, and P.A. Rudd. 2003. Contingent valuation and lost passive use: Damages from the Exxon Valdez oil spill. *Environmental and Resource Economics* 25: 257–286.

Carwardine, J., T. O'Connor, S. Legge, B. Mackey, H.P. Possingham, and T.G. Martin. 2012. Prioritizing threat management for biodiversity conservation. *Conservation Letters* 5: 196–204.

Caughley, G. 1994. Directions in conservation biology. *Journal of Animal Ecology* 63: 215–244.

Chape, S., M.D. Spalding, and M.D. Jenkins (Eds.) 2008. *The World's Protected Areas. Status, Values and Prospects in the 21st Century*. University of California Press, Berkeley.

Clements, F.E. 1916. *Plant Succession. An Analysis of the Development of Vegetation*. Carnegie Institute of Washington, Washington, DC.

Cody, M.L., and J.M. Diamond (Eds.). 1975. *Ecology and Evolution of Communities*. Harvard University Press, Cambridge, MA.

Cohen, J.E. 1995. *How Many People Can the Earth Support?* W.W. Norton & Company, New York, NY.

Cole, D.N., and L. Yung (Eds.). 2010. *Beyond Naturalness. Rethinking Park and Wilderness Stewardship in an Era of Rapid Change*. Island Press, Washington, DC.

Cook, J., D. Nuccitelli, S.A. Green, M. Richardson, B. Winkler, R. Painting, R. Way, P. Jacobs, and A. Skuce. 2013. Quantifying the consensus on anthropogenic global warming in the scientific literature. *Environmental Research Letters* 8 024024. doi:10.1088/1748-9326/8/2/024024.

Cooper, G.J. 2003. *The Science of the Struggle for Existence. On the Foundations of Ecology*. Cambridge University Press, Cambridge, UK.

Costanza, R. 2006. Nature: Ecosystems without commodifying them. *Nature* 443: 749.

Costanza, R., and M. Mageau. 1999. What is a healthy ecosystem? *Aquatic Ecology* 33: 105–115.

Costanza, R., R. d'Arge, R. de Groot, S. Farber, M. Grasso, B. Hannon, S. Naeem, K. Limburg, J. Paruelo, R.V. O'Neill, R. Raskin, P. Sutton, and M. van den Belt. 1997. The value of the world's ecosystem services and natural capital. *Nature* 387: 253–260.

Costanza, R., R de Groot, P. Sutton, S van der Ploeg, S.J. Anderson, I Kubiszewski, S. Farber, and R.K. Turner. 2014. Changes in the global value of ecosystem services. *Global Environmental Change* 26: 152–158.

Cramer, V. A., and R. J. Hobbs. 2002. Ecological consequences of altered hydrological regimes in fragmented ecosystems in southern Australia: Impacts and possible management responses. *Austral Ecology* 27: 546–564.

Creighton, M. 1891. Hodgson, John (1779–1845). in *Dictionary of National Biography*, Volume 27. (S. Lee, ed.). Smith, Elder & Co, London.

Crutzen, P. J., and E.F. Stoermer. 2000. The 'Anthropocene'. *Global Change Newsletter* 41: 17–18.

Cuddington, K., and B. Beisner (Eds.). 2005. *Ecological Paradigms Lost. Routes of Theory Change*. Elsevier, Amsterdam, The Netherlands.

Cukier, K., and V. Mayer-Schoenberger. 2013. The rise of Big Data: How it's changing the way we think about the world. *Foreign Affairs* 92(3): 28–40.

Cullen, R. 2012. Biodiversity protection prioritisation: A 25-year review. *Wildlife Research* 40(2): 108–116.

Curtis, J.T. 1956. The modification of mid-latitude grasslands and forests by man. pp. 721–736 in *Man's Role in Changing the Face of the Earth* (W.L. Thompson, Jr.,, ed.). University of Chicago Press, Chicago, IL.

Daily, G. 1997. *Nature's Services: Societal Dependence on Natural Ecosystems*. Island Press, Washington, DC.

Dale, V.H., K.L. Kline, J. Wiens, and J. Fargione. 2010. *Biofuels: Implications for Land Use and Biodiversity. Biofuels and Sustainability Reports*. Ecological Society of America, Washington, DC. Available at: http://www.esa.org/biofuelsreports.

Darling, F.F., and J.P. Milton (Eds.). 1966. *Future Environments of North America*. The Natural History Press, Garden City, NJ.

Davis, M.A., et al. 2011. Don't judge species on their origins. *Nature* 474: 153–154.

Day. R.H., S.M. Murphy, J.A. Wiens, G.D. Hayward, E.J. Harner, and L.N. Smith. 1997. Effects of the *Exxon Valdez* oil spill on habitat use by birds in Prince William Sound, Alaska. *Ecological Applications* 7: 593–613.

Deevey, E.S. (Ed.). 1972. *Growth by Intussusception. Ecological Essays in Honor of G. Evelyn Hutchinson*. Transactions of the Connecticut Academy of Arts and Sciences, New Haven, CT and Archon Books, Hamden, CT.

Delcourt, P.A., and H.R. Delcourt. 2008. *Prehistoric Native Americans and Ecological Change. Human Ecosystems in Eastern North America Since the Pleistocene*. Cambridge University Press, Cambridge, UK.

DeShazo, J.R., and J. Freeman. 2006. Congressional politics. pp. 68–71 in *The Endangered Species Act at Thirty*, Volume 1 (D.D. Goble, J.M. Scott, and F.W. David, eds.). Island Press, Washington, DC.

Diamond, J. 2005. *Collapse. How Societies Choose to Fail or to Succeed*. Viking, New York, NY.

Diamond, J., and T.J. Case (Eds). 1986. *Community Ecology*. Harper & Row, New York, NY.

Doak, D.F., V.V. Bakker, B.E. Goldstein, and B. Hale. 2014. What is the future of conservation? *Trends in Ecology and Evolution* 29: 77–81.

Donlan, J., et al. 2005. Re-wilding North America. *Nature* 436: 913–914.

Donlan, C.J., J. Berger, C.E. Bock, J.H. Bock, D.A. Burney, J.A. Estes, D. Foreman, P.S. Martin, G.W. Roemer, F.A. Smith, M.E. Soulé, and H.W. Greene. 2006. Pleistocene re-wilding: An optimistic agenda for twenty-first century conservation. *The American Naturalist* 168: 660–681.

Drury, W.H. Jr. 1998. *Chance and Change. Ecology for Conservationists*. University of California Press, Berkeley, CA.

Dudley, N. (Ed.). 2008. *Guidelines for Applying Protected Area Management Categories*. IUCN, Gland, Switzerland.

Ehrlich, P.R. 1968. *The Population Bomb*. Ballantine Books, New York, NY.

Elton, C.S. 1958. *The Ecology of Invasions by Animals and Plants*. Chapman and Hall, London, UK.

Errington, P.L. 1987. *A Question of Values*. Iowa State University Press, Ames, IA.

Estes, J.A., and J.F. Palmisano. 1974. Sea otters: Their role in structuring nearshore communities. *Science* 185: 1058–1060.

Estes, J.A., E.M. Danner, D.F. Doak, B. Konar, A.M. Springer, P.D. Steinberg, M.T. Tinker, and T.M. Williams. 2004. Complex trophic interactions in kelp forest ecosystems. *Bulletin of Marine Science* 74: 621–638.

Fenichel, E.P., and J.K. Abbott. 2014. Natural capital: From metaphor to measurement. *Journal of the Association of Environmental and Resource Economics* 1: 1–27.

Feynman, R.P. 1999. *The Pleasure of Finding Things Out*. Perseus Publishing, Cambridge, MA.

Finkelstein, M.E., D.F. Doak, D. George, J. Burnett, J. Brandt, M. Church, J. Grantham, and D.R. Smith. 2012. Lead poisoning and the deceptive recovery of the critically endangered California condor. *Proceedings of the National Academy of Science* 109: 11449–11454.

Firn, J., T.G. Martin, B. Walters, J. Hayes, S. Nichol. I. Chades, and J. Cadwardine. 2013. *Priority Threat Management of Invasive Plant Species in the Lake Eyre Basin*. CSIRO Climate Adaptation Flagship Working Paper No. 17, CSIRO and Queensland University of Technology. Available at: http://www.csiro.au/en/Organisation-Structure/Flagships/Climate-Adaptation-Flagship/CAF-working-papers.aspx.

Fitzpatrick, J.W. et al. 2005. Ivory-billed woodpecker (*Campephilus principalis*) persists in continental North America. *Science* 308: 1460–1462.

Flannery, T. 1994. *The Future Eaters*. Reed, Melbourne.

Ford, R.G., J.A. Wiens, D. Heinemann, and G.L. Hunt. 1982. Modelling the sensitivity of colonially breeding marine birds to oil spills: Guillemot and kittiwake populations on the Pribilof Islands, Bering Sea. *Journal of Applied Ecology* 19: 1–31.

Freyfogle, E.T. 2003. *The Land We Share. Private Property and the Common Good*. Island Press, Washington, DC.

Friedel, M.H., B.D. Foran, and D.M. Stafford Smith. 1990. Where the creeks run dry or ten feet high: Pastoral management in Australia. *Proceedings of the Ecological Society of Australia* 16: 185–194.

Fritts, T.H., and G.H. Rodda. 1998. The role of introduced species in the degradation of island ecosystems: A case history of Guam. *Annual Review of Ecology and Systematics* 29:113–140.

Game, E.T., J.A. Fitzsimons, G. Lipsett-Moore, and E. McDonald-Madden. 2013. Subjective risk assessment for planning conservation projects. *Environmental Research Letters* 8: 045027. (online) doi: 10.1088/1748-9326/8/4/045027.

Gammage, B. 2011. *The Biggest Estate on Earth. How Aborigines Made Australia*. Allen & Unwin, Sydney, Australia.

Gladwell, M. 2000. *The Tipping Point: How Little Things Can Make a Big Difference*. Little, Brown, New York, NY.

Gleason, H.A. 1926. The individualistic concept of the plant association. *Bulletin of the Torrey Botanical Club* 53: 1–20.

Golley, F.B. 1996. *A History of the Ecosystem Concept in Ecology*. Yale University Press, New Haven, CT.

Gotelli, N.J., and G.R. Graves. 1996. *Null Models in Ecology*. Smithsonian Institution Press, Washington, DC.

Gould, S.J. 2002. *The Structure of Evolutionary Theory*. Harvard University Press, Cambridge, MA.

Grafton, A. 1997. *The Footnote. A Curious History*. Harvard University Press, Cambridge, MA.

Grantham, T.E., and J.H. Viers. 2014. 100 years of California's water rights system: Patterns, trends and uncertainty. *Environmental Research Letters*. doi: 10.1088/1748-9326/9/8/084012.

Gregory, R., L. Failing, M. Harstopne, G. Long, T. McDaniels, and D. Ohlson. 2012. *Structured Decision Making. A Practical Guide to Environmental Management Choices*. Wiley-Blackwell, Chichester, UK.

Groffman, P.M., et al. 2006. Ecological thresholds: The key to successful environmental management or an important concept with no practical application? *Ecosystems* 9: 1–13.

Groves, C.R. 2003. *Drafting a Conservation Blueprint. A Practitioner's Guide to Planning for Biodiversity*. Island Press, Washington, DC.

Gunderson, L.H., and C.S. Holling (Eds.). 2002. *Panarchy. Understanding Transformations in Human and Natural Systems*. Island Press, Washington, DC.

Guntensbergen, G.R. (Ed.). 2014. *Application of Threshold Concepts in Natural Resource Decision Making*. Springer, New York, NY.

Haila, Y. 1994. Preserving ecological diversity in boreal forests: Ecological background, research, and management. *Annales Zoologica Fennica* 31: 203–217.

Hampton, S.E., C.A. Strasser, J.J. Tweksbury, W.K. Gram, A.E. Budden, A.I. Batcheller, C.S. Duke, and J.H. Porter. 2013. Big data and the future of ecology. *Frontiers in Ecology and the Environment* 11: 156–162.

Hanak, E., J. Lund, A. Dinar, B. Gray, R. Howitt, J. Mount, P. Moyle, and B. Thompson. 2011. *Managing California's Water: From Conflict to Reconcillation*. Public Policy Institute of California, San Francisco, CA.

Hannah, L., G. Midgley, S. Andelman, M. Araújo, G. Hughes, E. Martinez-Meyer, R. Pearson, and P. Williams. 2007. Protected area needs in a changing climate. *Frontiers in Ecology and the Environment* 5: 131–138.

Hansen, J., M. Sato, P. Kharecha, D. Beerling, R. Berner, V. Masson-Delmotte, M. Pagani, M. Raymo, D.L. Royer, and J.C. Zachos, 2008. Target atmospheric CO_2: Where should humanity aim? *Open Atmospheric Science Journal*, 2, 217–231, doi: 10.2174/1874282300802010217.

Hanski, I., and O.E. Gaggiotti. 2004. *Ecology, Genetics, and Evolution of Metapopulations*. Elsevier Academic Press, London, UK.

Hartley, C.I. 1922. *The Importance of Bird Life*. The Century Company, New York.

Hassell, M.P. 1998. The regulation of populations by density-dependent processes. pp. 29–51 in *Insect Populations in Theory and Practice* (J.P. Dempster and I.F.G. McLean, eds.). Kluwer Academic Publishers, Dordrecht, The Netherlands.

Heath, C., and D. Heath. 2007. *Made to Stick. Why Some Ideas Survive and Others Die*. Random House, New York.

Heinlein, R.A. 1961. *Stranger in a Strange Land*. Ace Books, New York, NY.

Hemmingway, E. 1952. *The Old Man and the Sea*. Charles Scribner's Sons, New York, NY.

Herring, S.C., M.P. Hoerling, T.C. Peterson, and P.A. Stott (Eds.). 2014. Explaining extreme events of 2013 from a climate perspective. *Bulletin of the American Meteorological Society* 95(9), S1–S96. Available at: http://www2.ametsoc.org/ams/assets/File/publications/BAMS_EEE_2013_Full_Report.pdf.

Hewitt, N., N. Klenk, A.L. Smith, D.R. Bazely, N. Yan, S. Wood, J.I. MacLellan, C. Lipsig-Mumme, and I. Henriques. 2011. Taking stock of the assisted migration debate. *Biological Conservation* 144: 2560–2572.

Hey, T., S. Tansley, and K. Tolle. 2009. *The Fourth Paradigm. Data-intensive Scientific Discovery.* Microsoft Research, Richmond, WA, USA.

Hiers, J.K., R.J. Mitchell, A. Barnett, J.R. Walters, M. Mack, B. Williams, and R. Sutter. 2012. The dynamic reference concept: Measuring restoration success in a rapidly changing no-analogue future. *Restoration Ecology* 30: 27–36.

Higgs, E. 2003. *Nature by Design. People, Natural Process, and Ecological Restoration.* MIT Press, Cambridge, MA.

Hirsh, J.E. 2003. *Shakespeare and the History of Solloquies.* Farleigh Dickinson University Press, Madison, NJ.

Hobbs, R.J. 2013. Grieving for the past and hoping for the future: Balancing polarizing perspectives in conservation and restoration. *Restoration Ecology* 21: 145–148.

Hobbs, R.J., E.S. Higgs, and C.M. Hall. 2013. *Novel Ecosystems: Intervening in the New Ecological World Order.* Wiley-Blackwell, Chichester, UK.

Hobbs, R.J., et al. 2014a. Managing the whole landscape: Historical, hybrid, and novel ecosystems. *Frontiers in Ecology and the Environment* 12: 557–564.

Hobbs, R.J., E.S. Higgs, and J.A. Harris. 2014b. Novel ecosystems: Concept or inconvenient reality? A response to Murcia et al. *Trends in Ecology and Evolution* 29: 645–646.

Hoegh-Guldberg, O., L. Hughes, S. McIntyre, D.B. Lindenmayer, C. Parmesam, H.P. Possingham, and C.D. Thomas. 2008. Assisted colonization and rapid climate change. *Science* 321: 345–346.

Holling, C.S., G. Peterson, P. Marples, J. Sendzimir, K. Redford, L. Gunderson, and D. Lambert. 1996. Self-organization in ecosystems: Lumpy geometries, periodicities, and morphologies. pp. 346–384 in *Global change and terrestrial ecosystems* (B. Walker and W. Steffen, eds.). Cambridge University Press, Cambridge, UK.

Hoose, P. 2004. *The Race to Save the Lord God Bird.* Farrar, Straus and Giroux, New York.

Horowitz, A. 2011. Will the e-book kill the footnote? *The New York Times* October 11, 2011.

Huff, D. 1954. *How to Lie with Statistics.* Norton, New York.

Hunter, M.L., Jr., K.H. Redford, and D.B. Lindemayer. 2014. The complementary niches of anthropocentric and biocentric conservationists. *Conservation Biology* 28: 641–645.

IPCC. 2014. *Fifth Assessment Synthesis Report, Summary for Policymakers.* Available at: http://www.ipcc.ch/pdf/assessment-report/ar5/syr/SYR_AR5_SPMcorr2.pdf.

IPCC. 2014. Summary for policymakers. pp. 1–32 in *Climate Change 2014: Impacts, Adaptation, and Vulnerability. Part A: Global and Sectoral Aspects. Contribution of Working Group II to the Fifth Assessment Report of the Intergovernmental Panel on Climate Change.* Cambridge University Press, Cambridge, UK.

IUCN/SSC. 2013. *Guidelines for Reintroductions and Other Conservation Translocations.* Version 1.0. IUCN Species Survival Commission, Gland, Switzerland.

Iwamura, T., K.A. Wilson, O, Venter, and H.P. Possingham. 2010. A climatic stability approach to prioritizing global conservation investments. *PLoS ONE* 5(11): e15103. doi: 10.1371/journal.pone.0015103.

Jachowski, D.S., and D.C. Kesler. 2009. Allowing extinction: Should we let species go? *Trends in Ecology and Evolution* 24: 180.

Jackson, S.T. 2012. Conservation and resource management in a changing world: Extending historical range of variation beyond the baseline. pp. 92–109 in *Historical Environmental*

Variation in Conservation and Natural Resource Management (J.A. Wiens, G.D. Hayward, H.D. Safford, and C.M. Giffen, eds.). Wiley-Blackwell, Chichester, UK.

Janzen, D.H. 1983. No park is an island: Increase in interference from outside as park size decreases. *Oikos* 41: 402–410.

Jenkins, C.N., S.L. Pimm, and L.N. Joppa. 2013. Global patterns of terrestrial vertebrate diversity and conservation. *Proceedings of the National Academy of Science USA*. doi.10.1073/pnas.1302251110.

Jongsomjit, D., D. Stralberg, T. Gardall, L. Salas, and J. Wiens. 2013. Between a rock and a hard place: The impacts of climate change and housing development on breeding birds in California. *Landscape Ecology* 28: 187–200.

Kareiva, P. 2011. Balancing the needs of people and nature. *Nature Conservancy Magazine*, Spring issue; Available at: http://www.nature.org/newsfeatures/magazine/part-two-balancing-the-needs-of-people-and-nature.xml.

Kareiva, P. 2014. New conservation: Setting the record straight and finding common ground. *Conservation Biology* 28: 634–636.

Kareiva, P., and M. Andersen. 1988. Spatial aspects of species interactions: The wedding of models and experiments. pp. 38–54 in *Community Ecology* (A. Hastings, ed.). Springer-Verlag, New York.

Kareiva, P., and M. Marvier. 2003. Conserving biodiversity coldspots. *American Scientist* 91: 344–351.

Kareiva, P., and M. Marvier. 2012. What is conservation science? *BioScience* 62:962–969.

Kareiva, P., S. Watts, R. McDonald, and T. Boucher. 2007. Domesticated nature: Shaping landscapes and ecosystems for human welfare. *Science* 316: 1866–1869.

Kareiva, P., M. Marvier, and R. Lalasz. 2012. Conservation in the Anthropocene: Beyond solitude and fragility. *Breakthrough Institute Journal*, Issue 2. Available at: http://thebreakthrough.org/index.php/journal/past-issues/issue-2/conservation-in-the-anthropocene.

Kelling, S., W.M. Hochachka, D. Fink, M. Riedewald, R. Caruana, G. Ballard, and G. Hooker. 2009. Data-intensive science: A new paradigm for biodiversity studies. *BioScience* 59: 613–620.

King, S. 2000. *On Writing: A Memoir of the Craft*. Scribner, New York, NY.

Kingsland, S.E. 1985. *Modeling Nature. Episodes in the History of Population Ecology*. University of Chicago Press, Chicago, IL.

Kingsland, S.E. 2005. *The Evolution of American Ecology 1890–2000*. The Johns Hopkins University Press, Baltimore, MD.

Kirby, K.J. 2012. A view from the past to the future. pp. 281–288 in *Historical Environmental Variation in Conservation and Natural Resource Management* (J.A. Wiens, G.D. Hayward, H.D. Safford, and C.M. Giffen, eds.). Wiley-Blackwell, Chichester, UK.

Kirby, K.A. 2014. "New conservation" as a moral imperative. *Conservation Biology* 28: 639–640.

Koch, G.W., P.M. Vitousek, W.L. Steffen, and B.H. Walker. 1995. Terrestrial transects for global change research. *Vegetatio* 121: 53–65.

Kolb, T.E., M.R. Wagner, and W.W. Covington. 1994. Concepts of forest health: Utilitarian and ecosystems perspectives. *Journal of Forestry* 92(7): 10–15.

Kricher, J. 2009. *The Balance of Nature. Ecology's Enduring Myth*. Princeton University Press, Princeton, NJ.

Kuhn, T.S. 1959. The essential tension: Tradition and innovation in scientific research. pp. 162–174 in *The Third (1959) University of Utah Research Conference on the Identification of Scientific Talent* (C.W. Taylor, ed.). University of Utah Press, Salt Lake City, UT; reprinted in Kuhn, T.S. 1977. *The Essential Tension. Selected Studies in Scientific Tradition and Change.* University of Chicago Press, Chicago, IL.

Kuhn, T.S. 1962, 1970. *The Structure of Scientific Revolutions.* University of Chicago Press, Chicago, IL. (2nd edition).

Lack, D. 1954. *The Natural Regulation of Animal Numbers.* Oxford University Press, Oxford, UK.

Lackey, R.T. 2001. Values, policy, and ecosystem health. *BioScience* 51:437–443.

Lackey, R.T. 2004. Normative science. *Fisheries* 29(7): 38–39.

Lackey, R.T. 2007. Science, scientists, and policy advocacy. *Conservation Biology* 21: 12–17.

Lakatos, I., and A. Musgrave (Eds.). 1970. *Criticism and the Growth of Knowledge.* Cambridge University Press, Cambridge, UK.

Le Guin, U. 1998. *Steering the Craft: Exercises and Discussions on Story Writing for the Lone Navigator or the Mutinous Crew.* Eighth Mountain Press, Portland, OR.

Learn, S. 2012. Water cutoff contributes to Klamath Basin bird deaths, highlights challenge facing crucial wildlife refuges. *The Oregonian,* April 5.

Leopold, A. 1933. *Game Management.* Charles Scribner's Sons, New York, NY.

Leopold, A. 1949. *A Sand County Almanac.* Oxford University Press, Oxford, UK.

Leopold, L.B. 1966. *Round River. From the Journals of Aldo Leopold.* Oxford University Press, Oxford, UK.

Leopold, A.S., S.A. Cain, D.M. Cottam, I.N. Gabrielson, and T.L. Kimball. 1963. Wildlife management in the national parks. *Transactions of the North American Wildlife and Natural Resources Conference* 28: 28–45.

Levin, S.A. 1992. The problem of pattern and scale in ecology. *Ecology* 73: 1943–1967.

Levin, P.S., and C. Möllmann. 2014. Marine ecosystem regime shifts: Challenges and opportunities for ecosystem-based management. *Philosophical Transactions of the Royal Society B* 370: 20130275. Available at: 10.1098/rstb.2013.0275.

Lewis, S.L., and M.A. Maslin. 2015. Defining the Anthropocene. *Nature* 519: 171–180.

Lindenmayer, D.B., and J. Fischer. 2006. *Habitat Fragmentation and Landscape Change.* Island Press, Washington, DC.

Lindenmayer, D.B., et al. 2012. Value of long-term ecological studies. *Austral Ecology* 37: 745–757.

Liu, H., S. Shah, and W. Jaing. 2004. On-line outlier detection and data cleaning. *Computers and Chemical Engineering* 28: 1635–1647.

Losos, J.B., and R.E. Ricklefs. 2010. *The Theory of Island Biogeography Revisited.* Princeton University Press, Princeton, NJ.

Lovejoy, T.E., and L. Hannah. 2005. *Climate Change and Biodiversity.* Yale University Press, New Haven, CT.

Low, T. 2001. *Freal Future. The Untold Story of Australia's Exotic Invaders.* Penguin Books, Ringwood, Victoria, Australia.

Ludwig, J.A., J.A. Wiens, and D.J. Tongway. 2000. A scaling rule for landscape patches and how it applies to conserving soil resources in savannas. *Ecosystems* 3: 84–97.

Lugo, A.E. 2008. Visible and invisible effects of hurricanes on forest ecosystems: An international review. *Austral Ecology* 33: 368–398.

Lund, J.R., E. Hanak, W.E. Fleenor, W.A. Bennett, R.E. Howitt, J.F. Mount, and P.B. Moyle. 2010. *Comparing Futures for the Sacramento-San Joaquin Delta.* University of California Press, Berkeley, CA.

MacArthur, R.H. 1971. Patterns of terrestrial bird communities. pp. 189–221 in *Avian Biology* (D.S. Farner and J.R. King, eds.). Academic Press, New York.

MacArthur, R.H. 1972. *Geographical Ecology. Patterns in the Distribution of Species*. Harper & Row, New York.

MacArthur, R.H., and E.O. Wilson. 1967. *The Theory of Island Biogeography*. Princeton University Press, Princeton, NJ.

Mace, G.M., A. Balmford, and J.R. Ginsberg (Eds.). 1998. *Conservation in a Changing World*. Cambridge University Press, Cambridge, UK.

Malthus, T.R. 1803. *An Essay on the Principle of Population; or, A View of its Past and Present Effects on Human Happiness; With an Enquiry into Our Prospects Respecting the Future Removal or Mitigation of the Evils which it Occasions* (Revised edition). Routledge/Thoemmes Press, London.

Mangel, M. 2008. Uncertainty in ecology: A retrospective and prospective. *Bulletin of the British Ecological Society* 39(1): 25–28.

Mann, C.C. 2006. *1491: New Revelations of the Americas Before Columbus*. Vintage Books, New York, NY.

Mann, C.C. 2011. *1493: Uncovering the New World Columbus Created*. Vintage Books, New York, NY.

Marris, E. 2011. *Rambunctious Garden. Saving Nature in a Post-wild World*. Bloomsbury, New York.

Marvier, M. 2014. New conservation is true conservation. *Conservation Biology* 28: 1–3.

Marvier, M., and P. Kareiva. 2014. The evidence and values underlying 'new conservation'. *Trends in Ecology and Evolution* 29: 131–132.

Marvier, M., J. Grant, and P. Kareiva. 2006. Nature: Poorest may see it as their economic rival. *Nature* 443: 749–750.

Max, D.T. 2014. Green is good. *The New Yorker* May 12: 54–63.

McCauley, D.J. 2006. Selling out on nature. *Nature* 443: 27–28.

McGrew, W.C., L.F. Marchant, and T. Nishida (Eds.). 1996. *Great Ape Societies*. Cambridge University Press, Cambridge, UK.

McIntosh, R.P. 1993. The continuum continued: John T. Curtis' influence on ecology. pp. 95–122 in *John T. Curtis. Fifty Years of Wisconsin Plant Ecology* (J.S. Fralish, R.P. McIntosh, and O.L. Loucks, eds.). Wisconsin Academy of Sciences, Arts & Letters, Madison, WI.

McLachlan, J.S., J.J. Hellmann, and M.W. Schwartz. 2007. A framework for debate of assisted migration in an era of climate change. *Conservation Biology* 21: 297–302.

McLuhan, M. 1964. *Understanding Media: The Extensions of Man*. McGraw-Hill, New York, NY.

Meadows, D.H., D.L. Meadows, J. Randers, and W.H. Behrens III,. 1972. *The Limits to Growth*. Universe Books, New York, NY.

Meadows, D.H., D.L. Meadows, and J. Randers. 1992. *Beyond the Limits*. Chelsea Green Publishing Company, Post Mills, VT.

Meadows, D., J. Randers, and D. Meadows. 2004. *Limits to Growth. The 30-year Update*. Chelsea Green Publishing Company, White River Junction, VT.

Melville, H. 1851. *Moby Dick*. Harper & Brothers, New York, NY.

Millennium Ecosystem Assessment. 2005. *Ecosystems and Human Well-being: Synthesis*. Island Press, Washington, DC.

Milly, P.C.D., J. Betancourt, M. Falkenmark, R.M. Hirsch, Z.W. Kundzewicz, D.P. Lettenmaier, and R.J. Stouffer. 2008. Stationarity is dead: Whither water management? *Science* 319: 573–574.

Mini, A.E., and R. LeValley. 2006. *Aleutian Cackling Goose Agricultural Depredation Plan for Del Norte County, California,*California Coastal Conservancy, San Francisco, CA.

Mini, A.E., D.C. Bachman, J. Cocke, K.M. Griggs, K.A. Spragens, and J.M. Black. 2011. Recovery of the Aleutian cackling goose *Branta hutchinsii leucopareia*: 10-year review and future prospects. *Wildfowl* 61: 3–29.

Mittermeier, R.A., P.R. Gil, M. Hoffman, J. Pilgram, T. Brooks, C.G. Mittermeier, J. Lamoreux, and G.A.B. Da Fonseca. 2005. *Hotspots Revisited. Earth's Biologically Richest and Most Endangered Terrestrial Ecoregions.* University of Chicago Press, Chicago, IL.

Moilanen, A., K.A. Wilson, and H.P. Possingham. 2009. *Spatial Conservation Prioritization. Quantitative Methods & Computational Tools.* Oxford University Press, Oxford.

Moore, K.D., and M.P. Nelson (Eds.). 2010. *Moral Ground. Ethical Action for a Planet in Peril.* Trinity University Press, San Antonio, TX.

Mora, C., and P.F. Sale. 2011. Ongoing global biodiversity loss and the need to move beyond protected areas: A review of the technical and practical shortcomings of protected areas on land and sea. *Marine Ecology Progress Series* 434: 251–266.

Moritz, C., J.L. Patton, C.J. Conroy, J.L. Parra, G.C. White, and S.R. Beissinger. 2008. Impact of a century of climate change on small-mammal communities in Yosemite National Park, USA. *Science* 322: 261–264.

Moyle, P.B. 2013. Novel aquatic ecosystems: The new reality for streams in California and other Mediterranean climate regions. *River Research and Applications*. doi: 10.1002/rra.2709.

Moyle, P.B., J.R. Lund, W.A. Bennett, and W.E. Fleenor. 2010. Habitat variability and complexity in the Upper San Francisco Estuary. *San Francisco Estuary & Watershed Science* 8(3). Available at: http://escholarship.org/uc/item/0kf0d32x.

Muir, J. 1912. *The Yosemite*. Century, New York.

Mulvaney, J., and J. Kamminga. 1999. *Prehistory of Australia*. Smithsonian Institution Press, Washington, DC.

Murcia, C., J. Aronson, G.H. Kattan, D. Moreno-Mateos, K. Dixon, and D. Simberloff. 2014. A critique of the 'novel ecosystem' concept. *Trends in Ecology and Evolution* 29: 548–553.

Murphy, D.D., and B.R. Noon. 1991. Coping with uncertainty in wildlife biology. *Journal of Wildlife Management* 55: 773–782.

Myers, N. 1988. Threatened biotas: "Hot Spots" in tropical forests. *The Environmentalist* 8: 1–20.

Næss, A. 1986. Intrinsic value: Will the defenders of nature please rise? pp. 504–515 in *Conservation Biology. The Science of Scarcity and Diversity* (M.E. Soulé, ed.). *Sinauer Associates,* Sunderland, MA.

Nassauer, J. I., M.V. Santelmann, and D. Scavia. 2007. *From the Corn Belt to the Gulf: Societal and Environmental Implications of Alternative Agricultural Futures.* Resources for the Future Press, Washington, DC.

National Academy of Sciences and The Royal Society. 2014. Climate Change Evidence & Causes. Available at: http://dels.nas.edu/resources/static-assets/exec-office-other/climate-change-full.pdf.

National Research Council. 2012a. *Sustainable Water and Environmental Management in the California Bay-Delta*. The National Academies Press, Washington, DC.

National Research Council. 2012b. *Sea-level Rise for the Coasts of California, Oregon, and Washington. Past, Present, and Future*. The National Academies Press, Washington, DC.

National Research Council. 2013. *An Ecosystem Services Approach to Assessing the Impacts of the* Deepwater Horizon *Oil Spill in the Gulf of Mexico*. The National Academies Press, Washington, DC.

National Research Council. 2014. *Can Earth's and Society's Systems Meet the Needs of 10 Billion People?: Summary of a Workshop*. The National Academies Press, Washington, DC.

Nelson, M.P., and J.A. Vucetich. 2009. On advocacy by environmental scientists: What, whether, why, and how. *Conservation Biology* 23: 1090–1101.

Nickles, T. (Ed.). 2003. *Thomas Kuhn*. Cambridge University Press, Cambridge, UK.

Norgaard, R.B. 2010. Ecosystem services: From eye-opening metaphor to complexity blinder. *Ecological Economics* 69: 1219–1227.

Norton, B. 2006. Toward a policy-relevant definition of biodiversity. pp. 49–58 in *The Endangered Species Act at Thirty. Conserving Biodiversity in Human-dominated Landscapes*, Volume 2 (J.M. Scott, D.D. Goble, and F.W. Davis, eds.). Island Press, Washington, DC.

Noss, R.F. 1996. Conservation or convenience? *Conservation Biology* 10:921–922.

Odum, E.P. 1953. *Fundamentals of Ecology*. W.B. Saunders Company, Philadelphia, PA.

Odum, E.P. 1969. The strategy of ecosystem development. *Science* 164: 262–270.

Oppenheimer, J.R. 1955. *The Open Mind*. Simon and Schuster, New York, NY, USA.

Oreskes, N., and E.M. Conway. 2010. *Merchants of Doubt*. Bloomsbury Press, New York.

Osborn, F. 1953. *The Limits of the Earth*. Little, Brown and Company, Boston, MA.

Parker, K.L. 1897. Australian Legendary Tales. Melville, Mullen & Slade, Melbourne. Available at: http://www.artistwd.com/joyzine/australia/dreaming/swans.php.

Parker, K.R., and J.A. Wiens. 2005. Assessing recovery following environmental accidents: Environmental variation, ecological assumptions, and strategies. *Ecological Applications* 15: 2037–2051.

Pastor, J., D.J. Mladenoff, Y. Haila, J. Bryant, and S. Payette. 1996. Biodiversity and ecosystem processes in boreal regions. pp. 33–69 in *Functional roles of biodiversity: a global perspective* (H.A. Mooney, J.H. Cushman, E. Medina, O.E. Sala, and E.-D. Schulze, eds.). John Wiley, New York, New York, USA.

Patten, M.A. 2015. Subspecies and the philosophy of science. *The Auk: Ornithological Advances* 132: 481–485.

Pedlar, J.H., D.W. McKenney, I. Aubin, T. Beardmore, J. Beaulieu, L. Iverson, G.A. O'Neill, R.S. Winder, and C. Ste-Marie. 2012. Placing forestry in the assisted migration debate. *BioScience* 62: 835–842.

Perera, A.H., C.A. Drew, and C.J. Johnson (Eds.). 2012. *Expert Knowledge and its Application in Landscape Ecology*. Springer, New York.

Pergams, O.R.W., and P.A. Zaradic. 2006. Is love of nature in the U.S. becoming love of electronic media? 16-year downtrend in national park visits explained by watching movies, playing video games, internet use, and oil prices. *Journal of Environmental Management* 80: 387–393.

Pergams, O.R.W., and P.A. Zaradic. 2008. Evidence for a fundamental and pervasive shift away from nature-based recreation. *Proceedings of the National Academy of Science* 105: 2295–2300.

Perring, M.P., and E.C. Ellis. 2013. The extent of novel ecosystems: Long in time and broad in space. pp. 66–80 in *Novel Ecosystems. Intervening in the New Ecological World Order* (R.J. Hobbs, E.S. Higgs, and C.M. Hall, eds.). Wiley-Blackwell, Chichester, UK.

Pew Research Center. 2015. Public and Scientists' Views on Science and Society. Available at: http://www.pewinternet.org/files/2015/01/PI_ScienceandSociety_Report_012915.pdf.

Pielke, Roger A., Jr. 2007. *The Honest Broker: Making Sense of Science in Policy and Politics*. Cambridge University Press, Cambridge, UK.

Pimm, S. 2014. Book review. *Biological Conservation* 180:151–152.

Pirsig, R.M. 1974. *Zen and the Art of Motorcycle Maintenance*. William Morrow, New York.

Pitelka, F.A. 1981. The condor case: An uphill struggle in a downhill crush. *Point Reyes Bird Observatory Newsletter* 53: 4–5.

Pitelka, L.F., and D.J. Raynal. 1989. Forest decline and acidic deposition. *Ecology* 70: 2–10.

Poirier, R. and D. Ostergren. 2002. Evicting people from nature: Indigenous land rights and national parks in Australia, Russia, and the United States. *Natural Resources Journal* 42: 331–351.

Pollack, H.N. 2003. *Uncertain Science … Uncertain World*. Cambridge University Press, Cambridge, UK.

Popper, K.R. 1959. *The Logic of Scientific Discovery*. Hutchinson & Co., London, UK (first published as *Logik der Forschung* in 1935 by Verlag von Julius Springer, Vienna.)

Portney, P.R. 1994. The contingent valuation debate: Why economists should care. *Journal of Economic Perspectives* 8: 3–17.

Rahiz, M., and M. New. 2013. 21st Century Drought Scenarios for the UK. *Water Resources Management* 27: 1039–1061.

Rapp, V. 2007. *Northwest Forest Plan—the first 10 years (1994–2003): First-decade results of the Northwest Forest Plan*. Gen. Tech. Rep. PNW-GTR-720. U.S. Department of Agriculture, Forest Service, Pacific Northwest Research Station, Portland, OR.

Reece, J.S., and R.F. Noss. 2014. Prioritizing species by conservation value and vulnerability: A new index applied to species threatened by sea-level rise and other risks in Florida. *Natural Areas Journal* 34: 31–45.

Reid, W.V. 2006. Nature: The many benefits of ecosystem services. *Nature* 443: 749.

Reiners, W.A., and K.L. Driese. 2004. *Transport Processes in Nature. Propagation of Ecological Influences through Environmental Space*. Cambridge University Press, Cambridge, UK.

Reisner, M. 1993. *Cadillac Desert*. Penguin Books, New York, NY. (Revised edition.)

Ricciardi, A., and D. Simberloff. 2009. Assisted colonization is not a viable conservation strategy. *Trends in Ecology and Evolution* 24: 248–253.

Richardson, D.M. (Ed.). 2011. *Fifty Years of Invasion Ecology. The Legacy of Charles Elton*. Wiley-Blackwell, Chichester, UK.

Ricketts, T.H., et al. 2005. Pinpointing and preventing imminent extinctions. *Proceedings of the National Academy of Science* 102: 18497–18501.

Ripley, S.D. 1981. Take the ultimate risk. *Point Reyes Bird Observatory Newsletter* 53: 1–3.

Robin, L., R. Heinsohn, and L. Joseph (Eds.). 2009. *Boom & Bust. Bird Stories for a Dry Country*. CSIRO Publishing, Collingwood, VIC, Australia.

Rohde, K. 2005. *Nonequilibrium Ecology*. Cambridge University Press, Cambridge, UK.

Rolston, H., III. 1994. *Conserving Natural Value*. Columbia University Press, New York NY.

Rolston, H. III. 2012 *A New Environmental Ethics. The Next Millennium for Life on Earth*. Routledge, New York, NY.

Roosevelt, T. 1902. *Hunting the Grisly and Other Sketches*. G.P. Putnam's Sons, New York.

Rosenzweig, M.L. 2003. *Win-win Ecology. How the Earth's Species can Survive in the Midst of Human Enterprise*. Oxford University Press, New York, NY.

Rosseau, D.L. 1992. Case studies in pathological science. *American Scientist* 80: 54–63.

Roth, J.E., K.L Mills, and W.J Sydeman. 2007. Chinook salmon (*Oncorhynchus tshawytscha*) —seabird covariation off central California and possible forecasting applications. *Canadian Journal of Fisheries and Aquatic Sciences* 64: 1080–1090.

Rout, T.M., E. McDonald-Madden, T.G. Martin, N.J. Mitchell, H.P. Possingham, and D.P. Armstrong. 2013. How to decide whether to move species threatened by climate change. *PLoS ONE* 8(10): e75814. doi: 10.1371/journal.pone.0075814.

Rubenstein, D.R., D.I. Rubenstein, P.W. Sherman, and T.A. Gavin. 2006. Pleistocene Park: Does re-wilding North America represent sound conservation for the 21st century? *Biological Conservation* 132: 232–238.

Ruckleshaus, M., and D. Darm. 2006. Science and implementation. pp. 104–126 in *The Endangered Species Act at Thirty. Conserving Biodiversity in Human-dominated Landscapes*, Volume 2 (J.M. Scott, D.D. Goble, and F.W. Davis, eds.). Island Press, Washington, DC.

Salafsky, S.R. 2015. Reproductive responses of an apex predator to changing climatic conditions in a variable forest environment. Ph.D. Dissertation, Colorado State University.

Scheffer, M., J. Bascompte, W.A. Brock, V. Brovkin, S.R. Carpenter, V. Dakos, H. Held, E.H. van Nes, M. Rietkerk, and G. Sugihara. 2009. Early-warning signals for critical transitions. *Nature* 461:53–59.

Schneider, S.H. 2009. *Science as a Contact Sport. Inside the Battle to Save Earth's Climate*. National Geographic Society, Washington, DC.

Schoener, T.W. 1983. Dr. Schoener replies. *American Scientist* 71: 235.

Schumacher, E.F. 1973. *Small is Beautiful: Economics as if People Mattered*. Harper & Row, New York.

Schwartz, M.W. et al. 2012. Managed relocation: Integrating the scientific, regulatory, and ethical challenges. *BioScience* 62: 732–743.

Scott, J.M., D.D. Goble, J.A. Wiens, D.S. Wilcove, M. Bean, and T. Male. 2005. Recovery of imperiled species under the Endangered Species Act: The need for a new approach. *Frontiers in Ecology and the Environment* 3: 383–389.

Scott, J.M., J.L. Rachlow, R.T. Lackey, A.B. Pidgorna, J.L. Aycrigg, G.R. Feldman, L.K. Svancara, D.A. Rupp, D.L. Stanish, and R.K. Steinhorst. 2007. Policy advocacy in science: Prevalence, perspectives, and implications for conservation biologists. *Conservation Biology* 21: 29–35.

Scott, J.M., D.D. Goble, A.M. Haines, J.A. Wiens, and M.C. Neel. 2010. Conservation-reliant species and the future of conservation. *Conservation Letters* 3:91–97.

Sears, P. 1935. *Deserts on the March*. University of Oklahoma Press, Norman, OK.

Seddon, P.J., A. Moehrenschlager, and J. Ewen. 2014. Reintroducing resurrected species: Selecting DeExtinction candidates. *Trends in Ecology and Evolution* 29: 140–147.

Sendak, M. 1963. *Where the Wild Things Are*. Harper & Row, New York, NY.

Shuford, W.D., and T. Gardali (Eds.). 2008. *California Bird Species of Special Concern: A Ranked Assessment of Species, Subspecies, and Distinct Populations of Birds of Immediate Conservation Concern in California*. Studies of Western Birds 1. Western Field Ornithologists, Camarillo, CA, and California Department of Fish and Game, Sacramento, CA.

Simberloff, D. 2015. "Novel Ecosystems" are a Trojan Horse for Conservation. *Ensia* January 21, 2015. Available at: http://ensia.com/voices/novel-ecosystems-are-a-trojan-horse-for-conservation/.

Simberloff, D., et al. 2011. Non-natives: 141 scientists object. *Nature* 475:36.

Simberloff, D., et al. 2012. Impacts of biological invasions: What's what and the way forward. *Trends in Ecology and Evolution* 28: 58–66.

Snow, C.P. 1963. *The Two Cultures: and a Second Look*. Cambridge University Press, Cambridge, UK.

Solbé, J. (Ed.). 2005. *Long-term Monitoring. Why, What, Where, When & How?*. Sherkin Island Marine Station, Sherkin Island, Co. Cork, Ireland.

Soulé, M. 2013. The "new conservation". *Conservation Biology* 27: 895–897.

Soulé, M. 2014. Also seeking common ground in conservation. *Conservation Biology* 28: 637–638.

Soulé, M., and R. Noss. 1998. Re-wilding and biodiversity: Complementary goals for continental conservation. *Wild Earth* 8: 18–28.
Spellerberg, I.F. 2005. *Monitoring Ecological Change.* Second Edition. Cambridge University Press, Cambridge, UK.
Stallcup, R. 1981. Farewell, skymaster. *Point Reyes Bird Observatory Newsletter* 53: 10.
Standish, R.J., et al. 2014. Resilience in ecology: Abstraction, distraction, or where the action is? *Biological Conservation* 177: 43–51.
Steadman, D.W. 2006. *Extinction and Biogeography of Tropical Pacific Birds.* University of Chicago Press, Chicago, IL.
Steinbeck, J. 1939. *The Grapes of Wrath.* Viking Press, New York, NY.
Stralberg, D., D. Jongsomjit, C.A. Howell, M.A. Snyder, J.D. Alexander, J.A. Wiens, and T.L. Root. 2009. Re-shuffling of species with climate disruption: A no-analog future for California birds? *PLoS ONE* 4(9): e6825. doi: 10.1371/journal.pone.0006825.
Strong, D.R., Jr., D. Simberloff, L.G. Abele, and A.B. Thistle (Eds.). 1984. *Ecological Communities: Conceptual Issues and the Evidence.* Princeton University Press, Princeton, NJ.
Sullivan, P.J., et al. 2006. Defining and implementing Best Available Science for fisheries and environmental science, policy, and management. *Fisheries* 31: 460–465.
Taleb, N.N. 2010. *The Black Swan: The Impact of the Highly Improbable.* Second edition. Random House, New York.
Tallis, H., J. Lubchenco, et al. 2014. Working together: A call for inclusive conservation. *Nature* 515: 27–28.
Tanner, J.T. 1942. *The Ivory-billed Woodpecker.* National Audubon Society, New York (reissued by Courier Dover Publications in 2012).
Terborgh, J., C. van Schaik, L. Davenport, and M. Rao (Eds.). 2002. *Making Parks Work. Strategies for Preserving Tropical Nature.* Island Press, Washington, DC.
Theobald, D. 2005. Landscape patterns of exurban growth in the USA from 1980 to 2020. *Ecology and Society* 10(1): 32. (online) Available at: http://www.ecologyandsociety.org/vol10/iss1/art32/.
Thomas, W.L. (Ed.). 1956. *Man's Role in Changing the Face of the Earth.* University of Chicago Press, Chicago, IL.
Thompson, A., and J. Bendik-Keymer (Eds.). 2012. *Ethical Adaptation to Climate Change. Human Virtues of the Future.* The MIT Press, Cambridge, MA.
Thompson, J., K.F. Lambert, D. Foster, M. Blumstein, E. Broadbent, and A.A. Zambrano. 2014. *Changes to the Land. Four Scenarios for the Future of the Massachusetts Landscape.* Harvard Forest, Petersham, MA. Available at: http://harvardforest.fas.harvard.edu/changes-to-the-land.
Tiffany, D.G. 2009. Economic and environmental impacts of U.S. corn ethanol production and use. *Federal Reserve Bank of St. Louis Regional Economic Development* 5: 42–58.
Tingley, M.W., W.B. Monahan, S.R. Beissinger, and C. Moritz. 2009. Birds track their Grinnellian niche through a century of climate change. *Proceeding of the National Academy of Science* 106 (Supplement 2): 19637–19643.
Turner, M.G., R.H. Gardner, V.H. Dale, and R.V. O'Neill. 1989. Predicting the spread of disturbance across heterogeneous landscapes. *Oikos* 55: 121–129.
Turner, B.L. II, W.C. Clark, R.W. Kates, J.F. Richards, J.T. Mathews, and W.B. Meyer. 1990. *The Earth as Transformed by Human Action.* Cambridge University Press, Cambridge, UK.
Udall, S. 1968. *1976: Agenda for Tomorrow.* Harcourt, Brace & World, New York, NY.
Valdez, R. 2013. Exploring our ancient roots. Genghis Kahn to Aldo Leopold: The origins of wildlife management. *The Wildlife Professional* 7(3): 50–53.

Van Horne, B., G.S. Olson, R.L. Schooley, J.G. Corn, and K.P. Burnham. 1997. Effects of drought and prolonged winter on Townsend's ground squirrel demography in shrubsteppe habitats. *Ecological Monographs* 67: 295–315.

Van Valen, L. 1973. A new evolutionary law. *Evolutionary Theory* 1: 1—30.

Veblen, T.T., and D.C. Lorenz. 1991. *The Colorado Front Range: A Century of Ecological Change*. University of Utah Press, Salt Lake City, UT.

Vogel, S. 2015. *Thinking Like a Mall. Environmental Philosophy after the End of Nature*. MIT Press, Cambridge, MA.

Voltaire. 1772. *La Begueule*. Chez Franc. Grasset et Comp.

Vucetich, J.A., J.T. Bruskotter, and M.P. Nelson. 2015. Evaluating whether nature's intrinsic value is an axiom of or an anathema to conservation. *Conservation Biology* 29: 321–332.

Waggoner, P.E. 1966. Weather modification and the living environment. pp. 87–98 in *Future Environments of North America* (F.F. Darling and J.P. Milton, eds.). The Natural History Press, Garden City, NJ.

Waldron, A., A.O. Mooers, D.C. Miller, N. Nibbelink, D. Redding, T.S. Kuhn, J.T. Roberts, and J.L. Gittleman. 2013. Targeting global conservation funding to limit immediate biodiversity declines. *Proceedings of the National Academy of Science USA* 110: 12144–12148.

Walker, B., and D. Salt. 2006. *Resilience Thinking. Sustaining Ecosystems and People in a Changing World*. Island Press, Washington, DC.

Walker, B., and D. Salt. 2012. *Resilience Practice. Building Capacity to Absorb Disturbance and Maintain Function*. Island Press, Washington, DC.

Walters, J.R., S.R. Derrickson, D.M. Fry, S.M. Haig. J.M. Marzluff, and J.M. Wunderle, Jr. 2010. Status of the California condor (*Gymnogyps californianus*) and efforts to achieve its recovery. *Auk* 127:969–1001.

Waples, R.S. 2006. Distinct population segments. pp. 127–149 in *The Endangered Species Act at Thirty. Conserving Biodiversity in Human-dominated Landscapes*, Volume 2 (J.M. Scott, D.D. Goble, and F.W. Davis, eds.). Island Press, Washington, DC.

Watson, J.E.M. 2014. Human responses to climate change will seriously impact biodiversity conservation: It's time we start planning for them. *Conservation Letters* 7: 1–2.

Watson, J.D., and F. Crick. 1953. A structure for deoxyribose nucleic acid. *Nature* 171: 37–738.

Weiher, E., and P. Keddy (Eds.). 1999. *Ecological Assembly Rules. Perspectives, Advances, Retreats*. Cambridge University Press, Cambridge, UK.

West, P., J. Igoe, and D. Brockington. 2006. Parks and peoples: The social impact of protected areas. *Annual Review of Anthropology* 35: 251–277.

Westoby, M. 1997. What does "ecology" mean? *Trends in Ecology and Evolution* 12: 166.

Whipple, A.A., R.M. Grossinger, D. Rankin, B. Stanford, and R.A. Askevold. 2012. *Sacramento-San Joaquin Delta Historical Ecology Investigation: Exploring Pattern and Process*. Prepared for the California Department of Fish and Game and Ecosystem Restoration Program. A Report of SFEI-ASC's Historical Ecology Program, SFEI-ASC Publication #672. San Francisco Estuary Institute-Aquatic Science Center, Richmond, CA.

White, P.S., and S.T.A. Pickett. 1985. Natural disturbance and patch dynamics: An introduction. pp. 3–13 in *The Ecology of Natural Disturbance and Patch Dynamics* (S.T.A. Pickett and P.S. White, eds.). Academic Press, Orlando, FL.

Whittaker, R.H. 1967. Gradient analysis of vegetation. *Biological Reviews* 42: 207–264.

Whittaker, R.H. 1969. Evolution of diversity in plant communities. pp. 178–196 in *Diversity and Stability in Ecological Systems* (G.M. Woodwell and H.H. Smith, eds.). Brookhaven Symposia in Biology Number 22. Brookhaven National Laboratory, Upton, NY.

Whittaker, R.H. 1975. *Communities and Ecosystems,* Second Edition. Macmillan, New York, NY, USA.

Wiens, J.A. 1966. On group selection and Wynne-Edwards' hypothesis. *American Scientist* 54: 273–287.

Wiens, J.A. 1969. An approach to the study of ecological relationships among grassland birds. *Ornithological Monographs* 8:1–93.

Wiens, J.A. 1973. Pattern and process in grassland bird communities. *Ecological Monographs* 43:237–270.

Wiens, J.A. 1974. Climatic instability and the "ecological saturation" of bird communities in North American grasslands. *Condor* 76: 385–400.

Wiens, J.A. 1977. On competition and variable environments. *American Scientist* 65: 590–597.

Wiens, J.A. 1981. Single-sample surveys of communities: Are the revealed patterns real? *American Naturalist* 117: 90–98.

Wiens, J.A. 1983. Interspecific competition. *American Scientist* 71: 234–235.

Wiens, J.A. 1989a. *The Ecology of Bird Communities. Volumes 1 and 2.* Cambridge University Press, Cambridge, UK.

Wiens, J.A. 1989b. Spatial scaling in ecology. *Functional Ecology* 3: 385–397.

Wiens, J.A. 1996. Oil, seabirds, and science. *BioScience* 46: 587–597.

Wiens, J.A. (Ed.). 2013a. *Oil in the Environment. Legacies and Lessons of the* Exxon Valdez *Oil Spill.* Cambridge University Press, Cambridge, UK.

Wiens, J.A. 2013b. Science and oil spills: The broad picture. pp. 423–445 in *Oil in the Environment: Legacies and Lessons of the* Exxon Valdez *Oil Spill* (J.A. Wiens, ed.). Cambridge University Press, Cambridge, UK.

Wiens, J.A., and T. Gardali. 2013. Conservation reliance among California's at-risk birds. *The Condor* 115: 1–15.

Wiens, J.A., and R.J. Hobbs. 2015. Integrating conservation and restoration in a changing world. *BioScience* 65: 302–312.

Wiens, J.A., and B.T. Milne. 1989. Scaling of "landscapes" in landscape ecology, or landscape ecology from a beetle's perspective. *Landscape Ecology* 3: 87–96.

Wiens, J.A., and K.R. Parker. 1995. Analyzing the effects of accidental environmental impacts: Approaches and assumptions. *Ecological Applications* 5: 1069–1083.

Wiens, J.A., N.E. Seavy, and D. Jongsomjit. 2011. Protected areas in climate space: What will the future bring? *Biological Conservation* 144: 2119–2125.

Wiens, J.A., G.D. Hayward, H.D. Safford, and C.M. Giffen (Eds.). 2012. *Historical Environmental Variation in Conservation and Natural Resource Management*. Wiley-Blackwell, Chichester, UK.

Wiens, J.D., R.G. Anthony, and E.D. Forsman. 2014. Competitive interactions and resource partitioning between northern spotted owls and barred owls in western Oregon. *Wildlife Monographs* 185: 1–50.

Wiese, F.K., and G.J. Robertson. 2004. Assessing seabird mortality from chronic oil discharges at sea. *The Journal of Wildlife Management* 68: 627–638.

Wilkinson, H.L. 1930. *The World's Population Problems and a White Australia*. P.S. King & Son, London, UK.

Williams, J.W., and S.T. Jackson. 2007. Novel climates, no-analog communities, and ecological surprises: past and future. *Frontiers in Ecology and the Environment* 5: 475–482.

Williams, D.M., and S. Knapp (Eds.). 2010. *Beyond Cladistics: The Branching of a Paradigm.* University of California Press, Berkeley, CA.

Wilson, E.O. (Ed.). 1988. *BioDiversity*. National Academy Press, Washington, DC.

Wilson, E.O. 1994. *Naturalist*. Island Press, Washington, DC.

Wilson, E.O. 1997. Introduction. pp. 1–3 in *Biodiversity II. Understanding and Protecting Our Biological Resources* (M.L. Reaka-Kudla, D.E. Wilson, and E.O. Wilson, eds.). Joseph Henry Press, Washington, DC.

Wilson, E.O. 2002. What is nature worth? *Wilson Quarterly* 26(1): 20–39.

Wilson, E.O. 2006. *The Creation. An Appeal to Save Life on Earth*. W.W. Norton & Company, New York, NY.

With, K.A. 2002. Using percolation theory to assess landscape connectivity and effects of habitat fragmentation. pp. 105–130 in *Applying Landscape Ecology in Biological Conservation* (K.J. Gutzwiller, ed.). Springer, New York, NY.

Wolfe, T. 1940 *You Can't Go Home Again*. Harper & Row, New York, NY.

Worster, D. 1994. *Nature's Economy: A History of Ecological Ideas*. Cambridge University Press, Cambridge, UK.

Wright, C.K., and M.C. Wimberly. 2013. Recent land use change in the Western Corn Belt threatens grasslands and wetlands. *Proceedings of the National Academy of Science* 110: 4134–4139.

Wu, J., and O.L. Loucks. 1995. From balance of nature to hierarchical patch dynamics: A paradigm shift in ecology. *Quarterly Review of Biology* 70: 439–466.

Wynne-Edwards, V.C. 1962. *Animal Dispersion in Relation to Social Behavior*. Oliver & Boyd, London.

Wynne-Edwards, V.C. 1986. *Evolution Through Group Selection*. Blackwell Scientific, Oxford.

Zimmerman, H. G., S. Bloem, and H. Klein. 2004. Biology, history, threat, surveillance and control of the cactus moth, *Cactoblastis cactorum*. *Joint FAO/IAEA Programme of Nuclear Techniques in Food and Agriculture*. IAEA, Vienna, Austria.

Scientific names of species mentioned in the text

African elephant (*Loxodonta africana*)
Akialoa (*Hemignathus ellisanus ellisanus*)
American chestnut (*Castanea dentata*)
American pika (*Ochotona princeps*)
Andean condor (*Vultur gryphus*)
Araripe manakin (*Antilophia bokermanni*)
Asian elephant (*Elephas maximus*)
Asian mongoose (*Herpestes javanicus*)
Atlantic cod (*Gadus morhua*)
Bark beetles (*Ips confusus*)
Barred owl (*Strix varia*)
Beech (*Fagus grandifolia*)
Bison (*Bison bison*)
Black-footed ferret (*Mustela nigripes*)
Black rhino (*Diceros bicornis*)
Black swan (*Cygnus atratus*)
Blue jay (*Cyanocitta cristata*)
Blue tit (*Cyanistes caeruleus*)
Blue whale (*Balaenoptera musculus*)
Bristlecone pine (*Pinus longaeva*)
Brown tree snake (*Boiga irregularis*)
Brown-headed cowbird (*Molothrus ater*)
Buffel grass (*Pennisetum ciliare*)
California clapper rail (*Rallus longirostris obsoletus*); now recognized as a separate species, Ridgway's rail (*Rallus obsoletus*)
California condor (*Gymnogyps californianus*)
California spotted owl (*Strix occidentalis occidentalis*)
Cane toad (*Rhinella marina*)
Cassin's auklet (*Ptychoramphus aleuticus*)

Ecological Challenges and Conservation Conundrums: Essays and Reflections for a Changing World, First Edition. John A. Wiens.

Cattle egret (*Bubulcus ibis*)
Cheatgrass (*Bromus tectorum*)
Cheetah (*Acinonyx jubatus*)
Chinook salmon (*Oncorhynchus tshawytscha*)
Delta smelt (*Hypomesus transpacificus*)
Dingo (*Canis lupus dingo*)
Eurasian lynx (*Lynx lynx*)
European bison (*Bison bonasus*)
Giant panda (*Ailuropoda melanoleuca*)
Giant sequoia (*Sequoiadendron giganteum*)
Golden toad (*Bufo periglenes*)
Hippopotamus (*Hippopotamus amphibius*)
House mouse (*Mus musculus*)
Ivory-billed woodpecker (*Campephilus principalis*)
Jack pine (*Pinus banksiana*)
Kākāwahie (*Paroreomyza flammea*)
Kelp (*Nereocystis luetkeana*)
Killer whale (*Orcinus orca*)
Kirtland's warbler (*Setophaga kirtlandii*)
Koala (*Phascolarctos cinereus*)
Krill (*Euphausia pacifica*)
Lupine (*Lupinus* spp.)
Mangrove (*Rhizophora* spp.)
Mexican spotted owl (*Strix occidentalis lucida*)
Monterey pine (*Pinus radiata*)
Nile perch (*Lates niloticus*)
Northern elephant seal (*Mirounga angustirostris*)
Northern goshawk (*Accipiter gentilis*)
Northern spotted owl (*Strix occidentalis caurina*)
ō'ō'ā'ā (*Moho braccatus*)
Oregon chub (*Oregonichthys crameri*)
Pallid sturgeon (*Scaphirhynchus albus*)
Parkinsonia (*Parkinsonia aculeata*)
Passenger pigeon (*Ectopistes migratorius*)
Peregrine falcon (*Falco peregrinus*)
Peruvian anchovetta (*Engraulis ringens*)
Polar bear (*Ursus maritimus*)
Prickley-pear cactus (*Opuntia stricta*)
Piñon pine (*Pinus edulis*)
Red-cockaded woodpecker (*Picoides borealis*)
Round goby (*Neogobius melanostomus*)
Sage sparrow (*Artemisiospiza nevadensis*)
Sagebrush (*Artemisia tridentata*)
Saltcedar (*Tamarix ramosissima*)

Saltcedar leaf beetle (*Diorhabda carinulata*)
Sea otter (*Enhydra lutra*)
Sea urchin (*Strongylocentrotus droebachiensis*)
Southern white rhino (*Ceratotherium simum simum*)
Southwestern willow flycatcher (*Empidonax traillii extimus*)
Striped bass (*Morone saxatilis*)
Taro (*Colocasia esculenta*)
Thylacine (*Thylacinus cynocephalus*)
Townsend's ground squirrel (*Urocitellus townsendii*)
Turkey oak (*Quercus cerris*)
Water buffalo (*Bubalus bubalis*)
Water hyacinth (*Eichhornia crassipes*)
Western black rhino (*Diceros bicornis longipes*)
White rhino (*Ceratotherium simum*)
Woolly mammoth (*Mammuthus primigenius*)

[illegible]
[illegible]
[illegible]
[illegible]
[illegible]
[illegible]
[illegible]

[illegible]
[illegible]
[illegible]
[illegible]
[illegible]
[illegible]
[illegible]
[illegible]

Index

Ecological Challenges and Conservation Conundrums: Essays and Reflections for a Changing World, First Edition. John A. Wiens.

www.ingramcontent.com/pod-product-compliance
Lightning Source LLC
Chambersburg PA
CBHW060101010325
22684CB00003B/10
9781118895108